THE NEW POLITICS OF SURVEILLANCE AND VISIBILITY

Edited by Kevin D. Haggerty and Richard V. Ericson

Since the terrorist attacks of September 2001, surveillance has been put forward as an essential tool in the 'war on terror,' with new technologies and policies offering police and military operatives enhanced opportunities for monitoring suspect populations. In addition, the last few years have seen consumer tastes become increasingly codified, with 'data mines' of demographic information such as postal codes and purchasing records. Surveillance has also recently emerged as a form of entertainment, with 'reality' shows becoming the dominant genre on network and cable television.

In *The New Politics of Surveillance and Visibility*, editors Kevin D. Haggerty and Richard V. Ericson bring together leading experts to analyse how society is organized through surveillance systems, technologies, and practices. They demonstrate how the new political uses of surveillance make visible that which was previously unknown, blur the boundaries between public and private, and alter processes of democratic accountability. This collection challenges conventional wisdom and advances new theoretical approaches on the subject of surveillance through a series of studies on its implications in such areas as policing, the military, commercial enterprise, mass media, and the health sciences.

KEVIN D. HAGGERTY is an assistant professor in the Department of Sociology and director of the Criminology Program at the University of Alberta.

RICHARD V. ERICSON is a professor with the Centre of Criminology at the University of Toronto.

The Green College Thematic Lecture Series provides leading-edge theory and research in new fields of interdisciplinary scholarship. Based on a lecture program and conferences held at Green College, University of British Columbia, each book brings together scholars from several disciplines to achieve a new synthesis in knowledge around an important theme. The series provides a unique opportunity for collarboration between outstanding Canadian scholars and their counterparts inter nationally, as they grapple with the most important issues facing the world today.

PREVIOUSLY PUBLISHED TITLES

Governing Modern Societies, edited by Richard V. Ericson and Nico Stehr (2000)

Risk and Morality, edited by Richard Ericson and Aaron Doyle (2003)

Re-alignments of Belonging: The Shifting Foundations of Modern Nation States, edited by Sima Godfrey and Frank Unger (2004)

Love, Hate and Fear in Canada's Cold War, edited by Richard Cavell (2004)

Multiple Lenses, Multiple Imaages: Perspectives on the Child Across Time, Space, and Disciplines, edited by Hillel Goelman, Sheila K. Marshall, and Sally Ross (2004)

The New Politics of Surveillance and Visibility, edited by Kevin D. Haggerty and Richard Ericson (2005)

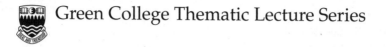
Green College Thematic Lecture Series

The New Politics of Surveillance and Visibility

Edited by
Kevin D. Haggerty and Richard V. Ericson

UNIVERSITY OF TORONTO PRESS
Toronto Buffalo London

© University of Toronto Press Incorporated 2006
Toronto Buffalo London
Printed in Canada

ISBN-13: 978-0-8020-3829-6 (cloth)
ISBN-10: 0-8020-3829-8 (cloth)

ISBN-13: 978-0-8020-4878-3 (paper)
ISBN-10: 0-8020-4878-1 (paper)

Printed on acid-free paper

Library and Archives Canada Cataloguing in Publication

The new politics of surveillance and visibility / edited by Kevin D. Haggerty
and Richard V. Ericson.

(Green College thematic lecture series)
ISBN-13: 978-0-8020-3829-6 (bound)
ISBN-13: 978-0-8020-4878-3 (pbk.)
ISBN-10: 0-8020-3829-8 (bound.)
ISBN-10: 0-8020-4878-1 (pbk.)

1. Electronic surveillance – Social aspects.   2. Privacy, Right of.   3. Social
control.   I. Haggerty, Kevin D.   II. Ericson, Richard V., 1948–   III. Series.

TK7882.E2N49 2006   303.3'3   C2005-906182-0

University of Toronto Press acknowledges the financial assistance to
its publishing program of the Canada Council for the Arts and the
Ontario Arts Council.

University of Toronto Press acknowledges the financial support for
its publishing activities of the Government of Canada through the
Book Publishing Industry Development Program (BPIDP).

University of Toronto Press acknowledges the financial support of the
University of Alberta for the publication of this volume.

Contents

Acknowledgments

This book arose out of a conference entitled 'The New Politics of Surveillance and Visibility' that was held at Green College, University of British Columbia, in May 2003. We are grateful to Green College, the University of Alberta, and the Social Sciences and Humanities Research Council of Canada for their generous financial support of the conference and publication. We would also like to thank Carolyn Anderson and Grace Dene of Green College for their help in organizing the conference, Rebecca Morrison for her excellent research assistance, and Allyson May for her skilful copy-editing. As always, Virgil Duff of the University of Toronto Press has been a most supportive and helpful sponsoring editor.

Kevin D. Haggerty and Richard V. Ericson

THE NEW POLITICS OF SURVEILLANCE AND VISIBILITY

1 The New Politics of Surveillance and Visibility

KEVIN D. HAGGERTY AND RICHARD V. ERICSON

Surveillance raises some of the most prominent social and political questions of our age. This volume brings together leading scholars of surveillance from the fields of political science, communications, media studies, anthropology, science studies, war studies, law, cultural studies, sociology, and criminology. Their contributions, written specifically for this volume, advance our theoretical understanding of surveillance through grounded investigations into its political dimensions and consequences.

Surveillance involves the collection and analysis of information about populations in order to govern their activities. This broad definition advances discussion about surveillance beyond the usual fixation on cameras and undercover operatives. While spies and cameras are important, they are only two manifestations of a much larger phenomenon.

The terrorist attacks of September 11, 2001 (hereafter 9/11) now inevitably shape any discussion of surveillance (Lyon 2003). While those events intensified anti-terrorist monitoring regimes, surveillance against terrorism is only one use of monitoring systems. Surveillance is now a general tool used to accomplish any number of institutional goals. The proliferation of surveillance in myriad contexts of everyday life suggests the need to examine the political consequences of such developments.

Rather than seek a single factor that is driving the expansion of surveillance, or detail one overriding political implication of such developments, the volume is concerned with demonstrating both the multiplicity of influences on surveillance and the complexity of the political implications of these developments. Contributors to this volume are concerned with the broad social remit of surveillance – as a tool of governance in military conflict, health, commerce, security and entertainment – and the new political responses it engenders.

Surveillance

Surveillance is a feature of modernity. The expansion of administrative surveillance was a key factor in the rise of the nation state (Giddens 1987). Surveillance was integral to the development of disciplinary power, modern subjectivities, and technologies of governance (Foucault 1977; 1991). Philosopher Gilles Deleuze, accentuating new technological developments, views surveillance as fundamental to a new order of global capitalism, which he terms the 'society of control' (1992). No single factor has caused this expansion of surveillance. It has not proliferated simply because it renders the state, capital, or power more effective. Rather, surveillance has been made to cohere with any number of institutional agendas, including rational governance, risk management, scientific progress, and military conquest.

Technological developments, while not determinative, have been profoundly important in the rise of new forms of surveillance. Computerization in particular has allowed for the routine processing and analysis of masses of electronic information. Computerized 'dataveillance' (Garfinkel 2000) facilitates integration of surveillance capabilities across institutions and technologies.

Surveillance technologies do not monitor people *qua* individuals, but instead operate through processes of disassembling and reassembling. People are broken down into a series of discrete informational flows which are stabilized and captured according to pre-established classificatory criteria. They are then transported to centralized locations to be reassembled and combined in ways that serve institutional agendas. Cumulatively, such information constitutes our 'data double,' our virtual/informational profiles that circulate in various computers and contexts of practical application.

The concept of 'surveillant assemblage' (Haggerty and Ericson 2000) points to the disconnected and semi-coordinated character of surveillance. No single Orwellian Big Brother oversees this massive monitory effort. Indeed, specific surveillance regimes typically include efforts to combine and coordinate different monitoring systems that have diverse capabilities and purposes. Part of the power of surveillance derives from the ability of institutional actors to integrate, combine, and coordinate various systems and components. Hence, while powerful institutions do not control the entire spectrum of surveillance, they are nonetheless relatively hegemonic in the surveillant assemblage to the extent that they can harness the surveillance efforts of otherwise disparate technologies and organizations.

Closed-circuit television (CCTV) is a telling example of the assemblage qualities of surveillance. While CCTV is often referred to as a single entity, it is actually comprised of multiple agendas and artifacts. Video cameras are the most rudimentary component of this assemblage. Cameras can be augmented by tilt and zoom capabilities, and can integrate digital recording devices to allow for post facto reconstructions. Some systems employ microphones to allow audio monitoring of distant locations. CCTV systems are occasionally enhanced by biometrics so that facial profiles can be compared to those stored on criminal and terrorist databases. Researchers are currently investigating the possibility of augmenting CCTV technologies with artificial intelligence that would allow the system itself to predict future criminal acts and automatically initiate a response to such algorithmically determined suspicious behaviour. The databases integrated into such systems will undoubtedly change and expand over time in accordance with perceptions of suitable enemies (Christie 1986). Moreover, CCTV can itself become integrated into a still larger human and technological assemblage designed to serve any number of purposes.

The proliferation of social visibility means that more people from more walks of life are now monitored. A comparison with Renaissance Europe is instructive. At that time populations knew almost nothing about the appearance or habits of their sovereign. This lack of familiarity allowed for the common game of dressing other members of court society in the sovereign's wardrobe, and then trying to determine if visiting dignitaries could identify the true monarch (Groebner 2001). Playing the same game today would be impossible. The mass media have made all members of society familiar with the appearance, personality, and foibles of our most powerful leaders. Mathiesen (1987) characterizes this as synopticism – the ability of the many to watch the few. Enhanced synoptic abilities are a direct result of the rise of the mass media (Meyrowitz 1985). The mass media have also helped foster a culture of celebrity where fame, or even notoriety, have become valuable in their own right. Increasing numbers of individuals seem eager to expose intimate details of their private lives. Evidence of this can be seen in the development of online diaries, reality TV, and web cameras in bedrooms, all of which speak to a reconfiguration of traditional notions of privacy.

Hierarchies of visibility are being levelled, as people from all social backgrounds are now under surveillance (Nock 1993). While surveillance has not eliminated social inequalities, certain groups no longer stand outside the practice of routine monitoring. Individuals at every location in the social hierarchy are now scrutinized, but at each level this monitoring is accomplished by different institutions with the aid of different technologies and for quite unique purposes. As a result, it is now possible to contemplate a group's characteristic surveillance profile, that is, the surveillance network in which it participates and that helps to define it.

While many commentators on surveillance are motivated by the political implications of these new developments in social visibility, there are myriad ways in which the 'politics of surveillance' can be understood. To date there has been no systemic effort to try and detail and conceptualize the different political axes of surveillance. This chapter fills this lacuna. As we set out below, the politics of surveillance includes both contestation over particular tactics and technologies of surveillance as well as the wide-ranging social consequences of monitoring practices. We distinguish between: 1) the unintended consequences of surveillance, 2) the stakeholder politics of surveillance, and 3) the politics of resistance. In practice, these three dimensions are usually related.

Unintended Consequences

The introduction of new surveillance technologies and regimes can alter social structures, practices, and opportunities in ways not envisioned by their advocates (Tenner 1996; Winner 1986). The prospect that new surveillance devices will evolve into tools of totalitarian control is perhaps the most familiar example of the unintended consequences of surveillance. More generally, analysts focus on *negative* unintended consequences, for example, how surveillance underpins regimes that erode privacy rights, create new forms of inequality, and lack mechanisms of accountability.

The expansion of audit as a tool of neo-liberal governance (Power 1997) illustrates how unintended consequences can arise from routine practices. Originally focused on financial criteria, auditing now encompasses various efforts to render institutions more transparent and accountable. This quest for visibility through surveillance has come at a cost. Auditing disproportionately values criteria that are amenable to being audited, often to the detriment of other outcomes that are less easy to measure. For example, crime rates for the police and standardized test scores in education are prominent auditing criteria that are only loosely connected with the diverse goals and accomplishments of these institutions. Auditing criteria can also distort organizational mandates, as the phenomena being measured is maximized at the expense of other desirable ends. Hence, the public political culture fixation on crime rates has encouraged police officials to subtly modify how they record events in hopes of positively influencing the final statistics. Educators worry that an undue emphasis on standardized test scores tempts teachers to 'teach the test' and neglect equally important but less scrutinized educational accomplishments.

The introduction of obstetric foetal monitoring technologies provides a quite different example of unintended consequences. These technologies make it easier for physicians to detect foetal anomalies by using sound waves to produce a foetal image. These images portray the foetus as a largely 'free floating' entity, a form of representation that exacerbates the tendency to disassociate the foetus from the mother. Foetal ultrasound has consequently contributed to the rise of 'foetal subjects' or 'foetal patients' with interests understood to be independent of, and occasionally at odds with, those of the mother. When deployed in the context of gendered cultural valorizations, foetal imaging technologies have also become an important part of a process whereby a disproportionate number of female foetuses are aborted (Mitchell 2001).

Stakeholder Politics

The stakeholder politics of surveillance involves efforts to influence the volume or configuration of surveillance. Some of the stakeholders in surveillance politics include elected officials, institutional representatives, privacy advocates, police, military personnel, and grassroots activists. The expansion of surveillance suggests that such claims-makers have been highly successful. At the same time, surveillance also proliferates through the everyday practices of institutions seeking security through visibility.

Oppositional groups – including ad hoc collations against particular surveillance initiatives as well as more formal privacy organizations –

employ political tactics to try to eliminate or limit surveillance practices. Some embrace media-savvy strategies, such as demonstrations or publicly releasing the readily available personal information of individuals who advocate on behalf of greater surveillance. The artistic community in particular has used various media to raise questions and cautions about the risks of surveillance (Levin, Frohne, and Weibel 2000). Troupes of activist 'surveillance camera players' now engage in public protest 'performances' before the very CCTV cameras that they oppose.

One cannot determine the stance that stakeholders take towards surveillance by simply positioning activists on the left or right of the political spectrum. Although it often appears that liberals oppose surveillance while conservatives are more predisposed to embrace it, the political demarcation is not that straightforward. Liberals occasionally champion greater surveillance, as is apparent in the demands for greater transparency of major social institutions such as the police, the media, the military, and corporations. At the same time, conservatives and libertarians often oppose new surveillance measures on the grounds that they increase state power and inhibit markets. Hence, organizations do not tend to have consistent political positions towards surveillance, concentrating instead on the specific governmental program with which a particular surveillance initiative is aligned.

The following sections detail some of the main discourses and attributes of the stakeholder politics of surveillance. We concentrate on the issues of privacy, effectiveness, technology, identity, error, and function creep. Cumulatively, these six dimensions provide an appreciation for the complexity of the politics of surveillance and the challenges posed to anti-surveillance strategies and discourses.

Privacy as a Legal Claim

The right to privacy is now the dominant legal and public discourse positioned against the proliferation of surveillance. That said, many advocates are quite pessimistic about the potential for privacy rights to effectively forestall the encroachment of surveillance. In a public lecture John Gilliom once compared the repeated turn to privacy rights as being akin to a pathological tendency to support an amiable and inoffensive elderly politician simply because we cannot conceive of an alternative – despite the fact that he chronically loses elections. David Brin is similarly resigned, suggesting that 'No matter how many laws are passed, it will prove quite impossible to legislate away the new surveillance tools and databases. They are here to stay' (1998: 9).

Privacy rights have helped modulate the excesses of surveillance, but numerous factors curtail their prospects for wider success. Part of the reason for the pessimism about privacy rights is that they are plagued by many of the problems of legal redress and rights discourse more generally (Bartholomew and Hunt 1990; Gotlieb 1996). Making a privacy claim entails translating subjective perceptions of violation into legal claims and categories. That in itself can be a difficult and frustrating exercise, as the very notion of privacy can seem alien to the people who arguably have the greatest need for such a right (Gilliom 2001). Legal privacy protections tend to be more circumscribed than the subjective experience of violation associated with new forms of surveillance. Where privacy rights have traditionally concentrated on the moment when information is acquired, citizens today seem increasingly anxious about how their personal data are combined and integrated with other pieces of information, and how they are then used. Hence, individual perceptions of the violations of surveillance are often inextricably linked with subjective assessments of the program of governance with which that surveillance is associated. Translating this sense of subjective violation into a legal privacy claim is very difficult, especially given the legal tendency to avoid embracing subjective notions of victimization.[1]

Several other factors limit the ability of legal privacy rights to constrain the drift towards ever greater surveillance. First, legal privacy rights are primarily focused on the state. Given the greater prominence of non-state agencies in conducting surveillance, this is a serious limitation. While some businesses have instituted 'fair information practices,' these are not widespread and have a limited record of success. Second, contestations over privacy rights usually pit individuals or small civil libertarian groups against well-funded institutions. Activists must consequently be selective about which battles they choose to fight, as legal remedies are expensive and notoriously slow. Not every potentially problematic surveillance development can be litigated. Finally, privacy protections have always existed uncomfortably alongside the recognition that officials often turn to informal networks to acquire

---

1 The rise of hate crimes legislation and sexual harassment provisions are notable examples of instances when the law is more willing to embrace subjective evaluations of criminal motivation and victimization.

ostensibly private information when they believe it is sufficiently important to do so.

Discussions about privacy rights often proceed as if privacy is itself a stable phenomena that must be protected from incursions or erosion. Such a conceptualization tends to downplay the historical variability and political contestation associated with the precise content of 'privacy.' Claims to privacy and secrecy are political efforts to restrict the ability of others to see or know specific things. One of the more intriguing developments in this regard concerns how powerful interests are now appealing to such rights. The synoptic capabilities of contemporary surveillance have produced a greater number of powerful individuals and institutions with an interest in avoiding new forms of scrutiny. As such, privacy rights that were originally envisioned as a means for individuals to secure a personal space free from state scrutiny are being reconfigured by corporate and state interests.

Corporations routinely appeal to legal privacy and secrecy protections. One of the more ironic of these involves efforts by firms that conduct massive commercial dataveillance to restrict the release of the market segment profiles that they derive from such information on the grounds that these are 'trade secrets.' The ongoing war on terror accentuates how the state is also concerned with carving out a sphere of privacy, even as it tries to render the actions of others more transparent. For example, the U.S. *Patriot Act* prohibits Internet service providers from disclosing the extent to which they have established governmental Internet monitoring measures. The U.S. administration has refused to meet Congressional demands for information regarding the implementation of the *Patriot Act*. The Pentagon regularly invokes claims to 'national security' to restrict public awareness of military matters, including the capabilities of its surveillance technologies. Hence, legal claims to privacy are being invoked as a means to render the actions of powerful interests more opaque at the same time that these same institutions are making the lives of others more transparent. Some see this trend towards non-reciprocal visibility as one of the greatest inequities in contemporary surveillance (Brin 1998).

It seems increasingly difficult for corporations and governments to maintain important secrets for any length of time in democratic societies. However, efforts to keep vital bits of information private are usually not designed to demarcate a sphere that will be hidden for all time, but are instead battles for time. The passage of time provides politicians with greater opportunities to chart new courses of action, distance them-

selves from unpalatable practices, and fashion a response to the inevitable release of state secrets. Businesses use time to capitalize on incre-
mental additions to knowledge before other companies become aware of
the 'secrets' of their corporate success. Hence, appeals to privacy by
powerful interests can ultimately be concerned with gaining time.

Privacy as a Social Fact

Where legal privacy rights are part of public contestations about surveillance, the social fact of privacy has been an important historical
precondition and impetus for the rise of new types of surveillance.

Our contemporary understandings of private spaces and selves can
be traced to some of the historical changes accelerated by the Industrial
Revolution, particularly the migration of rural populations into urban
centres. Freed from the familial and community networks that had
previously established their identities and reputations, the newly un-
recognizable masses exacerbated social problems associated with anonymity. New forms of credentialization, signification, and regulation
were developed to try to establish trustworthy identities (Groebner
2001; Nock 1993; Torpey 2000). Hence, some of the earliest bureaucratic
surveillance practices were a response to the social fact of privacy,
which was itself related to larger transformations in living arrangements, architectural forms, and cultural sensibilities.

Some scholars continue to highlight the problems of privacy, particularly the degree to which privacy can inhibit the development of desirable social policies (Etzioni 1999; Nock 1993). Others take a diametrically
opposite stance, pointing to the rise in surveillance to argue that privacy has been dangerously eroded (Garfinkel 2000; Whitaker 1999).
The contrast between these two positions can be reduced to an argument over whether we now have more or less privacy than in the past.
Unfortunately, framed in such a manner, the debate has become stag-
nant, largely because it is embedded in an increasingly outdated understanding of privacy. Each side of this dispute conceives of privacy in
quantitative terms; privacy is a thing of which we have more or less.
Such a formulation overlooks the vital point that new surveillance
practices have produced important qualitative changes in the experience of privacy. Privacy invasions now often *feel* different than they did
in the past. Generalizing broadly, where surveillance historically tended
to concentrate on monitoring discrete persons, today it often monitors
individuals as members of larger populations, disassembling them into

flows of information, chemicals, DNA, and so on. In the process, the subjective experience and perception of intrusion undergoes a qualitative transformation that remains difficult to articulate.

Discussions about privacy are also restricted by their reliance on spatial military metaphors. Hence, privacy is 'invaded,' 'intruded upon,' 'breached,' and 'violated.' While such metaphors have some rhetorical force, the fixation on images of 'invasion' draws attention away from one of the most important dynamics of contemporary privacy. In our day-to-day lives, privacy is not routinely 'invaded': it is not pried away from a resistant and apoplectic public. Instead, privacy is compromised by measured efforts to position individuals in contexts where they are apt to exchange various bits of personal data for a host of perks, efficiencies, and other benefits. Part of the ongoing politics of surveillance therefore does not involve efforts to 'capture' data, but to establish inducements and enticements at the precise threshold where individuals will willingly surrender their information. Surveillance becomes the cost of engaging in any number of desirable behaviours or participating in the institutions that make modern life possible.

The suggestion that surveillance erodes privacy implies that information that was once private has become widely available to the public. Again, this opposition between public and private does not do justice to the complexities of contemporary surveillance. New surveillance regimes often do not publicize information, but subject it to a form of re-privatization. As information about citizens becomes the property of the state or corporations, these institutions often jealously restrict access to such information. Indeed, institutional reluctance to share information is one of the great tensions in the surveillant assemblage, often complicating or inhibiting the integration of surveillance systems.

Effectiveness

New surveillance tools are not necessarily embraced because of their demonstrated effectiveness. The massive investment in CCTV cameras in the United Kingdom is a prominent example of a monitoring technology that was deployed long before any serious evaluation of its promised effects was conducted. Concerns about the 'effectiveness' of a surveillance technology involve a form of technical rational discourse that exists alongside more symbolic and emotional factors that contribute to the adoption of such tools.

When surveillance measures are opposed, the question of effectiveness quickly comes to the forefront. Opponents regularly claim that a proposed technology will not work, that it will not effectively identify criminals, consumers, or incoming missiles. Notwithstanding legitimate concerns about the effectiveness of many surveillance measures, addressing this issue is severely limited as a political strategy.

Technologies or practices are not 'effective' by virtue of some straightforward appeal to science. Instead, a determination of effectiveness is the net outcome of often contentious political struggles, including symbolic politics. Science is routinely invoked in such exchanges, often in a highly rhetorical fashion, with debates over effectiveness being predisposed to descend into battles between the experts. Given that such exchanges tend to involve contestations over statistics, methodology, and the interpretation of data, the public often has difficulty making independent sense of such arguments. This situation can benefit advocates of new monitoring practices, as scientific knowledge and technical expertise tend to be disproportionately aligned with such groups – not because they necessarily have truth on their side, but because the advocates for surveillance are often, but not exclusively, powerful economic and political interests that can afford to sponsor such studies. Opponents are often left scrambling to develop alternative interpretations of data produced according to the definitions, methodologies, and timelines of their adversaries.

Even when a particular technology is deemed to be ineffective, the matter seldom ends there. Instead, surveillance advocates often then return to their workshops, redoubling their efforts to ensure that their surveillance tool meets the new standard of effectiveness. The irony is that in order to meet the new criteria for 'effectiveness' the systems often have to become even more intrusive.

Technology

Soon after 9/11 the Bush administration proposed the ill-fated 'Operation TIPS' program which would have transformed thousands of government employees into de facto state informants. Russia is reestablishing a network of neighbourhood spies akin to the system used extensively by the KGB during the Soviet era (MacKinon 2003). These developments remind us that surveillance need not involve advanced technologies. Nevertheless, some of the most distinctive attributes of contemporary surveillance derive from their technological abilities to see more, at greater distances and in real time.

As commodities, surveillance devices are often supported by highly motivated companies that stand to reap a financial windfall if they are adopted. As a result many surveillance devices often resemble solutions in search of a problem.

Individual and institutional consumers can also be seduced by the commodity form of surveillance products (Haggerty 2003). Technological solutions have a strong cultural allure in Western societies and high technology has emerged as an important marker of progress and organizational reputation. Institutions can, therefore, be tempted to embrace technological solutions to address problems that might be better served through more mundane initiatives (Levy 2001). The recurrent embrace of a technological fix ignores how larger power structures tend to shape the actual content and application of technologies such that the liberatory potential of new technologies is regularly blunted or undermined. Moreover, any demonstrated successes of such techno-fixes in thwarting surveillance efforts usually marks the beginning of official efforts to criminalize, overwhelm, or bypass these new technologies.

Identity

The politics of identity as it pertains to surveillance has at least two dimensions. The first concerns the monitoring of pre-constituted social groupings; the second involves establishing new forms of identity. Both can severely alter opportunity structures and transform individual lives.

Although surveillance is now directed at all social groups, not everyone is monitored in the same way or for the same purposes. Different populations are subjected to different levels of scrutiny according to the logic of particular systems. Among the more contentious population categories used by surveillance systems are racial or ethnic identities.

It has been decades since J.H. Griffin's (1960) *Black Like Me* famously conveyed to a wider white American audience the routine discrimination and indignities directed at blacks in the American South. A cornerstone of the racialized system of domination Griffin documented was the use of visual scrutiny, particularly the 'hate stare' focused on blacks by southern whites as a means of intimidation. Such face-to-face intimidation continues, but in the intervening decades the monitoring of minority populations has become more technological and bureaucratic. As Fiske (1998) points out, to be young, male, and black in America is to be frequently reminded that you are being monitored as a potential criminal threat. In their study of CCTV in Britain, Norris and Armstrong

(1999) demonstrate how the racial stereotypes of camera operators result in young black men being subject to disproportionate and unwarranted levels of scrutiny.

Ongoing battles over 'racial profiling' have become a lightning rod for political concerns about institutional racism. Profiling increases the indignities and suspicions associated with intensive bureaucratic monitoring, subjecting certain individuals to longer delays, more (and more extensive) searches, and greater scrutiny of official documentation due to their perceived racial or ethnic identity. Profiling is a form of categorical suspicion that connects the operational requirements of surveillance systems with historical legacies of racism. The sheer volume of data produced by contemporary surveillance systems necessitates the development of techniques to single out particular transactions, incidents, or individuals for greater scrutiny. Profiling accomplishes this goal by establishing behavioural and identity markers that by themselves, or in combination, extract particular people for greater monitoring. The inclusion of racial variables on these profiles now means that being perceived to be black or Hispanic positions a citizen at greater risk of being perceived to be a drug courier, while appearing Arab or Muslim increases the likelihood of being dealt with as a terrorist.

Monitoring regimes also reproduce sexualized objectifications through the gendered voyeuristic gaze. New technologies allow such objectifications to be more intense and covert. A friend who formerly worked as a security officer in a downtown Toronto office complex recounts how the predominantly male CCTV camera operators passed the time by scrutinizing and judging the women they monitored on their screens, zooming in on legs, breasts, or other body parts that caught their fancy. Particularly titillating incidents were occasionally recorded, with videotapes being traded across shifts much like playing cards. Norris and Armstrong (1999) provide more systematic evidence of comparable voyeuristic monitoring of women by British CCTV operators.

The second major identity-related political dimension of surveillance concerns how surveillance can itself give rise to distinctive forms of subjectivity. Foucault made a comparable point in *Discipline and Punish* where he outlined how hierarchical disciplinary surveillance can fashion new subjectivities. Here we are concerned with a slightly different process related to the operation of the official categories used by monitoring systems. These systems often introduce identity categories that encourage new forms of self-identification. The rise of the census and other official counting exercises had similar effects. Censuses often

introduced entirely novel identity categories that bore little or no relationship to the history or self-identity of the groups being classified.
This was particularly apparent in colonial settings where arbitrary official identity categories subsumed groups of individuals with dramatically different histories and notions of group identity for the convenience
of the colonial administrators (Hawkins 2002). Ian Hacking refers to
this tendency for social groups to identify with such initially arbitrary
institutional categories as the processes of 'making up people' (Hacking
1986).

Struggles continue over the preferred identity categories to be used in
counting efforts. These struggles are often motivated by concerns about
the socio-political implications of employing specific racial, ethnic, and
sexual classifications. However, new surveillance systems also produce
a host of novel identity categories. This process is particularly apparent
with respect to the dataveillance conducted by the advertising industry,
which segments individuals into assorted market niches. Eager to increase the efficacy of their promotional efforts, advertisers want to
direct their promotions to prized market segments. Therefore they use
dataveillance to carve the population into a series of discrete groups
such as 'urban cowboys' or 'starter families.' Group membership is
determined by the degree to which these individuals share comparable
consumption patterns (Turow 1997). Where one is positioned in such
groupings increasingly structures the types of communications and
commercial opportunities a person receives. As the public becomes
increasingly attuned to such practices we are apt to see greater political
debate about the inequities of such practices as well as strategic efforts
to maximize where one is positioned within the new constellation of
market segments.

Error

If an important attribute of politics concerns how life opportunities are
structured, then we must be concerned with the play of error in surveillance systems. New forms of surveillance provide greater opportunities
for serious mistakes and establish more formalized error systems.

As surveillance and dataveillance are integrated into institutional
decision making, there is an attendant increase in the prospect for the
routine production of mistakes. Dataveillance relies on conscientious
and accurate data input by a widely dispersed and uncoordinated
network of individuals with varying degrees of commitment to this

task. Each keystroke contains possibilities for errors, some of which can have monumental consequences. Consumer advocates estimate that as many as 70 per cent of personal credit reports contain at least one error. Given that credit reports are only one database – or, more accurately, an assemblage of other databases – we can gain an appreciation of the sheer scope for potential errors produced by society-wide dataveillance. Simply verifying our personal information on all of the potentially relevant systems would be a monumental task, one that would be complicated by the fact that many of these databases have been structured to restrict our knowledge of their contents or even their existence.

The popular media are replete with horror stories about how data entry errors have produced disastrous consequences. Data errors in medical institutions have resulted in serious illness and even death. In his contribution to this volume (chapter 15), Gandy points out how a series of data entry errors by the Florida electoral boards may have helped to determine the outcome of the 2000 Presidential election. Security systems also produce a considerable volume of errors, although not all are as dramatic as the case of the two Welsh football fans deported from Belgium after being erroneously entered on a police 'hooligans' database. No longer able to travel in Europe, they had to undertake a six-year struggle simply to have their names removed from the Belgian database (Norris and Armstrong 1999: 221). This is an extreme but telling instance of what Brodeur and Leman-Langois (chapter 7 this volume) characterize as the inevitable rise of 'Big Bungler' as an attribute of the ascendancy of dataveillance.

The second type of surveillance error involves organizational decisions about when automated surveillance systems will subject individuals to greater scrutiny. Such systems employ algorithms that combine and analyse reams of data. Data thresholds are established such that if the right number or combination of variables emerges, the person or behaviour being analysed is directed to a human operative. Such tools are used in health, business, and social service, but are most familiar in the context of anti-terrorism surveillance. Biometric facial scanning systems, for example, map the human face using a series of discrete points. Authorities are alerted only when a face matching a predetermined number and combination of such points is identified.

None of these systems, however, establishes a perfect match. Officials must strike a compromise between a tolerable level of Type I versus Type II errors. If the system is structured towards producing a greater volume of Type I errors, then comparatively fewer combinations of data will be required to flag an individual for greater scrutiny. Such systems are designed to have a broad remit, extracting many individuals, most of whom will not belong to the class of individuals that is of concern. Alternatively, if the threshold is positioned at a level requiring a much larger number of data matches before individual cases are extracted, there is the greater likelihood that it will miss the occurrences that the system tries to capture. This is the systematic production of Type II errors. Where such lines are drawn in practice is a pragmatic political and organizational decision related to a series of social factors, most commonly the level of personnel available to supplement the surveillance technology.

Both random and structured surveillance errors have political implications for life opportunities, security, health, and the degree of routine indignities that citizens must endure. As yet, however, these issues are not widely politicized.

Function Creep

Critics tend to approach surveillance as a series of discrete laws or technologies. This often involves fixating on the moment when a new technology or piece of legislation is being introduced, in an effort to block or modulate such practices. Focusing on single-issue surveillance developments tends to miss how surveillance capacity is intensifying as a result of the combination and integration of systems. Surveillance power is expanding exponentially, with each new development potentially multiplying the existing surveillance capacity.

One of the most important dynamics shaping surveillance involves the process of 'function creep' (Innes 2001), whereby devices and laws justified for one purpose find new applications not originally part of their mandate. For example, cars now come factory-equipped with electronic black boxes designed as a safety feature to deploy airbags during a crash. The fact that these black boxes also record a car's speed, and whether the driver was wearing a seatbelt, has expanded the uses of such devices, with the police and lawyers now employing them for criminal investigations and civil litigation.

A blatant recent example of function creep has been the rapid development and expansion of DNA databases. Although the specifics of how these systems have evolved differ by nation and state, the general tendency is for DNA databases to be initially justified as a means to monitor a very small and particularly notorious set of criminals, usually convicted paedophiles or sex offenders. Very quickly, institutional players recognized the convenience and political expedience of recording the DNA of individuals convicted of a host of other crimes. Systems have consequently expanded to collect the DNA of larger classes of accused individuals. In some locales DNA can be collected from mere suspects, a potentially enormous group of people! Other jurisdictions have proposed using DNA collected for military or criminal justice identification purposes for pure research (Sankar 2001: 287). Politicians occasionally advocate the collection of DNA of every citizen for ill-defined health or security uses, as has occurred in Iceland with the support of pharmaceutical companies. At work here is a very familiar process whereby legal restrictions are loosened and political promises are ignored as new uses are envisioned for existing surveillance systems.

Arguably, function creep is one of the most important operational dynamics of contemporary surveillance. It is also notoriously difficult to transform into a coherent and successful stakeholder politics. Function creep tends to operate in a localized ad hoc and opportunistic fashion. New tools create a new environment of monitoring possibilities that were perhaps unanticipated by the original proponents of the system. Even activists can have difficulty imagining in advance precisely how a system will evolve. The expansion and combination of surveillance systems also tends to occur through unpublicized bureaucratic reforms that take place behind the scenes of public life. As such, they lack the moment of disclosure that helps to concentrate public interest. Finally, activists who do draw attention to the possibilities of systemic expansion meet with reassuring public promises that a new system will only be used for a limited set of pre-established and legally circumscribed purposes. The fact that such legal restrictions tend to be regularly relaxed, and political promises ignored, as new opportunities for surveillance efficiencies are discerned is a point that is often difficult to convey to the public.

Resistance

Resisting surveillance involves localized efforts to get by in the face of monitoring: to thwart a particular system, to live anonymously within its gaze, or to engage in any number of misdirection ploys. Critical scholars have eagerly documented some of the specifics of such resistance efforts (De Certeau 1984; Gilliom 2001; Moore and Haggerty 2001). Gary Marx (2003) identifies eleven general strategies whereby surveillance is resisted or subverted: switching, distorting, blocking, piggy-backing, discovery, avoidance, refusal, masking, breaking, co-operation, and counter-surveillance. All such efforts are based in an assumed familiarity with the protocols and optics of a particular sur-veillance system, knowledge that is used to defeat, deceive, or subvert the system.

The sheer ingenuity of many of these measures makes resistance an intrinsically appealing topic. Analysts identify with the resourcefulness of individuals who purchase clean urine to pass drug tests, or laugh at the black humour of individuals who drink typewriter correction fluid under the misguided assumption that it will help them pass a lie detector test. That said, we also risk romanticizing such behaviour, misrepresenting its political importance and turning a necessity into a virtue.

The criteria for evaluating the successes of resistance measures tend to be very local and individual. The victories of resistance are usually immediate and case-specific. In his contribution to this volume (chapter 5) John Gilliom observes how the introduction of photo radar cameras in Hawaii prompted drivers to smear mud on their licence plates and use special plastic plate-covers to block the cameras. To the extent that such efforts allow individuals to avoid receiving a ticket they are ex-amples of 'successful' resistance. Whether we applaud these successes will depend on whether we interpret them as a means to avoid unjusti-fiable intrusions, or as the acts of scofflaws dedicated to ensuring that they can drive dangerously.

Resistance is typically not motivated by a desire to eliminate or modify systems, but to evade their grasp. As such, it usually leaves the surveillance system intact, although resistance can become so wide-spread that specific surveillance initiatives are withdrawn. The fact that the Hawaiian authorities eliminated the photo radar system in the face of public resistance can be seen as a 'victory.' However, victory celebra-tions are sometimes tempered by a recognition that such resistance can ultimately and paradoxically culminate in more intrusive surveillance. As authorities become aware of the specifics of resistance they tend to modify technologies, laws, and bureaucratic procedures accordingly, often resulting in a cumulative escalation in surveillance. It is entirely likely that if the Hawaiian officials had decided to retain the photo-radar system, or if it is re-introduced, they would follow the lead of most jurisdictions and enforce existing laws that prohibit blocking li-cence plates and criminalize products that obstruct the cameras. Such actions would be examples of the general tendency for authorities to search out legal and technological resources to help make the world safe for the optics of surveillance systems (Ericson 1994).

While some of these new laws would undoubtedly be enforced, it is unlikely that such enforcement would occur in any coherent or consistent fashion. Instead, the legal frameworks that enable surveillance systems contribute to an environment that allows officials to draw selectively on regulations as they see fit in order to secure any number of desired outcomes. The net result of this enabling legal environment is further entrenchment of what many experience as arbitrary and capricious forms of governance. Hence an important political dimension of resistance concerns how it can foster a dynamic back and forth of evasion and official response that tends to ratchet up the overall level of surveillance and control.

Rather than glorify resistance to surveillance as a grassroots 'revolt against the gaze,' these practices are better appreciated as a predictable and telling consequence of the rise of surveillance. Greater familiarity with various resistance techniques is an inevitable by-product of the increased level of social monitoring. To the extent that surveillance is perceived to be unjust or stands in the way of desirable ends, more individuals will likely find themselves resisting surveillance in new and innovative ways.

Conclusion

Having reviewed some of the key dimensions of the politics of surveillance, we conclude with the question of whether the analytical category of 'surveillance' remains adequate to capture the wide range of technologies and practices that operate in its name.

All analytic categories are simplifying devices that suppress a world of specific differences under a broader rubric. Categorization entails an implicit claim that, irrespective of individual variations, the instances included under the category are sufficiently alike according to some set of criteria that they warrant being considered as part of the same grouping. In the case of the analytic category of 'surveillance,' however, the boundaries of that classification now risk being stretched beyond all recognition.

There is broad agreement that surveillance has expanded and in the process become more multifaceted and chaotic. Surveillance is accomplished by satellites, statistics, sensors, and spies. It has become a generalized tool used for some of the most socially laudable as well as condemnable ends. This multiplication of the aims, agendas, institutions, objects, and agents of surveillance has made it profoundly difficult to say anything about surveillance that is generally true across all, or even most, instances.

Nonetheless, we continue to encounter broad generalizations about 'surveillance.' Closer scrutiny reveals that such generalizations tend to be more narrowly concerned with surveillance as an instrument of enforced conformity. While this is undeniably a prominent aim of surveillance, it is only one of many. The political, social, and normative issues raised by surveillance as an instrument of enforced conformity often make little sense when applied to the broader spectrum of surveillance. One only need contemplate the very different issues raised by surveillance in the service of anti-terrorism efforts and surveillance as a key component of scientific inquiry to recognize this point. While surveillance is now ubiquitous, it is also diverse, multi-faceted, and employed in such a panoply of projects that it is almost impossible to speak coherently about 'surveillance' more generally.

Analysts therefore seem to be faced with one of two possibilities. They can continue to try to articulate statements about surveillance that are generally true across all, or most, instances. This articulation can only be accomplished by moving to ever higher levels of abstraction. Alternatively, they can adopt a stance towards surveillance that mirrors recent approaches to the conceptualization of power more generally. Following Foucault (1977; 1978), power is recognized as being neither inherently good nor evil, but as a tool to accomplish goals that meet particular interests. As such, analysts are encouraged to attend to how power is exercised, over whom, and towards what ends. A comparable approach towards surveillance would acknowledge that the proliferation of surveillance now means that it has few inherent attributes or universal forms. Consequently, analysts and activists must attend to the local contingencies of how surveillance regimes are coordinated and the ends they serve, focusing evaluations of surveillance at this low or mid-level of analysis.

Neither of these options is entirely satisfactory. Employing continually escalating levels of abstraction risks moving to such an intangible level of discourse that it becomes alienating. The second option of concentrating on the individual manifestations of surveillance and their relation to larger governmental projects has an intuitive appeal. However, in light of our understanding of the assemblage quality and dynamic of contemporary surveillance, focusing on single manifestations of surveillance risks overlooking how discrete tools and projects are combined and integrated into other programs and, in the process, are being transformed into something quite different. Undoubtedly, surveillance studies will continue to undertake generalizing and specifying projects, but both risk distorting our understanding of surveillance and its attendant politics.

REFERENCES

Bartholomew, A., and A. Hunt. 1990. 'What's Wrong with Rights?' *Law and Inequality* 9:1–58.

Brin, D. 1998. *The Transparent Society: Will Technology Force Us to Choose between Privacy and Freedom?* Reading, MA: Perseus Books.

Christie, N. 1986. 'Suitable Enemies.' In H. Bianchi and R. Van Swaaningen, eds., *Abolitionism: Towards a Non-Repressive Approach to Crime*, 42–54. Amsterdam: Free University Press.

De Certeau, M. 1984. '"Making Do": Uses and Tactics.' In *The Practice of Everyday Life*, 29–42. Berkeley: University of California Press.

Deleuze, G. 1992. 'Postscript on the Societies of Control.' *October* 59:3–7.

Ericson, R.V. 1994. 'The Decline of Innocence.' *University of British Columbia Law Review* 28:367–83.

Etzioni, A. 1999. *The Limits of Privacy*. New York: Basic Books.

Fiske, J. 1998. 'Surveilling the City: Whiteness, the Black Man and Democratic Totalitarianism.' *Theory, Culture and Society* 15:67–88.

Foucault, M. 1977. *Discipline and Punish: The Birth of the Prison*. New York: Vintage.

– 1978. *The History of Sexuality, Volume I: An Introduction*. New York: Vintage Books.

– 1991. 'Governmentality.' In G. Burchell, C. Gordon, and P. Miller, eds., *The Foucault Effect: Studies in Governmentality*, 87–104. Chicago: University of Chicago Press.

Garfinkel, S. 2000. *Database Nation: The Death of Privacy in the 21st Century*. Sebastopol: O'Reilly.

Giddens, A. 1987. *The Nation-State and Violence*. Cambridge: Polity.

Gilliom, J. 2001. *Overseers of the Poor: Surveillance, Resistance, and the Limits of Privacy*. Chicago: University of Chicago Press.

Gotlieb, C.C. 1996. 'Privacy: A Concept Whose Time Has Come and Gone.'
In D. Lyon and E. Zureik, eds., *Computers, Surveillance and Privacy*, 156–71.
Minnesota: University of Minneapolis Press.

Griffin, J.H. 1960. *Black Like Me*. Boston: Signet.

Groebner, V. 2001. 'Describing the Person, Reading the Signs in Late Medieval
and Renaissance Europe: Identity Papers, Vested Figures, and the Limits of
Identification, 1400–1600.' In J. Caplan and J. Torpey, eds., *Documenting
Individual Identity: The Development of State Practices in the Modern World*,
15–27. Princeton: Princeton University Press.

Hacking, I. 1986. 'Making Up People.' In T. Heller, M. Sosna, and D. Wellbery,
eds., *Reconstructing Individualism: Autonomy, Individuality and the Self in
Western Thought*, 222–36. Stanford: Stanford University Press.

Haggerty, K.D. 2003. 'From Risk to Precaution: The Rationalities of Personal
Crime Prevention.' In R.V. Ericson and A. Doyle, eds., *Risk and Morality*,
193–214. Toronto: University of Toronto Press.

Haggerty, K.D., and R.V. Ericson. 2000. 'The Surveillant Assemblage.' *British
Journal of Sociology* 51:605–22.

Haggerty, K.D., and A. Gazso. 2005. 'Seeing beyond the Ruins: Surveillance as
a Response to Terrorist Threats.' *Canadian Journal of Sociology* 30:169–87.

Hawkins, S. 2002. *Writing and Colonialism in Northern Ghana*. Toronto: University of Toronto Press.

Innes, M. 2001. 'Control Creep.' *Sociological Research Online* 6.

Levin, T.Y., U. Frohne, and P. Weibel, eds. 2000. *Ctrl [space]: Rhetorics of Surveillance from Bentham to Big Brother*. Cambridge, MA: MIT Press.

Levy, S. 2001. *Crypto: How the Code Rebels Beat the Government – Saving Privacy
in the Digital Age*. New York: Viking.

Lyon, D. 2003. *Surveillance After September 11*. London: Polity.

MacKinnon, M. 2003. 'You're Being Watched by "Big Babushka."' In *Globe and
Mail*, 22 Oct.: A3.

Marx, G. 2003. 'A Tack in the Shoe: Neutralizing and Resisting the New
Surveillance.' *Journal of Social Issues* 59:368–91.

Mathiesen, T. 1987. 'The Eagle and the Sun: On Panoptical Systems and Mass
Media in Modern Society.' In J. Lowman, R. Menzies, and T. Palys, eds.,
Transcarceration: Essays in the Sociology of Social Control, 559–75. Aldershot:
Gower.

Meyrowitz, J. 1985. *No Sense of Place: The Impact of Electronic Media on Social
Behaviour*. New York: Oxford University Press.

Mitchell, L.M. 2001. *Baby's First Picture: Ultrasound and the Politics of Fetal
Subjectivity*. Toronto: University of Toronto Press.

Moore, D., and K.D. Haggerty. 2001. 'Bring It On Home: Home Drug Testing and the Relocation of the War on Drugs.' *Social and Legal Studies* 10:377–95.

Nock, S.L. 1993. *The Costs of Privacy: Surveillance and Reputation in America.* New York: Aldine de Gruyter.

Norris, C., and G. Armstrong. 1999. *The Maximum Surveillance Society: The Rise of CCTV.* Oxford: Berg.

Power, M. 1997. *The Audit Society: Rituals of Verification.* Oxford: Oxford University Press.

Sankar, P. 2001. 'DNA–Typing: Galton's Eugenic Dream Realized?' In J. Caplan and J. Torpey, eds., *Documenting Individual Identity*, 273–90. Princeton: Princeton University Press.

Tenner, E. 1996. *Why Things Bite Back: Technology and the Revenge of Unintended Consequences.* New York: Vintage.

Torpey, J. 2000. *The Invention of the Passport: Surveillance, Citizenship and the State.* Cambridge: Cambridge University Press.

Turow, J. 1997. *Breaking Up America: Advertisers and the New Media Order.* Chicago: University of Chicago Press.

Whitaker, R. 1999. *The End of Privacy: How Total Surveillance is Becoming a Reality.* New York: New Press.

Winner, L. 1986. 'Do Artifacts Have Politics?' In *The Whale and the Reactor: A Search for Limits in the Age of High Technology*, 19–39. Chicago: University of Chicago Press.

PART ONE

Theorizing Surveillance and Visibility

In Part One David Lyon, William Bogard, Gary Marx, and John Gilliom explore some of the more pressing theoretical concerns pertaining to contemporary developments in surveillance. These include the relationship between surveillance and an apparently epochal shift in the nature of power, the popular appeal of surveillance, the prospects for resisting surveillance, and the complex set of factors that help shape our normative evaluation of monitoring practices.

David Lyon highlights the tension inherent in the public's repeated claim that it supports privacy, with the continuing embrace of new surveillance initiatives. He concentrates on the surveillance-related aspects of the official response to the 9/11 terrorist attacks, employing both sociological and psychoanalytic tools to demonstrate how surveillance has become a self-evident response to assorted social risks.

Fear drives the introduction of many new surveillance systems. September 11 demonstrated how specific fears can be exacerbated by the ability of new surveillance devices to produce dramatic visual evidence of horrible events and dangerous individuals. Outside of New York, the public experienced the attacks of 9/11 largely through televised news images. Only some of the total volume of such images were elevated to iconic status after having been filtered through institutional power structures. These images were then presented in the absence of historical context or political motivation. This helped foster the dominant public meanings of the attacks, which were channelled into a political and military response. Some of the most important of these responses have involved new surveillance mechanisms: popular media visualize threats which, in turn, prompt demands for yet more visualizing devices.

The media's ability to broadcast powerful images of rare events to a mass audience requires us to rethink the dominance of the panopticon model in surveillance studies. The Panopticon prison, as introduced by Jeremy Bentham and analysed by Michel Foucault, has come to stand for all systems of visibility that enable a few isolated watchers to scrutinize the behaviour of large groups. Architectural arrangements help maximize the possibility of this visibility while rendering its actual presence indeterminate. When combined with an explicitly articulated set of professional norms, such systems of visibility are understood to be an integral component in the operation of disciplinary power and the fashioning of modern subjectivities.

Lyon draws upon the work of Thomas Mathiesen to suggest that the emphasis on panoptic surveillance must be augmented by a recognition of synoptic forms of surveillance. Synopticism involves the ability of a large group of people to scrutinize the action of a few individuals – as exemplified by the 9/11 terrorist attacks. Synopticism is a function of the contemporary mass media that publicize the detailed actions of specific individuals, especially politicians and entertainment celebrities. The key dynamic is that the many are able to watch and judge the powerful few as seen through the eyes of television.

The prevalence of synopticism also draws attention to the types of enjoyment that people can derive from watching various behaviours and events. Individuals now expect to see an ever expanding range of 'back region' behaviours, a fact that has prompted some analysts to characterize Western nations as 'viewer societies.' To try to explain this culturally powerful practice of 'watching,' Lyon turns to the concept of 'scopophilia' as used in Lacanian psychoanalysis. Scopophilia refers to a love of looking that is characteristic of a particular stage of childhood development vital to our ability to see ourselves as others see us.

The current appeal of surveillance technologies is related to the sco-pophiliac viewer gaze. Viewing is now positioned as an intrinsically pleasant act, and the love of watching has ushered in yet more surveillance and monitoring. Evidence for this can be seen in the increasing public prominence of formerly private actions, combined with the cultural conception that the uninhibited scrutiny of others is acceptable.

William Bogard is also concerned about the applicability of Foucault's formulations on contemporary forms of surveillance. Bogard suggests that we are in the midst of an epochal or 'phase' shift in the nature of power in which technologies of surveillance and simulation are key. This shift is from a disciplinary society to a 'society of control.'

Bogard presents a fundamental reassessment of the continuing relevance of Foucauldian analyses to the contemporary dynamics of surveillance and control. For Foucault, disciplinary power was manifest in a series of bounded spaces such as the prison, school, or factory, where normalizing hierarchical monitoring systems were concentrated. The principle behind Foucault's analysis is therefore one of confinement and the division of territories that serve to cement power relations. Bogard argues that new practices and technologies of simulation, which are themselves situated in a larger context of global capitalism, have freed social control from specific territories or territorial logics. In the process control has become more inclusive. Discipline has reconstituted itself on a more deadly plane, having moved into cyberspace, mutated into simulation, and become hyperreal. Surveillance is no longer organized by centralized hierarchical observation or centralized powers, but is being replaced by more 'smooth' forms of simulated controls. The cumulative effect of this epochal change is that control has become more continuous, and hence more efficient and effective.

Bogard suggests that the society of control must be studied as a complex socio-technical machine that exerts control by decoding and deterritorializing subjectivity. To analyse this process he draws upon Deleuze and Guattari. These philosophers offer insights into how processes of destratification and restratification have become vital to the operation of surveillance systems.

Every territorial assemblage entails a characteristic code which operates like a filtering machine. At the heart of disciplinary regimes were coding practices that operated through a form of binary opposites that separated the normal from the abnormal, truth from falsity, sanity from madness. Power, in such a system, is on the side of the true, the real, and the normal. However, that system of coding is being replaced by a system that operates through the deconstruction of the binary basis of identity and the infinite production of differences. As such, the lines that used to separate various regimes and times of control are dissolving into a single form capable of infinite modulations. Rather than the bounded places of surveillance studied by Foucault, the emphasis today is on virtual realities produced by formal systems of modelling, which operate through digitized pre-programming. These virtual realities are not concerned with deterrence or normative regulation, but with perfecting a pre-emptive model of control where deviance is pre-programmed out of the system before it has the possibility to emerge. It is a model of control that dreams of doing away with the messy necessity of control, of letting the system itself assume control.

Disciplinary surveillance can be read as a form of machinic assemblage defined by its territoriality. Such a machinic assemblage is simultaneously a material arrangement of space and time, a mechanism for observation and ranking, a strategy of control, and an instrument for the distribution of desire and the production of truth. In the new society of control, discipline is destratified in terms of the emphasis on confinement, but restratified within the mode of information and simulation.

The degree to which an individual will find such monitoring intrusive is difficult to predict, and complicated by the sheer variety of forms that surveillance now assumes. In his contribution, Gary Marx suggests that such normative assessments are shaped by at least five factors:

1) inherent characteristics of the means of surveillance;
2) the application of the means;
3) the perceived legitimacy and nature of the goals of surveillance;
4) the structure of the setting in which surveillance is used; and
5) the actual content, kind, or form of data collected.

Marx concentrates on the last factor, how the content or form of data collected can inform our evaluation of the propriety or intrusiveness of surveillance. He presents a fine-grained analysis of the basic materials that surveillance systems might try and collect, accentuating the structural roots of our personal evaluations of surveillance.

Not all information about ourselves is perceived to be equally private. Marx draws attention to gradations in the apparent privacy of information by presenting these in the form of a series of concentric circles. Each step from the outermost circle towards the centre entails a movement towards information that is perceived to be more personal and personally damaging. At the outer edge is individual information which includes any data/category that can be attached to a person. Such data can be acquired from or imputed to a person, and is the least personal. Moving inward, the next circle consists of private information, which is defined by discretionary norms pertaining to its release. Such information is normally not available in the absence of specific circumstances that compel disclosure. The next innermost circle is intimate and/or sensitive information. Accessing such information can be useful to an opponent, and as such it is normally not known. The last two circles are information about 'unique' and 'core' identity. The concept of uniqueness implies that the information only pertains to a specific individual where the final, innermost circle of core identity involves information which tries to answer the question 'who is it?' As such, core information can involve details about personal names, location, parentage, identity papers, biometrics, and ancestry.

Various types of descriptive information can be connected to individuals. Marx suggests that this information can be broken down into:

a) individual identification;
b) shared identification;
c) geographical/locational information;
d) temporal information;
e) networks and relationships;
f) objects;
g) behavioural information;
h) beliefs, attitudes and emotions;
i) measurement characteristics; and
j) media references.

The multidimensional nature of personal information makes it difficult to generalize about how individuals will perceive privacy violations. Nonetheless, Marx proposes that we can discern some consistent patterns in public reactions to such violations, which will be more negative to the extent that the violation combines a series of variables. Hence, the most intrusive manifestation of surveillance would involve information about a core identity, a locatable person, information that is personal, intimate, sensitive, stigmatizing, strategically valuable, expensive, biological, naturalistic, predictive, that reveals deception, is attached to the person, and involves an enduring and unalterable documentary record. While it is unlikely that any specific instance of surveillance would meet all of these criteria, the greater the number of these factors present in any monitoring system, the more likely we are to experience such scrutiny as an intense violation.

Where Bogard sees the prospects for resistance and opposition being pre-emptively eliminated by a system of virtual hypercontrol, Gilliom takes a more pragmatic and cautiously optimistic stance towards the possibilities for resisting surveillance. Resistance occurs through partial and temporary victories against surveillance practices. Research on everyday resistance to surveillance provides insights into both the surveillant assemblage and public consciousness about such surveillance.

Gilliom turns to his research on the relationship of poor Appalachian women to Ohio's welfare system. Ohio introduced the Client Registry Information System – Enhanced, to collect more information about welfare claimants from interconnected databases. Sophisticated data-matching techniques were employed to identify potentially fraudulent claims. Focusing on the women's perceptions, language, and actions, Gilliom explores how a group of closely surveilled people understand the act of being monitored and the types of resistance they employ.

Different populations experience surveillance differently, giving rise to unique forms of resistance. One of the most important local variables in the Ohio system concerns the meager level of welfare provisions, which do not meet minimum subsistence requirements. The women Gilliom studied are structurally compelled to produce extra income, but doing so violates the welfare rules. Hence, the women's characteristic forms of resistance involve earning small amounts of money in the 'grey' cash economy which is more difficult for the authorities to monitor. Such resistance produces vital material benefits and also establishes a zone of relative autonomy in which the women can act somewhat free from the scrutiny of the surveillance system. However, doing so means that they are in constant danger of being apprehended and losing the welfare benefits on which their subsistence depends.

While the women are intimately aware of the degree of monitoring they are subject to, they do not appeal to abstract privacy rights to frame their experience of database surveillance. Privacy rights offer an opportunity to foster an oppositional consciousness, while also providing a framework around which more formal political opposition can be built. However, as the proliferation of surveillance makes privacy rights more important, their practical effects are increasingly limited. New surveillance measures do not fit into the existing privacy framework and the legal structure of privacy tends to push political struggles onto alien and alienating legal institutions. Moreover, privacy talk seems to foreclose more communal alternatives to the individualistic philosophical framework in which privacy is embedded. Most problematically, however, privacy no longer seems to be able to spark the public imagination.

These local and systemic failures of privacy rights accentuate the need for new concepts and vocabularies to understand the contemporary operation of surveillance. Rather than resist surveillance through abstract legal formulations, the Appalachian women situated their critique in a pragmatic series of claims about the necessity to meet daily requirements for themselves and their family. Gilliom believes that such grounded critiques of surveillance are an important way to supplement the abstract privacy formulations regularly invoked to resist surveillance.

Few comparable studies have sought to investigate the subjectivity and consciousness of individuals under surveillance. Gilliom urges us to abandon the general academic tendency to approach the surveilled as an undifferentiated mass and to recognize that surveillance is experienced differently by virtue of how people are situated and the attendant form of relationship they have with surveillance programs.

2 9/11, Synopticon, and Scopophilia: Watching and Being Watched

DAVID LYON[1]

Several persistent puzzles run through surveillance studies, and such puzzles are prominent in any discussion of the politics of surveillance and visibility. Among them, this one troubles many surveillance scholars and civil libertarians: why are surveillance technologies developed and deployed so promiscuously when, for example, people also claim to be concerned about privacy, and when credible evidence exists that these technologies often do not, or cannot, perform the functions required of them (even though they may perform other functions)?

It was clear before 9/11 that surveillance of many kinds was increasingly accepted as an everyday fact of life. Since then it has become even more obvious as numerous surveillance schemes, excepting only the most egregious, have been accepted apparently without a murmur, particularly by the American people. It is possible that on a simple calculus citizens accept that loss of privacy is the price to be paid for security – as the mass media have reiterated ad nauseam since 9/11 – but one wonders how these attitudes might change if the same citizens knew what exactly happened to their personal data as they circulate between distant databases. An implicit assumption seems to reign, that technological solutions are appropriate, and that they work, even if they did not on the morning of 9/11.

What makes surveillance *work* today? In the majority of situations, searchable databases are now central to its operation (Lessig 1999). Personal data are thus being processed for various purposes, whether to secure airports from guerilla fighters (a better term than the ill-

---

1 This chapter includes and elaborates on some passages from chapter 1 of my *Surveillance after September 11* (Polity/Blackwell 2003).

defined 'terrorist'; see Downey and Murdock 2003) or to intercept messages from violent conspirators. Does such personal data processing equate with the normalizing discipline of the panoptic, operating through the unseen gaze to produce compliance based on uncertainty and fear? Or is it the subtle power of categorizing social groups, discriminating between them for different treatment?

Many other puzzles confront those who would understand surveillance today, but the ones I highlight here are still significant and at the heart of any contemporary analysis of surveillance power. In what follows, I wish to explore these questions through the prism of the events of 9/11 and their aftermath. I put the accent on *literal* watching, rather than the more *literary* use of the word 'watching' as a trope for some form of dataveillance. But I hope that the analysis of literal watching may throw some light on the more metaphorical kinds. At the same time, the treatment is sociological in the sense that empirical constraint and theoretical explanation are in view, even though this is an initial exploration.

My hypothesis is that surveillance – which at its social and etymological core is about watching – is easily accepted because all sorts of watching have become commonplace within a 'viewer society,' encouraged by the culture of TV and cinema. As things once 'private' have become open to the 'public gaze' of many, and as intimate and once-sequestered areas of life are 'screened,' so it seems of less and less consequence that this or that bit of once-protected personal data is disclosed. However, two other processes are also set in train, which are illustrated by 9/11. One is that the gaze of the many, fixed on the few, may foster some rather specific interpretations of the world. In the case of 9/11, the TV gaze permitted the development of a context-free narrative about American victims of totally unexpected foreign violence. The other is that this narrative, once accepted, becomes the means of legitimizing other kinds of official 'watching' (for 'terrorists' in this case) of the many by the few. However, this latter watching is not merely the occasional foray into specific segments of suspects' lives, but a systematic watching that 'screens' and sorts the watched into categories in order to determine who gets what kind of treatment.

9/11 and Surveillance: From Screen to Screening

There are several specific ways in which the events of 9/11 relate to surveillance and the politics of visibility. While the timing of the attack on the World Trade Center (WTC) seems to have been determined by the large numbers of workers likely to be in the buildings around 9:00 A.M., and the morning departure of fuel-heavy flights from airports on the eastern seaboard, one can also assume that it was designed to achieve maximum media coverage. Millions of people watched the horror unfold both during and, increasingly, after the attacks. The time delay between the assaults on the first and second of the Twin Towers ensured that television crews had time to get equipment in place for the spectacle to be viewed with fullest effect. This is a prime example of what has been called the 'viewer society' or the 'synopticon' (Mathiesen 1997) where the many watch the few (cf. Meyrowitz 1985). In this case, the audience was huge, and the opportunity was exploited by repeats of the approaching jetliner, the crash into the tower, the explosion, the flames, the smoke, the pitiful people jumping to their death from windows, and the eventual terrifying collapse. Violence and mass media operated in perverse harmony.

These images, in their concentrated and horribly spell-binding form, are themselves what happened on 9/11. Apart from those directly involved in Manhattan as victims, escapees, emergency workers, and bystanders, the events were 'experienced' or 'known' through mass media. Cut off from any antecedent history of American economic and military activities around the world, or of the growing resentment and hatred directed against the United States by a number of nations, the attacks were apparently spontaneous, utterly unexpected, and unpredictable. This made them all the more heinous in their bloody toll and their callous calculation. This is the media spectacle that was 9/11. With no crash into the White House, and only grainy shots of the Pentagon attack, it was the WTC that maintained its position as the key symbol of that day.

The media spectacle of 9/11 stimulated crucial public opinion effects of sympathy, anger, fear, and the quest for retribution, effects which turn out to play a valuable role in justifying political and military responses. Without 9/11, many legal and technical measures, long-cherished dreams of some politicians and technocrats, would never have appeared plausible or workable. But TV-assisted fear and outrage are powerful weapons for changing the course of policy. Those – the majority in the global north – who watched those TV images had their consciousness etched not just with death and destruction, but with a particular view of the world. This view brooks little contradiction, and permits little debate.

Not only did the images circulate at the time, and in the immediate months that followed, they were also repackaged as documentaries that circulated around the world, and were reinforced by new 'one year after' film and television accounts in September 2002. Many images and accounts of 9/11 were also available in newspapers, magazines, books, and of course online. An unprecedented number of web-hits were recorded in relation to 9/11. So in three related media in particular – television, cinema, and the Internet – 9/11 became a major event, providing viewers with many different angles on the attacks. While most accounts within the United States were reverential, stressing American victimhood, some made outside the country were branded as 'anti-American.'

Many had questions, of course. The question about how the attacks could have occurred was prominent, and at first posed primarily in technical terms: why were U.S. security and intelligence services unaware of the threat? The response was a determination 'not to let it happen again' – again, it was posed in primarily technical terms, but also involved wide-ranging legal and policy initiatives. In fact, the order of these is reversed. Hastily formulated legislation came first, while new technologies, though much discussed, will continue to be implemented for several years to come. The response to what was seen on the screen, I suggest, is to seek other items to be shown on other kinds of screens. Indeed, a process of *screening*, for example, racial profiling, of the many by the few is in some important ways consequent on the *screening*, in the TV sense, of the few for the many.

One could object that this is too simple, a verbal *trompe d'oeil* or an exaggeration. After all, it is not only screening by police and security forces that counts today, but a general screening of everyone, by everyone. This is of course another aim of Homeland Security, to enlist all in the task of identifying and alerting authorities to potential 'terrorists.' Hotline information abounds at state borders and municipal offices in the United States. But the categories used – such as 'Middle-Eastern' or 'Arab-Muslim' – derive directly from official, media-amplified definitions, which buttress the overall state security effort. People who seem to fit such categories are the primary targets in the 'enhanced culture of suspicion' (the term is from O'Neill 2002).

Many are also involved in the other kind of screening. In a world where millions of households possess video cameras, TV cameras do not have a monopoly of screened images (see chapter 8 in this volume). The many also watch the many via the mediation of camera and screen besides what is shown on TV. Again, however, larger questions of power have to be asked. The major media conglomerates monopolize official images, and even the celebrated case of the bystander video of the Rodney King beating by police in Los Angeles does not on its own show that the cameras of 'little brothers' have somehow achieved a political clout commensurate with that of the guardians of law and order. While in principle everyone can indeed watch everyone else in mediated ways – screening the results – some forms of watching carry more weight than others.

To return to the main thrust of the argument, we can safely note that prominent among the technical responses to 9/11 is the reinforcement of surveillance. With better systems in place, their promoters insist, the risks of another attack could be drastically reduced. With better means of identifying, classifying, profiling, assessing, and tracking the population, the chances of preventing future attacks may be increased. Such means are provided by surveillance systems, suitably automated to allow a few inspectors to watch many people. This is not Thomas Mathiesen's 'synopticon,' where many watch the few, but the 'panopticon' in which the few watch the many. Yet as Mathiesen argues, the two systems work together, the one strengthening the other.

Let us review, for a moment, the range of responses to 9/11 that justify thinking of it as a surveillance watershed. These are most apparent in the United States, but they relate to global trends and in some important respects – such as border requirements – demand reciprocal responses from other countries as well. The biggest and best-known relate to intelligence gathering and policing. This includes the Pentagon's notoriously named 'Total Information Awareness' scheme (amended from 'Total' to 'Terrorist' and ultimately dropped in the face of public opposition); the CAPPS II initiative for 'pre-screening' of airline passengers at airports; the slowly developing plan that will bring together at a federal level the state records on drivers to create a de facto national ID card system; plus other initiatives to tighten border crossings, intercept telephone, fax, and Internet communications, and to encourage ordinary citizens to become the 'eyes and ears' of intelligence and law-enforcement authorities.

What these indicate is a reinforcing and augmenting of existing trends, particularily towards algorithmic surveillance, using searchable databases. In each case, new, higher levels of integration are sought, as corporate and government databases are increasingly merged, permitting cross-checking over a broader informational terrain. New software and systems are being designed or retooled from other purposes (customer relationship marketing for terrorist targeting, for instance; see Pridmore and Lyon 2003) for identification, monitoring, tracking, checking, mining, predicting and, purportedly, pre-empting terrorist activity. The most prominent initiatives are technological. They promise to bring more details about behaviours, proclivities, transactional traces, and apparently trivial characteristics (such as gait, used with CCTV) into the purview of the powers-that-be than ever before.

Although Jeremy Bentham ultimately failed to persuade potential customers to buy his 1791 plan for the Panopticon prison, Michel Foucault was right to use the Panopticon as a model of modern discipline. Foucault supposed that the system of spectacle, tied tightly to punishment from medieval times, was being *replaced* in the modern world by subtler means of social control, exemplified in the self-disciplining routines of the Panopticon. Against this, Mathiesen suggests that the two continue to operate together. Indeed, they do so in mutually reinforcing ways. The many watching the few does not *give way* to the few watching the many. Rather, both occur simultaneously, and both depend increasingly on similar electronic communication technologies. On 9/11, specific persons perpetrated an unprecedented attack, which could be viewed almost anywhere. Those acts, abstracted from previous events, generated visceral fear. They were then used to justify panic regimes and stereotypes, which in turn were fed into the freshly augmented surveillance systems, giving them both their rationale and their coded content.

Even if it seems like a truism, it is still worth noting that mediated watching has become a key feature of contemporary societies, sometimes to an obsessive degree (Meyrowitz 1985). The popularity of so-called reality TV (and Big Brother!) is some measure of this, as is the proliferation of webcams depicting mundane activities inside people's homes, and the continuing significance of cinema. Both the watching of the few by the many and the many by the few are fuelled by a desire to watch, or, perhaps, what film theorists such as Christian Metz (1982) call 'scopophilia,' or the love of looking. In both cases scopophilia may be seen (in one aspect at least) as a sort of voyeurism that reduces the rights of the watched. While reality TV show heroes may have given consent to mass viewing, others who appear on screen – such as the victims of 9/11 – have not. Equally, many who are under the panoptic gaze are not informed or have not consented to having their personal lives exposed to view.

In terms of the basic case of 9/11, our starting questions may be reviewed as follows: why are surveillance technologies adopted when people also claim to be concerned about privacy and when little evidence exists that they work for the purposes stipulated? The screen images of victims justify the 'screening' and 'imaging' of others. The mass media play an important role in legitimizing new surveillance technologies. But do people know how today's surveillance works? This is a vital research question to which we as yet have few answers. Here the ambivalence of the panopticon is important. In some contexts, knowledge of surveillance (such as cameras in the shopping mall) can have the direct, desired, deterrent effect, whereas in others, surveillance operates more subtly as social sorting and digital discrimination (Lyon 2003b). After 9/11, both aspects are expanding, but one suspects that fewer people are aware of the latter than of the former. They may agree that 'intrusions into privacy' are necessary for 'security,' knowing only *that* someone has data on them, not what is *done with* those data.

Why is watching others so culturally significant at the start of the twenty-first century? Is there a broader frame in which all-pervasive surveillance technologies makes sense? I shall suggest that Meyrowitz's work on the social consequences of electronic media and Mathiesen's 'viewer society' yield some important clues. While they focus on television, others have shown how the mutating twentieth-century cinema throws light on this phenomenon by highlighting the role of the panoptic gaze and synoptic consumption. Internet media seem to play a parallel role.

Why do people permit themselves to be watched? The case of 9/11 does not get us into this question yet, except obliquely. If the TV version of 9/11 is a kind of voyeurism, then the victims did not permit anything: they were unaware that their behaviours and their bodies would be seen repeatedly. Voyeurism is justified as journalism, which depends heavily on images. Those who permit themselves to be watched generally do so because of some perceived benefit to them, or, perhaps, simply because they enjoy being watched. They are rewarded for a TV appearance and feel safer with security cameras overhead.

The Case for Synopticon

Thomas Mathiesen argues that while Foucault contributed to contemporary understandings of surveillance through his work on the panopticon, his analysis overlooks, among other things, the opposite and simultaneous process of 'synopticism.' The few may well watch the many, as they do in surveillance situations of constantly increasing magnitude, but this does not mean that the many no longer watch the few, as Foucault suggested in his analysis of the demise of public executions and other punitive spectacles. Indeed, the same communication and information technologies today permit an unprecedented watching of the few by the many – mainly through television – as well as an unprecedented watching of many by the few through visual surveillance and dataveillance of various kinds.

Mathiesen acknowledges the continued role of the panoptic, whose normalizing gaze produces self-controlling subjects fitted for democratic capitalist society. He observes that the gaze extends beyond prisons to track those who are released from them; that computerized surveillance uses categorical suspicion in predicting future criminal acts; and that other agencies – such as medical, psychiatric, or schooling agencies – use classificatory and diagnostic techniques that are panoptic and carceral. But he expresses surprise that Foucault would totally neglect the mass media, and television in particular. Had he considered their effect, avers Mathiesen, his 'whole image of society as far as surveillance goes' (1997: 219) would have changed. Mathiesen's comments apply to the computer as well. Curiously enough, Foucault worked with a basic blind spot as far as any twentieth-century technological devices and systems are concerned. The same communication and information technologies that are missing in any 'synoptic' account are also absent from the account of the panoptic in modern times.

Following from this, Mathiesen notes the parallels between the synoptic and panoptic. First, they have developed through exactly the same time period, from 1800 to 2000. Second, they are both potential means of power, present since ancient times. The census and tax registration were panoptic precursors, while spectacles organized by the powerful were synoptic precursors – both were technologically upgraded and intensified in the later twentieth century. Third, the two systems interact intimately and may even be fused with each other. Mathiesen gives the example of early prison chapels that were simultaneously panoptic (the preacher would see all prisoners) *and* synoptic (the prisoners could see only the preacher). Orwell's telescreen performed a similar two-way function. The television has similar capacities, says Mathiesen, in that viewers may watch while themselves being watched by, for example, advertisers who track their purchases or companies who check their ability to pay.

Today, such examples could be multiplied with regard to the more sophisticated panoptic-synoptic tendencies of television and the networked media of the Internet (see chapters 11 and 12 in this volume). In the case of television, set-top cable or satellite boxes may contain software that tracks channel-surfing patterns and creates digital profiles of each person in the household using the remote control. These become the basis for targeted advertising and other personalized treatment and solicitations. Of course, the popularity and acceptability of set-top TV boxes has waxed and waned since their first appearance, but some argue that a new generation of such devices is likely to make better headway. The price paid for more channel choices is more data mining and profiling of the viewer (Manjoo 2003 and chapter 12 in this volume). Similarly the Internet presents parallel synoptic-panoptic processes. The Internet is a gateway to a wealth of diverse information and the more it is used as such, the more companies providing the information also engage in tracking, data mining, and profiling of surfers (Lyon 2002).

Mathiesen shows that mass media personalities tend to be regarded as reliable sources of information, and that only selected non-professionals (usually male, and from elite institutions) typically appear on television. Add to this the widespread role of gatekeeping 'information professionals' and the powerful influence of the mass media becomes clear. In these ways mass media are a power resource that actually make a difference. They are involved in the discipline or control that is implied in surveillance, which is seen less in individual effects than in general patterns. The panoptic may have an effect on behaviours – as McCarthyism led communists to be more circumspect in their speech and activities but not in their commitments in the 1950s – but it does not today touch the *soul*, suggests Mathiesen. Instead this is the task of the mass media, which can have a more direct effect on consciousness. It offers a world paradigm which fits current situations neatly, and is highly attractive to consumers who are positioned as 'choosers' rather than 'creators.'

Mathiesen's final point is that the synoptic and panoptic mutually feed on each other (1997: 231). While the old panopticon wishes to live under the cover of the synopticon, the latter makes extensive use of highly desirable news from the panopticon – prisoners, escapes, murders, and, today, terrorists. Mathiesen does not note this, but a double purging also goes on. Just as the panoptic purges the person of everything except the behavioural trace, so the synoptic performs a similar reduction. The larger-than-life criminal (or terrorist) is repackaged for consumption as a spectacle within a panic regime, a stereotype cut off from antecedent events, social ties, or political motivation. So much for the demise of the spectacle! In fact, hints Mathiesen, the opportunity is created for the creation of public opinion sympathetic to increased panoptical controls. He could hardly have better foreseen the relationship between the flames in the Twin Towers and the Pentagon's 'Total Information Awareness' integrated data-mining schemes (Lyon 2003a).

The Case(s) for Panopticon

The relation of the panopticon to surveillance has been thoroughly rehearsed. Indeed, it must be the most discussed and debated theoretical concept (Lyon 1993; Boyne 2000). The panoptic urge is to make everything visible; it is the desire and the drive towards a total gaze, to fix the body through technique and to generate regimes of self-discipline through uncertainty. The inmate cannot evade the eye of the inspector because the prison cell is open to view and back-lit so nothing can be hidden. At the same time, the inspector's eye cannot be seen by inmates, because it is shielded with a system of blinds. You never know when the inspector's eye will be on you so you modify your visible behaviour in order to avoid negative sanctions. This is the soul-training characteristic of disciplining power.

Early studies suggested that electronic technologies made possible the perfection of the panopticon in ways Bentham could never even have dreamed. Such technologies complete the panoptic project both by bringing more behaviours to light, and by rendering the surveillance apparatus more opaque. Thus Shoshana Zuboff (1988), Nancy Reichman (1987), and Frank Webster and Kevin Robbins (1986) once hinted that to live in the later twentieth century was to live in an 'electronic panopticon.' Each study had a very good point, that contemporary life is made visible in a multitude of ways, and that these have profound implications for life-chances as well as for self-discipline, but it was a point that was vulnerable to overstatement. Some subsequent studies continue the tradition of generalizing the panoptic effects of electronic surveillance, rather than focusing on its most salient features. Of the latter, I would argue that one-way observation (the unseen 'inspector'), classification, and training in self-discipline are three of the most significant features of electronic systems.

Later studies have queried many aspects of this thesis. One important question concerns which aspect of the panopticon is really at issue: the unseen observer, or the phenetic drive to classify? Or is there something in Foucault's work which is more salient to current developments? This might involve his analysis of the treatment of plague victims who experienced power as differentiation, segmentation, and training or his discussion on lepers who were segmented, confined, and exiled (Green 1999; Norris 2003). Alternatively, is there something about electronic technologies that qualifies their databases as 'discourses' and would thus connect them more fully with Foucault's concept of biopower (Poster 1996)? Each of the studies cited hints at a different modus operandi for the panopticon. It works either by fear and uncertainty about the unseen observer, by classifying populations for ease of management, or discursively, by constructing subjects. These three may also overlap.

Yet other panopticon studies suggest that the whole Benthamite enterprise is over, rendered redundant by other means of social management and control. Zygmunt Bauman (2000: 85) for example, takes Mathiesen to be saying that just as Foucault saw the new regimes of panoptic self-discipline to have supplanted the spectacle, so now it is the synopticon that pushes the panopticon aside. Bauman's post-panoptical era focuses rightly on the extraterritorial aspects of power: those 'operating the levers of power on which the fate of the less volatile partners in the relationship depends can escape beyond reach – into sheer inaccessibility' (2000: 11). Post-9/11 secrecy has served to encourage this development, one might add. But Bauman still maintains that the synopticon now does the panopticon's work, through seduction and enticement rather than coercion. He seems unaware of the extent to which, for instance, consumer desires are created through database marketing, a realm in which the phenetic is writ large. Haggerty and Ericson (2000) also state that it is time to leave the privileged panopticon concept behind, along with Big Brother, in favour of the 'surveillant assemblage,' within which the panoptic is just one dimension.

The debate is likely to continue for some time, but it does seem to me that several things may be said about it. One, it is possible to put the totalizing and anti-humanist aspects of Foucault's treatment of the panopticon on one side and consider the relevance of the self-discipline system in some significant settings, such as CCTV. Moreover, in contexts like this, it is still necessary to specify what sorts of things the panopticon is said to accomplish. For Norris and Armstrong, for example, it is the classificatory power of CCTV which makes it panoptic

(1999: 219). It is the *discriminatory* eye of video surveillance which, like the Panopticon prison's carefully classified and segregated sections, yields at least part of its power.

Setting the concept of the panopticon aside altogether, in favour of another, seems to me to be a mistake.[2] Too many useful insights would be lost from surveillance studies. The panoptic operates alongside the assemblage and the synopticon, without being simply displaced by them. I take the points made by those who wish to drop the panopticon, that control is achieved by means of subtle regimes of seduction in the world of consumer desires, or that rigid, top-down, centralized surveillance schemes have partially liquefied into pulsating flows. But it does seem that, at the very least, the creation of compliance through fear of uncertainty and the classificatory impulse of 'making up people' (see Hacking 1986) so central to the panopticon ought to be retained.

A fine example of different surveillance operations 'operating alongside' is offered by Norris and Armstrong, who show that television has proved to be one of the biggest allies of CCTV. As I suggest is the case with 9/11, there is a 'television love affair' with CCTV in the United Kingdom (Norris and Armstrong 1999: 67f). Television and CCTV are both visual media that observe and appear to have been made for each other. 'Add one ingredient, crime, and you have the perfect marriage. A marriage that can blur the distinction between entertainment and news, between documentary and spectacle, and between voyeurism and current affairs.' British television shows such as *Crime Beat* and *Eye Spy* use CCTV footage in their presentations, and they achieve high ratings. But the CCTV regimes are also reinforced, because they can be 'seen' to be helping to deter crime, or at least to enable police to be deployed to apprehend suspects, or failing that to enlist the help of ordinary viewers to catch 'this man' seen on film (see chapter 8 in this volume). While television use of CCTV footage reveals little or nothing about how CCTV really operates, it ensures that the budgets for new camera systems will continue to be found.

My argument thus far is that the concept of the panopticon ought to be retained in surveillance studies, not just for its focus on the unseen observer (beloved by those who see the potential for electronic fulfilment of Bentham's dream), but for its emphasis on classification and categorization – also amenable to coding within the algorithms of software-

---

2 Haggerty and Ericson do not exclude the panopticon from debate. Soul-training, classification, and one-way visibility are all still significant, from their point of view.

based sorting, along with 'soul training.' But with Mathiesen, I argue that the panopticon does not operate on its own. Other metaphors and concepts, including especially the synopticon, help us to understand the broader context within which the panopticon works. In the case of 9/11 the evidence seems strong that there is a reciprocal, mutually reinforcing relation between the synopticon and the panopticon. The many watching the few gives grounds for legitimating the watching of the many by the few.

But the puzzle remains about watching itself. Why does this take particular forms at the start of the twenty-first century, and why has watching of some specific kinds become so compelling and yet so commonplace? It is more than the presence of reality TV and surveillance movies that constitute cultures in the global north as 'viewer societies.' Among the many answers that might be given to this question, ones from the psychoanalytic stable have not featured much in the surveillance studies literature. I now turn to one of these – the notion of scopophilia – to discover if it can offer any leverage on this question.

Viewer Societies and Scopophilia

I begin with the question of why people permit others to watch and, at least by implication, to be watched. What is it about watching others that is so compelling and enjoyable? Whence the fascination with seeing the intimate lives of others on, say, 'Big Brother' shows, and why the willingness to join Neighbourhood Watch schemes and snoop on the folks next door? Many are, apparently, willing to be watched, both literally and figuratively, and some even actively seek the experience. Teenage stars of the shopping mall video surveillance system provide one example, and in the more figurative, non-visual, sense, some consumers seem willing to have their purchases tracked as a reflection of the identity they build for themselves through their clothing and accessories. This watching-and-being-watched appears to be a phenomenon that has existed for some time in its modern form, at least as it was explored in post-war movies such as *The Third Man*. Mathiesen's 'viewer society' thesis, while it provides a welcome corrective to the overdetermining panopticon, does not go far enough in considering the culture of viewing in ways that Meyrowitz (1985) does. Neither author tries to explain the very process of 'viewing' itself, within the 'viewer society.'

If aspects of late- or post-modernity may be thought of in terms of 'viewer society,' then as much attention should be accorded to viewing as to being viewed. In a sense this raises fresh question about the body. If the pre-modern concern was how a body should *live*, and modernity was dominated by the question of how a body is *known* (which, of course, lies behind surveillance in many of its forms), late- or post-modern culture shifts the question to how the body *looks*. There is an apparent shift from the analytic to the aesthetic. The question of why people permit themselves to be watched may be paralleled by the question of why people want to watch. One possible answer that bears some thought emerges from film theory and psychoanalysis, 'scopophilia' and its deviant accomplice, 'voyeurism.' The cinema has probed and played with the notion of surveillance, voyeurism, and the power of the gaze for many years – classically in Hitchcock's *Rear Window* (1954), where the neurotic need to surveil is explored. The spectator in the cinema, or in the television lounge, is offered a context within which watching is allowed, right into life areas commonly considered 'private' or out of bounds.

Scopophilia means, at its simplest, the love of looking. The origins of the concept lie in Freudian and especially Lacanian psychoanalysis. Jacques Lacan followed Freud in seeing the experience of infancy as determinative for developing a sense of self, but tightened this by reconsidering Freud's ideas on narcissism. The infant is at first merged with the maternal body, where all desires are met, but then discovers, devastatingly, that s/he has a separate identity. But the infant may obtain a sense of bodily unity through the recognition of its image in a 'mirror' (Lacan 1977), or feedback from others. Unfortunately the mirror offers only a mirage of coherence: it lies. The act of seeing ourselves as others see us is fragmenting, destructive of our former unity. The infant imagines that he or she is stable and unified when the psychical space is actually splintered and physical movements are uncoordinated. Lacan's story gets worse, as alienation deepens and the subject seems to have little chance for reflection, autonomy, or transformation (Elliot 1992: 138–46). The key point is that the capacity to look, and to see ourselves as others see us, is crucial for identity formation.

Applied to the cinema or, more broadly, to the viewer society, scopophilia has been translated as the predominantly male gaze of Hollywood that depersonalizes women, turning them into objects to be looked at, rather than subjects with their own voices and subjectivities. In a classic article on visual pleasure and narrative cinema, Laura Mulvey says that three kinds of looking are present in the cinema: the look of the camera recording events, the audience looking at the screen, and the looks between characters (Mulvey 1975). The audience watching reproduces the lively curiosity of infants deriving pleasure from what they see: 'Looking itself is a source of pleasure, just as ... there's pleasure in being looked at.' But there is a darker side to this cinematic looking.

For Mulvey, the infant may be trained and normalized in his viewing habits. But the love of looking may become fixated into a perversion, found in peeping Toms and obsessive voyeurism. In the latter cases, viewers gain satisfaction from 'watching, in an active controlling sense, an objectified other.' This puts a decidedly negative – and, of course, feminist – spin on scopophilia, even though it also acknowledges that there may be modes of 'good looking' (see Schuurman 1994) as well. Mulvey's case is well made, and is strong precisely because she makes clear the grounds of critique. Following Lacan himself will not help here because for him it is the symbolic realm in general that dominates the individual, not specific political or ideological formations. Nor will much aid come from postmodern critics of Lacan, such as Gilles Deleuze and Felix Guattari (1977), unless you hold, with them, that 'impersonalized flows of schizoid desire can herald a radical transformation of society' (Elliot 1996: 190).

It is not too much of a stretch to suggest that part of the enthusiasm for adopting new surveillance technologies, especially after 9/11, relates to the fact that in the global north (and possibly elsewhere too) the *voyeur gaze* is a commonplace aspect of contemporary culture (Denzin 1995). This does not explain the political economy of surveillance in neo-liberal societies in which social defence technologies fit with forms of exclusion and privilege (Garland 2001; Rose 1999). Not does it explain why particular techniques might be adopted in certain settings, or why technical solutions would be sought at all to cope with problems – like Islamist 'terrorism' – that appear to be deeply social and cultural in origin. But it might help to explain why surveillance, both seeing and being seen, is accepted as a viable mode of social ordering, management, and control. It fits the cultural paradigm, the basic orientation to the world, of Mathiesen's viewer society and Meyrowitz's media-saturated world. And again, the feedback loop is self-reinforcing. The more that can be seen, the more we want to see. If the techniques are available, the will is there for deeper mining of the data.

This kind of analysis squares with Norman Denzin's approach to viewer society, which he calls 'the cinematic society' (Denzin 1995). This is a world in which the 'voyeur's gaze' has become increasingly central, to the postmodern point that 'the gaze is openly acknowledged, and its presence everywhere, including the living room, is treated as commonplace' (Denzin 1995: 9). The voyeur takes many forms – from medical checks to sociological surveys to peeping Toms to the tourist gaze – but in each s/he helps to keep alive the sense of 'private' sphere – so important to the cultural logics of capitalism and democracy – by invading it. The voyeur's views about what s/he is up to altered with time during the twentieth century, starting with fairly clear scientific realist epistemologies and ending with uncertainty about truth claims and cause and effect. The voyeur was gradually incorporated into Hollywood's productions, with treatments of the illicit gaze, the use of surveillance technologies, and other techniques. This occurs within a broader frame, says Denzin, in which film carries out ideological tasks for capitalism in its various stages – local, monopoly, and multina-tional-consumerist. Hence the ubiquitous, up-front, domesticated gaze of postmodern times, in which the audience is drawn into the voyeur's activities.

Equally, in the realm of TV, a world which once was 'private', or 'backstage' (to use Goffman's classic terms) has only moved into the 'public' domain. As Joshua Meyrowitz puts it, 'electronic media make public a whole spectrum of information once confined to private inter-actions' (1985: 95). Meyrowitz sees this as an Innisian 'bias' away from the tendency of print media to favour the 'front region.' The public/ private distinction is thus blurred decisively in an age of TV such that what was once seen as private becomes public. The paradox here of course is that today's surveillance is mainly about behavioural traces and bits of data, that will be used for sorting – not merely moments of 'private' life that have become 'public.' This also connects with Daniel Solove's (2002) questions about which metaphor is right for surveil-lance – should it really have privacy as its obverse side?

If the viewer society and the voyeur gaze are indeed as significant as these theorists suggest, then the expansion of surveillance into the twenty-first century makes a lot of cultural, as well as economic and political, sense. Gary T. Marx (1996) observes that scholars should pay more attention to popular media as a means of understanding surveil-lance and it seems to me that this is more true than ever today. In Mulvey's and Denzin's hands, the notion of voyeur is used to under-stand some crucial cultural trends, but also, importantly, as the basis for critique. Mass media–saturated populations in the global north have become increasingly inured to the notion that uninhibited watching of others is acceptable, even though, as Denzin observes, this simultaneously helps to keep alive the belief in privacy as a valued condition. Scopophilia, while it may start out innocently enough as the love of looking, is all too easily perverted into an obsessive and controlling gaze that objectifies the image under observation. While Mulvey's specific concern is the male gaze, and Denzin enlarges this to encompass questions of ethnic complexity, it is not hard to see how a critique of the kinds of gaze consequent on 9/11, on any marginalized groups, may find a rationale here.

Conclusion

The attacks of 9/11 and their aftermath represent a watershed in surveillance. Existing surveillance practices and processes are being ratcheted up in a rapidly rising spiral. They may be understood not merely as an outgrowth of technological trajectories, or neo-liberal patterns of political economy, but in relation to cultural trends that render surveillance progressively more commonplace, unexceptional, and even desirable. Those cultural trends may be seen clearly through the lens of the panopticon-synopticon relationship, which is reciprocal and mutually reinforcing.

This is emphatically not an argument for the displacement of the panoptic by the synoptic, but rather a plea that the two be considered together. The synoptic helps justify the panoptic, which in turn provides some of its most telling images. The panoptic must be retained in surveillance studies because it helps to highlight a key aspect of surveillance today – social sorting and digital discrimination by means of searchable databases. Such searchable databases are basic to the intensified integration of surveillance following 9/11. If ordinary citizens and consumers understood this aspect of surveillance, they may well be less sanguine about having 'nothing to hide' or about exchanging their 'privacy' for more 'security.'

At the same time, some deeper understanding of the viewer society (with its panoptic-synoptic features) is sought in psychoanalytic and social theories of film and television. Scopophilia, and in particular the voyeur gaze, seem to have become culturally central in late or postmodern times, which helps to explain further why companies and governments seem to have so little trouble selling and installing surveillance technologies. The fact that there are at least alleged benefits to be gained from parting with personal data – not to mention the obverse of the love of looking at others, which is the love of being looked at – means again that many do not object to being surveilled.

Part of the appeal of the notion of the voyeur gaze is that it is a critical concept. It strongly pushes us towards an appreciation for the need for normative approaches to surveillance. These would aim to limit the current operations of such practices where the power to objectify or categorize operates without reference to any ethical codes. Unfortunately, in a world where *aesthetic* concerns of how we may look at others tends to dominate, we may have to search elsewhere for ethical resources to deal with the modern obsession to 'know' the body in abstract ways. It seems a long way off to a world where we have eliminated to objectification of persons and the digital discrimination against the weak. This does not make it less worth hoping for and working towards.

REFERENCES

Bauman, Z. 2000. *Liquid Modernity*. Cambridge: Polity Press; Malden: Blackwell.
Boyne, R. 2000. 'Post-panopticism.' *Economy and Society* 29 (2): 285–307
Deleuze, G., and F. Guattari. 1977. *Anti-Oedipus: Capitalism and Schizophrenia*. New York: Viking.
Denzin, N. 1995. *The Cinematic Society: The Voyeur's Gaze*. London: Sage.
Downey, J., and G. Murdock. 2003. 'The Counter-Revolution in Military Affairs: The Globalization of Guerilla Warfare.' In D. Thussu and D. Freeman, eds., *War and the Media: Reporting Conflict*, 70–86. London: Sage.
Elliot, A. 1992. *Social Theory and Psychoanalysis in Transition: Self and Society from Freud to Kristeva*. Oxford: Blackwell.
– 1996. 'Psychoanalysis and Social Theory.' In B. Turner, ed., *The Blackwell Companion to Social Theory*. Oxford: Blackwell.
Garland, D. 2001. *The Culture of Control*. Chicago: University of Chicago Press.
Green, S. 1999. 'A Plague on the Panopticon: Surveillance and Power in the Global Information Economy.' *Information, Communication, and Society* 2 (1): 26–44.
Hacking, I. 1986. 'Making Up People.' In T. Heller, M. Sosna, and D. Wellbery, eds., *Reconstructing Individualism: Autonomy, Individuality and the Self in Western Thought*. Stanford: Stanford University Press.

Haggerty, K., and Ericson, R. 2000. 'The Surveillant Assemblage.' *British Journal of Sociology* 51 (4): 605–22.

Lacan, J. 1977. *Écrits: A Selection*. London: Tavistock.

Lessig, L. 1999. *Code and Other Laws of Cyberspace*. New York: Basic Books.

Lyon, D. 1993. 'An Electronic Panopticon? A Sociological Critique of Surveillance Theory.' *Sociological Review* 41 (4): 653–78.

– 2002. 'Surveillance in Cyberspace: The Internet, Personal Data, and Social Control.' *Queen's Quarterly* 109 (3): 345–56.

– 2003a. *Surveillance after September 11*. Cambridge: Polity Press; Malden: Basil Blackwell.

– 2003b. ed. *Surveillance as Social Sorting: Privacy, Risk, and Digital Discrimination*. London and New York: Routledge.

Manjoo, F. 2003. 'Your TV Is Watching You.' http://www.*Salon.com*. 8 May.

Marx, G.T. 1996. 'Electronic Eye in the Sky: Some Reflections on the New Surveillance and Popular Culture.' In D. Lyon and E. Zureik, eds., *Computers, Surveillance and Privacy*. Minneapolis: University of Minnesota Press.

Mathiesen, T. 1997. 'The Viewer Society: Michel Foucault's 'Panopticon' Revisited.' *Theoretical Criminology* 1 (2): 215–34.

Metz, C. 1982. *The Imaginary Signifier: Psychoanalysis and the Cinema*. Bloomington: Indiana University Press.

Meyrowitz, J. 1985. *No Sense of Place: The Impact of Electronic Media on Social Behavior*. New York and Oxford: Oxford University Press.

Mulvey, L. 1975. 'Visual Pleasure and Narrative Cinema.' *Screen* 16 (3): 6–18.

Norris, C. 2003. 'From Personal to Digital: CCTV, the Panopticon, and the Technological Mediation of Suspicion and Social Control.' In Lyon, ed., *Surveillance as Social Sorting*.

Norris, C., and G. Armstrong. 1999. *The Maximum Surveillance Society: The Rise of CCTV*. Oxford and New York: Berg.

O'Neill, O. 2002. *A Question of Trust*. Cambridge: Cambridge University Press.

Poster, M. 'Databases as Discourse; or Electronic Interpellations.' In D. Lyon and E. Zureik, eds., *Computers, Surveillance and Privacy*, 175–92. Minneapolis: University of Minnesota Press.

Pridmore, J., and D. Lyon. 2003. 'Customer Relationship Management as Surveillance: Software for Social Sorting in the Twenty-First Century.' Paper presented at the Canadian Sociology and Anthropology meetings, June, Halifax, Nova Scotia.

Reichman, N. 1987. 'The Widening Webs of Surveillance.' In C. Shearing and P. Stenning eds., *Private Policing*, 247–65. Thousand Oaks, CA: Sage

Rose, N. 1999. *Powers of Freedom*. Cambridge: Cambridge University Press.

Schuurman, P. 1994. 'Good Lookin'.' MA thesis, Queen's University, Kingston.

Solove, D. 2002. 'Privacy and Power: Computer Databases and Metaphors for Information Privacy.' *Stanford Law Review* 53 (6): 1393–1462.

Webster, F., and K. Robbins. 1986. *Information Technology: A Luddite Analysis.* Norwood, NJ: Ablex.

Zuboff, S. 1988. *In the Age of the Smart Machine: The Future of Work and Power.* New York: Basic.

3 Welcome to the Society of Control: The Simulation of Surveillance Revisited

WILLIAM BOGARD

I

Here are two illustrations of the simulation of surveillance, one from an Internet advertisement for Spector, a piece of software that monitors and records people's computer use, the second from an editorial outlining the dangers of a plan by the U.S. Defense Department, called Total Information Awareness (TIA), to create a universal database that will supposedly thwart terrorists and enhance 'homeland security.'

Spector

Imagine a surveillance camera pointed directly at your monitor, filming away everything that is done on your Macintosh. That is the idea behind the number one selling Internet Monitoring and Surveillance software, Spector.

Spector works by automatically taking periodic screen shots of a Power PC-based Macintosh and saves those screenshots to a local or network drive for later viewing. Screen shots can be taken as often as every few seconds, or as infrequently as once every few minutes.

Spector is ideal for consumers and corporations alike. Consumers now have the ability to see exactly what their children or spouse do on their computer when they cannot be around. Corporations and Educational Institutions can now make sure their employees and students are using their computers appropriately.

Recognizing that Internet filtering software is inadequate and inconvenient, Spectorsoft decided that the best way to put parents and teachers in control is to allow them to see exactly what kids do on the computer by recording their actions. With Spector, a parent/teacher sees everything the child sees. If a child tries to access a checking account, or visits adult-oriented web sites, or is approached by a stranger on the Internet, the parent/teacher will be able to see that by playing back the recorded screens.

'Internet filters don't solve the problem. They fail to filter out all the bad stuff, and they prevent users from doing completely legitimate tasks by producing far too many false positives,' adds Fowler.

'In addition, filtering programs require constant updates, and that is extremely inconvenient. Spector doesn't try to stop the user from doing anything. Instead, it records their actions. That places the issue of responsibility directly on the user. When a child or employee knows their actions may be recorded and viewed at a later point in time, they will be much more likely to avoid inappropriate activity.'

In addition to recording by taking screen snapshots, Spector also records every keystroke typed. With Spector's detailed and automatic snapshot recordings, one can see all e-mails, chat conversations, instant messages and web sites visited.

Users have been raving about the simplicity and accuracy of Spector: 'Within 36 hours of installing Spector I had enough evidence to go to the police. It turns out that our daughter was caught up in a sexual relationship with her 37-year-old Middle School teacher. The man was arrested, pled guilty, was sentenced and barred for life from teaching. None of this would have been possible without the evidence that we obtained using your Spector software' writes Bob Watkins of Tennessee. http://www.spectorsoft.com/products/Spector_Macintosh/index.html

Homeland Security

November 14, 2002

You Are a Suspect
Copyright The New York Times Company

[By, of all people, WILLIAM SAFIRE (former speechwriter for Richard Nixon)]

WASHINGTON. If the Homeland Security Act is not amended before passage, here is what will happen to you:

Every purchase you make with a credit card, every magazine subscription you buy and medical prescription you fill, every Web site you visit and e-mail you send or receive, every academic grade you receive, every bank deposit you make, every trip you book and every event you attend – all these transactions and communications will go into what the Defense Department describes as 'a virtual, centralized grand database.'

To this computerized dossier on your private life from commercial sources, add every piece of information that government has about your passport application, driver's license and bridge toll records, judicial and divorce records, complaints from nosy neighbors to the F.B.I., your life-time paper trail plus the latest hidden camera surveillance – and you have the supersnoop's dream: a 'Total Information Awareness' about every U.S. citizen.

This is not some far-out Orwellian scenario. It is what will happen to your personal freedom in the next few weeks if John Poindexter gets the unprecedented power he seeks. Remember Poindexter? Brilliant man, first in his class at the Naval Academy, later earned a doctorate in physics, rose to national security adviser under President Ronald Reagan. He had this brilliant idea of secretly selling missiles to Iran to pay ransom for hostages, and with the illicit proceeds to illegally support contras in Nicaragua.

A jury convicted Poindexter in 1990 on five felony counts of misleading Congress and making false statements, but an appeals court overturned the verdict because Congress had given him immunity for his testimony. He famously asserted, 'The buck stops here,' arguing that the White House staff, and not the president, was responsible for fateful decisions that might prove embarrassing.

This ring-knocking master of deceit is back again with a plan even more scandalous than Iran-contra. He heads the 'Information Awareness Office' in the otherwise excellent Defense Advanced Research Projects Agency, which spawned the Internet and stealth aircraft technology. Poindexter is now realizing his 20-year dream: getting the 'data-mining' power to snoop on every public and private act of every American.

Even the hastily passed U.S.A. Patriot Act, which widened the scope of the Foreign Intelligence Surveillance Act and weakened 15 privacy laws, raised requirements for the government to report secret eavesdropping to Congress and the courts. But Poindexter's assault on individual privacy rides roughshod over such oversight. He is determined to break down the wall between commercial snooping and secret government intrusion. The disgraced admiral dismisses such necessary differentiation as bureau-cratic 'stovepiping.' And he has been given a $200 million budget to create computer dossiers on 300 million Americans.

When George W. Bush was running for president, he stood foursquare in defense of each person's medical, financial and communications privacy. But Poindexter, whose contempt for the restraints of oversight drew the Reagan administration into its most serious blunder, is still operating on the presumption that on such a sweeping theft of privacy rights, the buck ends with him and not with the president.

This time, however, he has been seizing power in the open. In the past week John Markoff of The Times, followed by Robert O'Harrow of The Washington Post, have revealed the extent of Poindexter's operation, but editorialists have not grasped its undermining of the Freedom of Information Act.

Political awareness can overcome 'Total Information Awareness,' the combined force of commercial and government snooping. In a similar overreach, Attorney General Ashcroft tried his Terrorism Information and Prevention System (TIPS), but public outrage at the use of gossips and postal workers as snoops caused the House to shoot it down. The Senate should now do the same to this other exploitation of fear.

The Latin motto over Poindexter's new Pentagon office reads 'Scientia Est Potentia' – 'knowledge is power.' Exactly: the government's infinite knowledge about you is its power over you. 'We're just as concerned as the next person with protecting privacy,' this brilliant mind blandly assured The Post. A jury found he spoke falsely before.

II

All the old debates are raised by these two cyber-stories:[1] the disappearance of privacy, the cover-up and betrayal of secrets, the excesses of police power, Big Brother, Brave New World. Despite their continuing and even increasing relevance today, in a 9/11 world, there is much more going on here than these tired discussions indicate. We are, it seems, in the midst of a major transition, underway for years now, from *disciplinary societies*, which were organized principally, although not exclusively, by *technologies of confinement*, to what Deleuze has called 'societies of control,' or what I referred to less eloquently several years ago as 'hypercontrol in telematic societies' (Deleuze 1995: 167–82; Bogard

1 A sign of the times: our second example is already dated. Having been stung by the publicity surrounding Total Information Awareness, the U.S. Congress has slashed, although not eliminated, its funding. Whether or not the specific program survives, however, is a moot point, since its logic is what matters, and what is immanent to the forms of state command and control today.

1996). Less and less do we see social control technologies bound to specific territories, or governed by conventional territorial logics. The old institutions of confinement – prisons, schools, hospitals, workplaces, families – are breaking down, as they say, and new forms of control are emerging that are 'deterritorialized' and 'decoded,' in a word, *destratified*. The exclusionary techniques perfected in prisons, schools, and hospitals from the nineteenth to the mid-twentieth century and described so well by Foucault are old hat, as he himself recognized. State and military power are going the way of power in the workplace and on the street, indifferent to location and time, to longitude and latitude, and to constraints of scale (Foucault 1979). Control is now an inclusive, continuous, and virtual function, traversing every level and sequence of events, simultaneously molecular and planetary, no longer limited by walls or schedules. It is *disarticulated* – no longer organized by a principle of hierarchical or stratified observation, nor by a centralized power or set of rigidly segmented operations. Confinement henceforth is abandoned in favour of simulated controls that work with far more smoothness than the old strategies of spatial and temporal division (Poster 1990; Der Derian 1992).

In one sense, this development marks a shift from material to immaterial forms of coercion, set in motion by the new technologies of information and communication management (cf. Zuboff 1988: 219ff.). Disciplinary societies, of course, are not without immaterial controls. Statistical summaries and comparisons, numerical formulations, and a whole differential calculus of bodies and spaces were essential concomitants of discipline in the nineteenth century. Territories in the modern era, beginning in that age, became *coded as spaces of probability*, and confinement not just a matter of brute restraint but of organizing a whole social machinery in probabilistic terms. How is the collective body to be deployed for optimal effort? How are its movements to be synchronized, serialized, and standardized? Discipline became a science and art of correlations, of the rational connection of forces and functions; statistics became its means and compository of knowledge. Together, they produced 'efficient and effective' control across a host of interconnected institutions, always trying to improve performance and raise the 'confidence level' of various outcomes, to force a multiplicity to function maximally as a unit. In the factory, in school, and at home statistics combined with discipline to regulate the problematic relation between the possible and the real. How could training join with precise observation and measurement to convert what is merely latent or possible into a manifest function? With the advent of modern statistics that relation became formalized as a calculus of probability. The real was reconceptualized as a degree of likelihood, and possibility, once the most open of concepts, was rethought in terms of normal curves, tests of significance, and rules of inference.

The nature of control, however, is changing. Control is no longer merely a question of probability or efficiency. Statistical control, no doubt, is still very much with us. But it is no longer mired in the problem of the possible and the real. Who cares what is possible, or what efficient means exist to realize it, when one commands the virtual? Virtual realities are the order today, and reality is not a matter of statistical inference but of pure deduction, its truth less an outcome of controlled experimentation than of formal modelling, pre-programming, and digitalization. Control today is more about scanning data for deviations from simulation models than patrolling territories. Territorial control – old-fashioned surveillance – is only the final step in a series of prior operations that now take place on a purely axiomatic level (Hardt and Negri 2000: 326-7). It is the simulated crime – the virtual and not the 'possible' crime – that drives policing today. It is the model of delinquency, not its 'reality,' that pre-structures the field of monitoring and intervention.

The goal of information and communication management technologies is simply to control as perfectly and seamlessly as possible all conceivable outcomes *in advance*. This is the logic behind data mining, profiling, cloning, scenario engineering, sim-training, and the like: to substitute proactive measures for the old reactionary regimes of spatial and temporal division. While they still utilize statistics, the projected line of such measures nonetheless is to eliminate the dialectic of the real and the possible, or the probable and the unlikely, and with it the whole discourse of efficiency. The smoothest form of control, according to this logic, is not merely 'efficient,' it is 'prefficient,' that is, it eliminates problems *before* they emerge, absolutely, before they even have the chance to *become* problems. This is hypercontrol, an ultimate resolution to the problem of efficiency, with all the techno-determinist, totalitarian, racist, imperial images associated with that phrase. It is the *pre-emptive strike*, to use the terms of the Bush doctrine for combatting terrorism: reaction precedes reacting, precession of reaction, finality of reaction.

On the other hand, and without the slightest contradiction, with hypercontrol, nothing is ever final. These systems of proactive reaction, for all their finalities, produce a state of existential and not simply statistical uncertainty. We can imagine the indeterminacy for the inmate surrounding the exercise of panoptic power in the twenty-first century perfected to the point of its own disappearance. At first glance, such uncertainty is something like the conditions of 'ostensible acquittal' and 'indefinite postponement' Kafka describes in *The Trial* (Kafka 1968: 156–60). But it is even beyond this. In that tale, one never knows the crime of which one stands accused, or when one might be arrested and charged. The current situation rather points to an uncertainty that extends to the *possibility of crime itself*. It is easy to imagine Kafka in a not-so-distant world where crime is perfectly pre-empted and guilt and innocence have lost their meaning, where everything and nothing is a crime. This too is part of the logic of hypercontrol. Henceforth, crime does not exist in virtue of discipline, which seeks to know and control it (Foucault 1979), but in virtue of simulation, which derives it from a model and thus renders its existence undecideable (Baudrillard 1996: 1-8).

We cannot fall back on easy metaphors to explain these developments. This is not Big Brother. In a world already scoured of problems, who needs an omnipresent watcher? And it is not Brave New World either. The new controls do not work on the level of pleasure or pain, but on the *plane of desire*. There is nothing revolutionary about control that operates through pleasure: any psychologist can make a rat run a maze. Only desire, Deleuze and Guattari say, is revolutionary, and the unparalleled intensification of technical control we are witnessing today is nothing short of a revolutionary movement within – and against – desire (Deleuze and Guattari 1983: 222–62; Guattari 1995: 204ff.). In fact, it is so revolutionary that it eliminates the problem of control itself; at least, that is, its imaginary. 'Societies of control' have dreams of a time when they can get out of the control business altogether, when the ancient 'war between technology and desire,' as Sandy Stone (1995) puts it, is over and everything goes on automatic. This is the dream behind Spector and Poindexter's Total Information Awareness program, indeed the whole global program to simulate surveillance. Not to watch, not to have to react, not to police, not even to measure or correlate, but to sit back and let the system, itself a product of desire, indeed of a kind of delirium, take command.

Let's Begin Again

It used be that when you went to work, you were not in school anymore, or that when you went home, you were no longer at work

(Deleuze 1988: 40). Now, the lines that once served to divide the various regions and times of control from one another are dissolving into a single form capable of infinite modulations. Deleuze and Guattari call this form an 'abstract machine,' while Hardt and Negri refer to it as the increasing abstraction of panoptic mechanisms within global empire (2000: 330). Deleuze (1995) described the old logic of confinement versus the new logic of control societies this way: 'The various placements or sites of confinement through which individuals pass are independent variables: we're supposed to start all over again each time, and although all these sites have a common language, it's analogical. The various forms of control, on the other hand, are inseparable variations, forming a system of varying geometry whose language is digital (though not necessarily binary). Confinements are molds, while controls are a modulation, like a self-transmuting molding continually changing from one moment to the next ...' (178–9).

If we consider them from the perspective of sign systems rather than territories, disciplinary societies, Deleuze writes, have two poles: *signatures* that stand for individuals, and *numbers recorded in registers* that stand for the places of individuals in a mass (and form the basis of statistical correlations). There is no contradiction or opposition between these poles. Disciplinary societies exercise power in both ways, by individuation and massification. In control societies, however, this duality collapses in favour of a single system capable of finely modulated adjustments. To take a musical example, modulation changes the key signature and various registers of a composition but not its formal structure or the internal relation of its parts. Signatures and numbers, Deleuze claims, are replaced with *passwords*, which determine whether or not you have access to information. Passwords in turn are *codes*, and codes are the new language of control in digital systems. The switch to digital forms of control involves a massive abstraction and a corresponding homogenization, the disappearance of both individuals and masses into packets of information, into bits and signals and spectra. 'Individuals become "dividuals" [that is, subject to control at multiple levels of the organization of the individual], and masses become samples, data, markets, or "banks"' (180). The new forms of control inaugurate not simply changes in the extension but *intensive* transformations of disciplinary procedures. They represent a 'phase shift' in the history of the exercise of power, in the same sense that Foucault described the historical transformation from sovereign to disciplinary power (1979, 1980a). We could make an analogy between ice changing into water and what happens when disciplinary societies become societies of control. Discipline becomes liquid: it flows into every hole, fills every crack, and leaves nowhere to hide.

In fact, the matter is more complicated, as Deleuze himself would admit. What distinguishes control societies from disciplinary societies is not really the use of codes, which historically exist in all societies, but rather the specific form of *decoding and recoding* that control societies initiate. Baudrillard (1983, 1995) has characterized the shift as the passage from systems of representation to systems of simulation, or as the metamorphosis of reality into hyperreality. What is decoded in this passage is the relation of equivalence between the sign and its object, or alternately, the difference between truth and falsity, reality and fiction, and so on. We do not have to accept Baudrillard's at times fantastic conclusions to acknowledge that a fundamental recoding of relations of knowledge and power – and pleasure/desire, to complete the triad constituted by Foucault – are underway in information societies (Foucault 1980a). What this recoding amounts to in the first instance is a destruction of the reality principle upon which modern forms of both knowledge and power base themselves and its reconstitution as hyperreality. Knowledge becomes information, power becomes display. Ultimately, however, this resolves into a fundamental *disarticulation of subjectivity*, that is, of the status of the modern subject, its relations of identity and alterity, its connections to the law, and so forth. The subject is recreated as a 'virtual subject,' such as we find on computer networks.

We must study control society in exactly the same way that Deleuze and Guattari analysed capital (1983: 222ff.), as a complex socio-technical machine that exerts control through decoding and deterritorializing subjectivity. Where this was previously the function of surveillance, and more broadly, discipline, today increasingly it is the function of simulation, in particular, the production of simulated or hybrid subjects occupying simulated spaces, and targets of simulated forms of control. Modulated and infinitely 'modulatable' subjectivities are indifferent to institutional setting (home, work, play, school), to time and place, and to all the outdated strategies of modern state power. Let me explain.

In their recent book *Empire*, Hardt and Negri, who draw heavily on Deleuze and Guattari, list some of the features of these new systems of control (2000: 22–30). They are decentralized yet global in scope. They are nonlinear. They are increasingly immaterial and invest immaterial forms of production such as knowledge production, publicity and communications, service work, and so on. Rather than fixing identities in hierarchies and systems of exclusion, they operate through the proliferation and management of multiple and hybrid identities. For Foucault, a defining feature of discipline in the production of the modern subject was its recoding of multiplicities and differences to binary oppositions: self and other, normal and abnormal, sane and mad, healthy and sick, delinquent and non-delinquent (Foucault 1965, 1975, 1979, 1980a; cf. Canguilhem 1978; Deleuze 1988: 23ff.). Biunivocalism and the resulting essentialization of identities are the hallmarks of modern power. Societies of control, in contrast, exercise power precisely through the production of differences and the radical deconstruction of the binary basis of identity. What matters most in postmodern forms of control is the absolute fluidity of identity, the disappearance of the line between self and other, the seamless integration of bodies and information systems. The global production of capital increasingly demands flexible, modulated subjectivities, receptive to the appeals of mass marketing, the swings of opinion polls, and the decentring of management practices. It also demands rapid mobility of global workforces, and the obliteration of the distinction between labour time and leisure time, work and play, factory, school and home (Zuboff 1988; Hochschild 1997). These requirements are increasingly accomplished through the digitalization of control. Rejecting the claim of some postmodernist theories that difference, play, and hybridity are liberatory in themselves and can be opposed to the modernist production of essentialist identities, Hardt and Negri have argued that postmodern capital has promoted these qualities to further systematic control at the global level (2000: 142). Empire no longer functions, in other words, to suppress differences, but to produce and micro-manage differences at both the level of content and expression.

III

It is most useful to think about hypercontrol in terms of the destratification of social control. This may sound contradictory, because we normally associate social control with the production of strata, that is, with the constitution of hierarchies of power (race/class/gender, etc.), categorical exclusions (normal/abnormal, mad/sane, healthy/sick), and so on. In one sense, it is quite true that nothing has changed: control still operates to create and maintain systems of unequal power and value, even more so than in the past. As Deleuze and Guattari say, destratification always entails *restratification* (1987: 54). The destratification of control we are observing with regard to disciplinary societies is perfectly compatible with its reestablishment on a different level, or within the same level in different ways. This is why there can be no universal definition of control, no single form. Its elements change, its historical function varies, in response to changes in its milieu that themselves involve other kinds of control. There is no 'dialectic of control' (see, e.g., Giddens 1983: 39), only shifting between states of more or less control, contests among types of control, states of relative order and disorder. Change or 'becoming' is always a movement between less and more articulated states. There is no escape from (re)stratification. In the words of Artaud, who hated all fixed or imposed forms, it is a kind of condemnation, 'the judgment of God' (Deleuze and Guattari 1987: 150; Artaud 1988). One resists its commands, but it always finds new ways to speak, to 'express' itself, always institutes other modes of restriction or limitation. Discipline may no longer depend on confinement. It is destratified today, but it restratifies within the mode of information, and it demands a new practice of resistance. In the same way, the panoptic model no longer governs surveillance, but restratifies it within the mode of simulation. To fully understand the transmutation of discipline and surveillance into hypercontrol and simulation we need to frame the question of social control within the broader problem of stratification.

Deleuze and Guattari define strata as phenomena of 'thickening' in a 'plasma field' (1987: 502). For 'thickening' we can substitute the concept of 'articulation' – both refer to a change of *content* (the manner in which unattached elementary particles are selected and connected), and to a change of expression (how that content becomes rigid, develops fixed functions, etc.) (40, 502). To articulate is to stratify. The plasma field – which Deleuze and Guattari sometimes refer to as the 'Body Without Organs' or, with a somewhat different and broader connotation, the 'Plane of Consistency' – consists of 1) unformed, unstable matters and flows, 2) free intensities (energy levels), and 3) singularities (threshold events or bifucators). It is *disarticulated*, that is, destratified, lacking both the connections and successions of elements that form the content of a stratum, as well as the functional relations between parts that constitute its particular expressive qualities. Stratification is the movement from less to more articulated states, from less to more organized milieus, and always involves the loss of degrees of freedom. Think, for instance, of expression in formal languages, as progressive articulation eliminates useless phonetic, syntactic, and semantic elements. De Landa (1997:

227ff.) has analysed such linguistic processes in terms of creolization and pidginization, while Laporte (2000) has likened the stratification of expression in language to a process of sanitation, specifically, the elimination of shit.

Following tradition, Deleuze and Guattari distinguish three kinds of stratification: physico-chemical, organic, and social or alloplastic, referring to the capacity to bring about modifications in the external world (1987: 502). Of course, our concern is primarily with the last kind, particularly as it relates to technologies or 'machinic assemblages' of control. The three kinds, however, are inevitably found mixed together. Stratification in general, as a process of thickening or articulation, involves 1) giving form to matter, 2) imprisoning intensities, and 3) locking singularities into systems of redundancy involving repetition and succession. Foucault's work on the prison provides an excellent example of social stratification in these terms. It is a machine whose 'articulations' comprise social, organic, and physical elements. Developing a microphysics of space and time, discipline sorts bodies and the gestures they are capable of making into homogeneous collections that serve formal functions. It maximizes and concentrates the collective body's energies for the completion of functional tasks such as work through repetition, drill, exercise, and so forth (1979: 135ff.). More recent authors drawing on Foucault (e.g., Gandy 1993; Lyon 1994, 2001) have emphasized the sorting and categorization functions of surveillance as a concomitant of disciplinary practice. Surveillance is one of those assemblages that acts as a 'surface' of stratification, a machine where element-particles are tested and some are selected for inclusion in the stratum, others rejected. What are the 'elements,' the content, of social strata? Not race, class, and gender, which are levels of expression, second-order phenomena. At the level of content, the elementary particles of social strata are body parts, partial movements, skin colours, scars, bits of hair, vocal sounds, DNA, semen, marks, that is, almost any 'dividual' item amenable to selection, repetition, or useful connection in the production of individuals.

Deleuze and Guattari (1987: 40-1) emphasize that all strata, including alloplastic strata, are 'double articulations' (see also De Landa 1997: 60). The first articulation involves a process of 'sedimentation,' which involves the sorting of particles into similarly composed layers and the imposition of a statistical order of connections and successions among those particles. They call this first articulation a 'connective synthesis,' and it concerns the production of a stratum's content, or alternatively, its primary coding. Deleuze and Guattari refer to 'statistical connection' in the sense of quantum relations, and not at all in the molar sense of statistical relations between already constituted strata, what they sometimes call 'overcoding' (1987: 62, 219). Again, discipline provides an example: Foucault refers to training the body in certain skills in this way, as a connection/coding of small movements. Here surveillance is a means of sorting particular quanta. The second articulation is a process of 'folding,' or what De Landa (1997: 60) has called 'cementing,' which establishes functional structures from collected elements and constructs molar compounds or expression. With regard to discipline, this articulation refers to the production of what are commonly, if inadequately, known as macro-institutional structures, that is, integrated functions and systems of rank. Surveillance, at this level, is concerned predominantly with the regulation of molar divisions of class, race, gender, and so on. Deleuze and Guattari sometimes call the second articulation the 'disjunctive synthesis' (1983: 12–3). It forms the rigid segments we often associate with formal structures, such as bureaucracies, as well as 'distributes centers of power and overcodes aggregates' (1987: 210–13). The distinction between the two articulations is real, not merely conceptual. In practice, however, they are always mixed up together. State power, for instance, operates at both levels of stratification, content and expression. It is interested in the molecular as well as molar organization of society, and it polices not only class, race, and gender, but body fluids and chemistry as well.

Social stratification, for Deleuze and Guattari, is a 'machinic assemblage,' or the product of a machinic assemblage. I must defer a discussion of this complex idea here, except to say that 'machinic assemblages,' in their sense, are more than just technical equipment, but include systems of knowledge and relations of affect (Guattari 1990, 1995; Guattari 1996: 236). They produce, or if you will, acquire, a subjectivity. The panopticon – as a material arrangement of space and time, a machinery of observation and ranking, a strategy of control, a means of gathering information, and an instrument for the distribution of desire and the production of truth – is an important kind of machinic assemblage, part architecture, part philosophy and design, part pleasure and dream. When we refer to surveillance as a machinery of social stratification and control, it is also always in this expanded sense of the machine, that is, a collection of tools, engineering plans, infused with a kind of passion around which a certain collective body and subjectivity develops. For Foucault (1979), such a machine is immanent to the production of 'docile bodies' (level of content) and 'delinquents' (level of expression).

For Deleuze and Guattari, every machinic assemblage is territorial, and the first rule for analysing assemblages is to discover what territoriality they envelope, for there always is one. The territory of the surveillance assemblage is essentially the field of actions and passions, upon which it imposes form and serves as mechanism for their distribution into relatively homogeneous 'layers.' Foucault (1979), describes this in terms of technologies of examination, normalizing judgment, and so on. But the general principle behind his analysis remains one of confinement, the division of space-time into affective and somatic territories that cement relations of power and produce subjectivity. One must be in this place at this hour, one must be visible and open to inspection within these prescribed zones, one may not enter or leave this area without the right credentials: different sets of rules governing each territory, and rules that separate one territory from another. The body itself becomes a territory upon which relations of power are exercised. The body is mapped out, its lines of force and resistance carefully recorded. We are all familiar with this story. Surveillance and discipline at bottom enforce a territorial principle: the forces that control the terrain, the knowledge of its boundaries, its high points and centres, its relief, its exposed regions and blind spots, control what unfolds there, the movements of its populations, the flows of materials and concepts.

Every territorial assemblage, insofar as it imposes a form, also entails a code, that is, a rule of selection or strategy of repetition. *We cannot separate territorialization and coding,* they always go together in the production of strata (Deleuze and Guattari 1987: 41). Coding, like territorialization, occurs at both the level of content and the level of expression. Languages, for example, are coded both as connections/successions of vocal elements and as grammatical rules. With regard to content and expression, codes are like filtering machines. They separate noise from information, or nonsense from sense. They 'deduct' free elements in their milieu and arrange them in graded orders. De Landa draws the analogy of a stream that deposits variably sized particles of sediment in relatively homogeneous layers on its bed, which later become compressed into strata, so that the moving water acts as a sorting mechanism (1997: 60).

Thanks to Foucault, we are familiar with the coding schemes, the diagrams and 'abstract machines' at the heart of disciplinary regimes. They are the 'engineering' schemas that separate normal from abnormal populations, truth from falsity, reality from illusion, sanity from madness. In all cases, the code organizes a territory of control and divides one population from another, or compounds forces to produce integrated functions. Inevitably, in modern Western societies, the code is formulated to place power firmly on the side of truth, normality, and the real. Foucault sometimes refers to technologies of the self, the obligation to speak the truth about oneself (Foucault, Lotringer et al. 1997). We have already seen how Deleuze sees a movement from the language of signatures and numbers in societies of confinement to the language of codes in societies of control. But it is more accurate to say that one set of codes has been exchanged for another, in a movement that involves a fundamental decoding of one system of control and its recoding into another, from signatures and numbers to passwords and models, from signs of identification and systems of registration to informated profiles and genetic markers.

Deleuze and Guattari in fact make this very case in *Anti-Oedipus* (1983: 222ff.). There, capital is depicted as an immense decoding (and deterritorializing) machine. Every movement – flows of money, of resources, of body parts, of waste, of signs, of art, of power – must be decoded and recoded to serve the *axiomatic* of capital, which beyond the simple production of surplus value involves the destratification of identities and destabilization of systems of signification. Indeed, it requires the destruction of its own axiomatic in order to reduplicate it on ever higher, more inclusive levels. *It is no different with the disciplinary assemblage.* It is willing to sacrifice its own principle in order to reconstitute itself on a more deadly and smooth plane, and at an even lower degree of intensity. For Deleuze and Guattari, this is capital as a cancerous body without organs, the zero degree of its own death that its experiments in the control of production and work always aim for, including the annihilation of living labour and its reincarnation in 'cyborg work' (Deleuze and Guattari 1987: 163; Bogard 1996: 98ff.). Discipline moves into cyberspace, it dematerializes; surveillance mutates into simulation, it becomes hyperreal; the hyperreal is deterritorialized, it is decoded space.

IV

When I wrote *The Simulation of Surveillance*, my biggest mistake was to give too much weight to the imaginary of surveillance, a move inspired, in part, by reading Baudrillard. I argued simulation was the 'dream-logic' or 'imaginary (pataphysical) solution' of the surveillance machine: flawless control, control in advance, and thus in effect the end of control (1996: 23). If surveillance was a strategy of visibility, then simulation was perfect exposure. If surveillance was continuous observation, then simulation was the fantasy of vicarious experience: not only can I see inside your head, but I can have the same perceptions and experiences as you. If surveillance was the 'recording machine,' then simulation was the illusion of perfect reproduction: in Spector, the image it recorded is an exact reproduction of the image produced.

To be sure, Baudrillard saw simulation not quite as an imaginary, but as the hyperrealization of the image, or the disappearance of what separates the image and the real, and that is an entirely different matter. The image, he says, conceals its 'murder of the real' as it itself becomes 'more real than real' (the 'simulacrum is real'). It is the 'perfect crime' because the new reality, the same in every detail, replaces the old, and no one knows the difference (Baudrillard 1996: 1–8).

My book described the simulation of surveillance as a logic of control that is materializing before our very eyes in postmodern society, ever more totalizing, intensive, and thoroughgoing. The notion of control in advance was meant to emphasize its pre-emptive aspect; Baudrillard called it simply the code, the substitution of signs of the real for the real itself (Baudrillard 1983). If simulation is the perfect crime, the disguised murder of the real, than the simulation of surveillance is the perfect police, viz., absolute control over the production of reality. Perfect crime and absolute control all wrapped up in one. This is hypercontrol, more controlling than control, and I simply meant the policing logic that inspires technologies like Spector, or the homeland security measures that John Poindexter dreams up for the U.S. government, or a host of other current measures designed to push surveillance to its limit – indeed, to move it beyond its limit, which is territorial and implies, as we have seen, a logic of confinement. Dream up a world where surveillance is perfect, where it operates as a constant and complete background to daily life. The new surveillance no longer targets bodies per se – messy, unpredictable things – but the information bodies produce or harbour about themselves, contained on their hard drives or in their genes. There are no secrets here; all codes conform to one code that decodes them all. The police do not wait for a crime to happen, because they already have staged it, using profiles or genetic indicators. This imaginary is even more poisonous, if possible, than Orwell's, because in it control itself finally disappears into a pure operationality against which all resistance is impossible.

As I have indicated, I now think that these developments in control are more usefully described through a model of destratification and restratification, or deterritorialization and reterritorialization, decoding and recoding. The simulation of surveillance is not a dream or imaginary state of perfect control. It involves real and imperfect strategies of extending social control beyond systems of confinement, deterritorializing the space of enclosure, allowing enclosure to operate, as it were, 'at a distance,' or rather without regard to distance, the model of *telematic or virtual confinement*. The classic example is the monitoring bracelets worn by sex offenders to track their movements outside the prison, but it is easy to think of even more radical devices such as genetic mitigations and implants. To be sure, what we are witnessing is a kind of dematerialization of control, but although it has an imaginary logic, it is a fully positive movement in relation to the positive forms of surveillance and discipline it is in the process of replacing.

It is in the realm of decoding technologies that the simulation of surveillance truly transports us into a new logic of control. Confinement has a code as well as a territory, and it is this code that is broken by simulation. It is the very idea of confinement, as a means of control, its 'reality principle,' that is at stake here. The old notion of confinement entailed a restriction on movement, a limitation to one place, and a separation of that place from spaces of free movement. Today, the system itself demands nomadic flows of populations and resources and therefore porous borders. Foucault (1979) had already suggested in *Discipline and Punish* that confinement need not be physical, that confinement is not even necessary for discipline (cf. also Deleuze 1988: 42). Although he does not exactly use these terms, the sense of this work is that the production of the 'modern soul' involves a kind of self-policing, without the need for walls. At the end of the book, in the story of the child of Mettray, Foucault portrays the individual who comes to love his confinement (1979: 293). But already the limitation to one place is inoperative here, and Mettray, symbol of the carceral society, could be anywhere. A kind of decoding of the model of imprisonment, of confinement and the carceral, was already at work at the very beginning of disciplinary societies in the form of an interiorization of control and of self-monitoring. Today, societies of control are embarked on the next stage of this process, which involves the cancellation of the interior/exterior duality and, as Deleuze has noted, the replacement of the individual self as a locus of social control with the form of the modulated 'dividual': no longer a unified self, but a kind of fractal subjectivity, endlessly divisible, and upon which control can be exercised at will in any context and for any purpose. A dividual can be any partial object or event, at any scale of organization, human or otherwise.

> We don't have to stray into science fiction to find a control mechanism that can fix the position of any element at any given moment – an animal in a game reserve, a man in business (electronic tagging). Felix Guattari has imagined a town where anyone can leave their flat, their street, their neighborhood, using their (dividual) electronic card that opens this or that barrier, but the card may also be rejected on a particular day, or between certain times of day; it doesn't depend on the barrier but on the computer that is making sure everyone is in a permissible place, and effecting a universal modulation. (Deleuze 1988: 182)

None of these changes in the nature of control suggest that the prison, or the carceral more generally, is disappearing. They are perfectly consistent with more prisons and even higher rates of incarceration. What is changing is the *diagram* of relations of power, that is, how control is engineered. While barriers and enclosures certainly still constitute the most visible aspects of control, there is no doubt that computerization and biotechnologies are increasingly virtualizing the space of punishment. It is not difficult to imagine a day when crime control is simply a matter of comparing genes to a model of genetic normality. Today, of course, it has become normal at any barrier – transactional, financial, political, educational, residential – to present one's dividualizing mark – password, DNA, retina, face, whatever – for comparison to a soon to be 'virtually centralized' database of models. One need not be aware of any of these developments, as they are being accomplished automatically, without the least disruption to the normal flow of events. It is not the prisons that will disappear. Soon it will simply be the people (or parts of people!) that do not quite measure up to the model that will disappear, even before they are born, before they are formed. For in the end, this is what the decoding and deterritorialization of control, the simulation of surveillance, come down to: not individualization, or even self-monitoring/self-confinement, but the pre-formation of molecules according to models that allow for their easy experimentation and quick recombination. We know these things are on the horizon or even closer. They are imagined in science fiction films like *Gattaca*, where 'invalids' are those with natural parents and the wrong genes, and the literature on cyborgs (Haraway 1990).[2] Hardt and Negri refer to them as the global functions of 'biopolitical production,' Poster as the Superpanopticon, I call them as hypercontrol, and so on (Poster 1990; Hardt and Negri 2000). We are all talking about the same things in different registers, the information and simulation revolutions in control. It is almost enough to make one nostalgic for the old systems of discipline and surveillance.

V

It is easy to present the transformation of control in black-and-white terms. The destratification of confinement, the deterritorialization of enclosed spaces, and the simulation of surveillance suggests a kind of one-way process. In fact, the picture is tremendously complicated. In Deleuze and Guattari, we have seen, stratification, destratification, and restratification are entirely relative terms. Nothing is ever destratified without its elements immediately being restratified in other ways and on other levels. De Landa has suggested that the model of stratification is itself too linear and deterministic and has proposed a notion of 'meshworks' to account for systems of control that are nonlinear and self-organizing (De Landa 1997: 62). Such systems evolve 'autocatalytically' and are nonhierarchical and decentralized. They are composed of heterogenous elements and relations rather than the relatively homogeneous layers that comprise strata, which is not to say that they cannot give rise to stratified structures. There is a great deal of merit in such an approach. Machinic assemblages have rhizomatic properties, they are multiplicities and have fractal branchings ('any point can, and must, connect to any other'; Deleuze and Guattari 1987: 7). Surveillance-simulation assemblages are complicated arrays of sensors, storage and recording mechanisms, databases, channels, screens, but also models,

---

2 In *Gattaca*, 'invalids' are contrasted to 'valids' whose parents assured their genetic perfection before they were born. In the film, invalids occupy the bottom of the stratification ladder and do all the unpleasant tasks of society. Here, typical of Hollywood, we are still given a portrayal of a future as technologies of truth and hierarchies of power. One can just as easily claim that invalids represent the destratified segment of society, below the social ladder altogether, a pool of waste materials waiting to be recycled into useful products. One could also imagine a society where invalids are done away with altogether, unpleasant work is handled by machines, and genetic engineering is a purely formal operation with no relation to truth and falsity, validity and invalidity.

statistical probabilities, and feedback loops. They embody desire and dreams, theatrics and disguise, lines of flight and capture.

In Deleuze and Guattari, machinic assemblages are already presented as something different than strata (1987: 503). They have a side that 'faces the strata' – in that sense, they operate on the edges or faces of strata as apparatuses of capture, that is, as machines that collect and sort the elements that will become a stratum's content. This is their function of territorialization and coding. On the other hand, machinic assemblages are 'cutting edges of deterritorialization and decoding.' They are the movements 'by which one leaves the territory' and are composed of 'lines of flight' or resistances to capture (508). They operate on *surfaces*, between strata (epistratic phenomena), or between a stratum and a 'body without organs' (unformed matter, non-formal functions).[3] This is why Hardt and Negri have likened them to 'smoothing machines,' and have written about the development of global capital and its variable machineries of control as a process of smoothing. Smoothing is destratification, the loss of substantive and functional coherence as, for example, when a hard metallic object is melted down into liquid form (Bogard 1996, 2000; Guattari 1995; Hardt and Negri 2000: 332). It involves a shift in the mode of control from transcendent operator to immanent force:

> Capital ... demands not a transcendent power but a mechanism of control that resides on the plane of immanence. Through the social development of capital, the mechanisms of modern sovereignty – the processes of coding, overcoding, and recoding that imposed a transcendent order over a bounded and segmented terrain – are progressively replaced by an axiomatic; that is, a set of equations and relationships that determines and combines variables and coefficients immediately and equally across various terrains without reference to prior and fixed definitions or terms. The primary characteristic of such an axiomatic is that relations are prior to terms. In other words, with an axiomatic system, postulates 'are not propositions that can be true or false, since they contain relatively indeterminate variables. Only when we give these variable particular values ... do the postulates become propositions, true or false ...' Capital operates through just such an axiomatic of propositional functions ... [It] tends toward a smooth space defined by uncoded flows, flexibility, continued modulation, and tendential equalization. (Hardt and Negri 2000: 326–7)

3 On the difficult notion of surfaces, as both material and abstract features of strata, cf. Avrum Stroll (1988).

Foucault described with great subtlety how the transcendent order of sovereign power was replaced by the immanent order of discipline (Foucault 1979, 1980a, 1980b). With the advent of societies of control, the immanent order, articulated through the deployment (*dispositif*) of a series of stages of abstraction – what we call hypercontrol or simulation – finally loses its institutional walls. Again, Hardt and Negri write:

> We can say that the dispositif (translated as mechanism, apparatus, or deployment) is the general strategy that stands behind the immanent and actual exercise of discipline. Carceral logic, for example, is the unified dispositif that oversees – and is thus abstracted and distinct from – the multiplicity of prison practices. At a second level of abstraction, the diagram enables the deployments of the disciplinary dispositif. For example, the carceral architecture of the Panopticon, which makes inmates constantly visible to a central point of power, is the diagram or virtual design that is actualized in the various disciplinary dispositifs. Finally, the institutions themselves instantiate the diagram in particular and concrete social forms as well. The prison ... does not rule its inmates the way a sovereign commands its subjects. It creates a space ... in which inmates discipline themselves ... Sovereignty has become virtual (but it is for that no less real), and it is actualized always and everywhere through the exercise of discipline.
>
> Today the collapse of the walls that delimited the institutions and the smoothing of social striation are symptoms of the flattening of these vertical instances toward the horizontality of the circuits of control. The passage to the society of control does not in any way mean the end of discipline. In fact, the immanent exercise of discipline ... is extended even more generally in the society of control. What has changed is that, along with the collapse of institutions, the disciplinary dispositifs have become less limited and bounded spatially in the social field. (2000: 330)

I have been writing as if surveillance represents the territorial dimension of control assemblages, and simulation the deterritorialized aspect, but in fact that matter is much more complex. In practice, surveillance and simulation, although really distinct assemblages, always mutually implicate and complicate each other's operations and development. Monitoring and recording inevitably involve an element of stealth, and modelling depends upon the systematic collection and distillation of information. It would be a mistake to posit a simple linear relation between surveillance and simulation.

When Deleuze remarks that the diagram of confinement is giving way to the society of control, although he sees this as a historical development, he does not intend us to view the matter as a simple replacement of one form of power by another. Relations of power in every society are a matter of mixed constitution; control always operates as a function of relative degrees of territorialization and deterritorialization, coding and decoding. Every 'apparatus of capture' (surveillance) is crossed by 'lines of flight' (simulation), and everything that flees, at some point, becomes a trap.

The scope of surveillance, of course, like discipline, is far wider today than in the past. The powers of monitoring and recording have expanded exponentially in postmodern societies, to the point where virtually every space, interior and exterior, has become a space of observation (Dürrenmatt 1988). There is no denying that this expansion produces intense effects of stratification. Everything is exhaustively classified and categorized, exclusions are more detailed, space and time are more rigorously mapped and divided. At the same time, control is more decentred and dematerialized, exercised increasingly in virtual space-time, well in advance of the operation of assemblages of surveillance. Just as Foucault claims that power cannot be understood apart from resistance, the relation between surveillance and simulation is not one of contradiction, but of implication and complication. The task in any analysis of social control is to determine in as detailed a way as possible the concrete mechanisms of stratification and destratification at work, and how they transform the traditional practices of institutional confinement.

Ultimately, as Guattari has written, this is a problem in the production of desire and subjectivity (1996: 193–203). What is at stake in the emergence of the society of control is an exceptionally complicated redeployment of global relations of affect and identity. It is clear that whereas the strategy of discipline was to channel a multiplicity of affects into a single stream and to produce unified institutional identities, to separate normal and abnormal, sane and mad, true and false, today the task of control is to destratify desire and the subject, to multiply channels of affect and promote the emergence of hybrid subjects, to free information from its connections to signification and truth, and to virtualize relations of power, all within the axiomatic of capital. The simulation of surveillance is the effort to convert the revolutionary force of desire into the carefully regulated production of pleasures – mass media, computers, marketing, gaming – and the bounds of essentialist identity and experience into a bestiary of grotesque hybrid forms – cyborgs, mutants, emoticons – receptive to the commands of global production. One can only imagine such an unimaginable project will fail.

REFERENCES

Artaud, A. 1988. *Antonin Artaud: Selected Writings*. Berkeley: University of California Press.
Baudrillard, J. 1983. *Simulations*. New York: Semiotext(e).
– 1995. *Simulacra and Simulation*. Ann Arbor: University of Michigan Press.
– 1996. *The Perfect Crime*. New York: Verso.
– 2001. *Impossible Exchange*. New York: Verso.
Bogard, W. 1996. *The Simulation of Surveillance: Hyper-control in Telematic Societies*. New York: Cambridge University Press.
– 2000. 'Smoothing Machines and the Constitution of Society.' *Cultural Studies* 14 (2): 269–95.
Canguilhem, G. 1978. *On the Normal and the Pathological*. Boston: Reidel.
De Landa, M. 1997. *A Thousand Years of Nonlinear History*. New York: Zone Books.
Deleuze, G. 1988. *Foucault*. Minneapolis: University of Minnesota Press.
– 1995. *Negotiations, 1972–1990*. New York: Columbia University Press.
– 1997. *Essays Critical and Clinical*. Minneapolis: University of Minnesota Press.
Deleuze, G., and F. Guattari. 1983. *Anti-Oedipus: Capitalism and Schizophrenia*. Minneapolis: University of Minnesota Press.
– 1987. *A Thousand Plateaus: Capitalism and Schizophrenia*. Minneapolis: University of Minnesota Press.
– 1994. *What is Philosophy?* New York: Columbia University Press.
Der Derian, J. 1992. *Antidiplomacy: Spies, Terror, Speed, and War*. Cambridge, MA: Blackwell.
Dürrenmatt, F. 1988. *The Assignment (or, On the Observing of the Observer of the Observers)*. New York: Random House.
Foucault, M. 1965. *Madness and Civilization: A History of Insanity in the Age of Reason*. New York: Pantheon Books.
– 1975. *The Birth of the Clinic; An Archaeology of Medical Perception*. New York: Vintage Books.
– 1979. *Discipline and Punish: The Birth of the Prison*. New York: Vintage Books.
– 1980a. *The History of Sexuality*. New York: Vintage Books.

– 1980b. *Power/Knowledge: Selected Interviews and Other Writings, 1972–1977*. Brighton, Sussex: Harvester Press.

Foucault, M., and S. Lotringer, et al. 1977. *The Politics of Truth*. New York: Semiotext(e).

Gandy, O. 1993. *The Panoptic Sort: A Political Economy of Personal Information*. Boulder, CO: Westview Press.

Giddens, A. 1983. *Central Problems in Social Theory*. London: Macmillan Press.

Guattari, F. 1984. *Molecular Revolution: Psychiatry and Politics*. New York: Penguin.

– 1990. 'On Machines.' *Journal of Philosophy and the Visual Arts* 6:8–12.

– 1995. *Chaosmosis: An Ethico-Aesthetic Paradigm*. Bloomington: Indiana University Press.

– 1996. *The Guattari Reader*. Ed. G. Genesko. Cambridge: Blackwell.

– 2000. *The Three Ecologies*. London: Athlone Press.

Haraway, D. 1990. *Simians, Cyborgs, and Women: The Reinvention of Nature*. London: Free Association

Hardt, M., and A. Negri. *Empire*. 2000. Cambridge: Harvard University Press.

Hochschild, A.R. 1997. *The Time Bind: When Work Becomes Home and Home Becomes Work*. New York: Metropolitan Books.

Kafka, F. 1968. *The Trial*. New York: Schocken Book.

Laporte, D. 2000. *The History of Shit*. Cambridge, MA: MIT Press.

Lyon, D. 1994. *The Electronic Eye: The Rise of Surveillance Society*. Minneapolis: University of Minnesota Press.

– 2001. *Surveillance Society: Monitoring Everyday Life*. Philadelphia: Open University Press.

Marx, G. 1988. *Undercover: Police Surveillance in America*. Berkeley: University of California Press.

Negri, A. 1999. *Insurgencies: Constituent Power and the Modern State*. Minneapolis: University of Minnesota Press.

Poster, M. 1990. *The Mode of Information: Poststructuralism and Social Context*. Chicago: University of Chicago Press.

Stroll, A. 1988. *Surfaces*. Minneapolis: University of Minnesota Press.

Stone, A.R. 1995. *The War of Desire and Technology at the Close of the Mechanical Age*. Cambridge, MA: MIT Press.

Zuboff, S. 1988. *In the Age of the Smart Machine: The Future of Work and Power*. New York: Basic Books.

4 Varieties of Personal Information as Influences on Attitudes towards Surveillance*

GARY T. MARX

For it is a serious thing to have been watched. We all radiate something curiously intimate when we believe ourselves to be alone.

E.M. Forster, *Where Angels Fear to Tread*

'You ought to have some papers to show who you are,' the police officer advised me.

'I do not need any paper. I know who I am,' I said.

'Maybe so. Other people are also interested in knowing who you are.'

B. Traven, *Death Ship*

Social organization is increasingly based on technologies which, with laserlike focus and spongelike absorbency, root out and give meaning to personal data. Fortunately there is now a growing literature looking at broad causes, forms, processes, policies, and consequences of our move towards a surveillance society (see, e.g., Allen 2003; Lyon 2001; Garfinkle 2000; Brin 1998; Ericson and Haggerty 1997; and Etzioni 1999). In contrast, several decades ago emerging large-scale computer systems drew only modest social science attention.

The majority of writers take it for granted that there is trouble, as the *Music Man* warned, in River City. Sometimes this is explicitly stated, but more often it is an underlying theme or subtext – contemporary surveillance is viewed as something to be sceptical and suspicious of and social critics sound the alarm. The extensive and intensive noting of 'every move you make, every breath you take' is seen as a major

* A longer version of this paper with extensive endnotes is at garymarx.net.

element in the destruction of the traditionally human in an increasingly engineered, fail-safe, risk-adverse society. In such a society technology serves to further inequality and domination in ways that may not be visible (either because of transparency – they are seen through and folded into routine activities or because of opaqueness – they are obscured). This is often achieved under the guise of doing the opposite and with a heavy dose of unintended consequences. There are also technophiles, often nesting in engineering and computer science environments, who error on the other side, uncritically and optimistically welcoming the new surveillance amidst the challenges and risks of the twenty-first century.

As a citizen concerned with calling public attention to the unequal playing fields of social control technology, whether involving undercover police and informing, computer matching and profiling, drug testing, electronic location and work monitoring, or new communications, I have walked among the strident (see for example the articles on garymarx.net). Yet as a social scientist partial to the interpretive approach and aware of the richness and complexity of social reality, I have done so reluctantly and with some degree of role conflict. In this paper I combine these interests in analysing some of the structural roots of concern over surveillance involving personal information.

I seek an empirical, analytic, and moral ecology (or geography or mapping) of surveillance (see Altheide 1995; Solove 2002; Marx 2002). Of particular interest are data which are involuntary collected and recorded, whether through intrusive and invasive methods – prying out what is normally withheld, or using technology to give meaning to what the individual offers (e.g., appearance, emissions unrecognized by the unaided senses, or behaviour which traditionally was ignored, unrecorded, and/or uncollated such as economic transactions, communication, and geographical mobility). Also of interest are data gathered through deception or by unduly seductive or coercive methods.

Here I take the role of the somewhat disinterested outside observer, rather than the more interested subject – whether as surveillance agent or subject. I seek to uncover the factors that help shape surveillance practices and our evaluation of them. I do this by identifying characteristics of the technologies, applications, goals, settings, and data collected. This paper considers the last topic – the properties of the basic material surveillance technologies may gather.

I adopt a structural approach with caution, if not trepidation, since as Hamlet claimed, 'nothing's either true or false but thinking makes it so' (at least from the point of view of the beholder). Meaning lies not in the behaviour, but in perception and interpretation. Yet rather than being idiosyncratic and random, the latter are largely context dependent. The challenge is to link the presence and evaluation of surveillance to the structural characteristics of the situation.

The best-known normative perspective on this involves the Principles of Fair Information Practice developed in 1973 by the Health Education and Welfare Department (http://www.epic.org/privacy/code_fair_info.html). These are important but not sufficient, since they say little about the data collection itself, including the means used and how they are applied, the goals sought, the setting, and the kind of data collected. Nor do they adequately deal with issues of identification (including the move to merge distinct identifiers and to create a single identifier for each individual across settings), anonymity, pseudo-anonymity, and authentication.

At least five related analytic categories can be applied to the evaluation and practice of surveillance. The first involves *the inherent characteristics of the means* used (e.g., are they bodily invasive, do they extend the senses, are they covert, what degree of validity do they possess). The second is the actual *application of the means*, including the collection of the data and its subsequent treatment – is the procedure competently and fairly applied, and are trust and confidentiality of the collected data sustained, is there adequate security, are undesirable consequences minimized or otherwise mediated? Traditional data protection principles primarily apply to this second element. A third factor considers the *legitimacy* and *nature of the goals* the surveillance tool is used for (e.g., for the protection of health as against voyeurism). A fourth factor involves the *structure of the setting* in which the surveillance is used (e.g., reciprocal versus non-reciprocal, familial versus non-familial). A final factor conditioning evaluations of surveillance is the actual *content* or the *kind* or *form* of data gathered – which is the subject of this article.

Kinds of Data

Beyond issues concerning means, goals, and contexts there is the question of what type of data is involved. I emphasize *content* here, although the related issue of *form* with respect to the method of data collection and presentation is also briefly considered. What are the major kinds of data that surveillance may gather and what characteristics unite and separate them? When we speak of surveillance just what is it that is surveilled? What is surveillance information? How does the 'what' condition evaluations of surveillance?

In the West we place particular emphasis on the sanctity of the borders around the person, the body, the family, and the home, and on the protection of information gathered in certain professional relationships such as those involving religion, health, and law.

There is enormous variation in the morality and frequency of surveillance behaviours. At the extremes this is easy to see. For example, contrast the discovery behaviour of the Watergate burglars with that of lifeguards. Much domestic surveillance falls in a greyer area calling for a more fine-grained analysis, but even at the extremes it is useful to be able to be more specific about why a form is acceptable or unacceptable.

We can push the seemingly self-evident and ask what broad cultural assumptions inform evaluations of the substance of surveillance, even while acknowledging that there will be local variation. Let us consider some questions illustrating the 'kinds of data' issue.

With respect to form, under normal conditions presumably to be secretly video- and audiotaped is more revealing and a greater violation than either of those alone. But what if just one or the other was involved? Holding the kind of behaviour constant, is it a greater violation to be secretly videotaped (with no audio) in a space presumed to be private, or to be audiotaped with no video? How do these compare to a secret narrative account written up after the fact by a participant (infiltrator), or by an eavesdropper listening behind a door? What about comparing these when the recording is not secret?

Is it a greater invasion if (holding the permission issue apart) a video is shown to a medical class of a psychiatric interview in which the patient is anonymous, or is it a greater invasion if, instead, the class receives only a written narrative account, but in this case the patient is identified by name or other details? How does the presence or absence of various aspects of 'identifiability' such as face, voice, name, and composite details effect evaluations (e.g., face blocked but voice unaltered, face revealed but voice altered)? Why would these issues not arise if this was an instructional video on the perfect form of a world-class athlete?

Is there any difference between a camera on a public road noting only licence plate numbers and one that also captures the image of the person in the car?

When purchasing something with cash, why does it seem inappropriate for a merchant to ask for the customer's home phone number? Why do some local merchants refuse to accept calls with Caller-Id? Why should information about my hobbies be requested when I seek a warranty for my toaster? How does unwarranted access to an unlisted phone number or home address compare to discovering a restricted e-mail or postal box address? How does (or should?) the leaking of the fact that a CEO candidate was in a mental hospital differ from the leaking of a report based on multiple biological and social indicators that the candidate has a high probability of developing heart disease within ten years?

In my junior high school student athletic performance was assessed by ten measures. The top 10 per cent received athletic letters and their scores were posted. Would it be appropriate to list all scores? Or consider contemporary cases in which schools have been prohibited from publicly listing the names of honour students.

On an airplane trip does it matter if the stranger sitting next to you is silent or instead begins asking you for, or revealing, personal information? Can the kind of unsolicited information offered or asked for by strangers be scaled from acceptable to unacceptable? What are the expectations here, even for close relationships? What tacit organizing principles are embedded? Contrast questions or revelations about sports teams or popular singers with those involving employment, politics, religion, family, sexual orientation, and health.

Most of the above cases deal with taking information from the person but similar questions and latent patterning may be identified when unrequested information or stimulation comes to the person as well. For example, contrast attitudes towards a sales appeal delivered by text, orally, on a cell phone, over a land line phone, over a computer as spam, by regular mail, and by a door to door salesperson. Factors conditioning attitudes towards such unrequested information include the ease with which it can be ignored, the degree of intrusion into an ostensibly 'protected' space, and any direct costs to the person – such as having to pay for a cell phone message.

With respect to content, what are the major kinds of information that can be known about a person and how is the seeking and taking of these evaluated by the culture? As noted this is distinct from, if related to, the context and how the data are (and should *be*) treated according to policies such as the Principles of Fair Information Practice, once collected and used.

It is difficult of course to talk about the meaning of personal information apart from its context. Such contextual factors can include the differences between a discussion of political beliefs with a friend versus with a feigned friend serving as a government informer, revealing one's social security number for employment versus having it be required as a driver's licence number, or drug testing bus drivers as against all seventh graders who wish to participate in extra-curricular activities.

Rather than seeing the personal or private as inherent properties, they are more usefully viewed in relation to particular persons, roles, and contexts which may be fluid. Let us also differentiate empirical and normative meanings of private.

The word private has two meanings, which often confuses discussion of these issues. Private can have an empirical meaning – referring to the actual state of knowledge about individual information. Unlike apparent age, gender, skin colour, and so forth (aspects which are usually visible in face-to-face settings) or readily available information such as numbers in a telephone book, much information about the individual is not known by others. Such information is 'private' in being not known, rather than being 'public.' But beyond the actual condition of being known or unknown, private refers to privacy norms about the appropriateness of this empirical status. Is it right or wrong that various others have, or do not have, information about the person?

We must ask private for whom and under what conditions? A doctor's knowledge of a patient's condition is information the patient shares. It ceases to be private as unknown to the doctor even though that knowing does not obviate the doctor's obligation to treat it confidentially. Knowledge that one's partner has a tattoo in places not usually warmed by the sun is different from that knowledge obtained by an unknown voyeur with a hidden camera.

In objectifying and treating types of information more abstractly as I do in this paper, it is vital not to lose sight of the centrality of the empirical settings. However, granted the centrality of situational or contextual matters to the topic, some conceptual mileage still lies in considering the 'objective' attributes of various kinds of information about persons. This can be seen by holding the context constant and imagining different kinds of information. Thus as part of treatment a doctor may know that a patient is HIV positive, has arthritis, no religious preference, is a vegetarian, and a Chicago Cubs fan. Merely because the doctor is aware of the above does not mean that the various kinds of information have become equivalent. Losing (or giving up) control of information does not make it impersonal or eliminate the normative aspects.

All private information about an individual is in one sense personal, but all personal information does not involve expectations of privacy, nor when it does are those expectations held to the same degree. Within contexts it is possible to make comparative statements across kinds of information and information may retain its normative status, regardless of whether or not it is known.

There is no easy answer to the question 'what is personal information?' or to how it connects to perceived assaults on our sense of dignity, respect for the individual, privacy, and intimacy. However, even giving due consideration to the contextual basis of meaning and avoiding the shoals of relativism as well as reification, it is possible to talk of information as being more or less personal. There is a cultural patterning to behaviours and judgments about kinds of personal information.

Surveillance involving information that is at the core of the individual and more 'personal' is likely to be seen as more damaging than that involving more superficial matters. But what is that core and what radiates from it? How does the kind of information involved connect to that core?

Scientific explanation and moral evaluation require understanding what personal information is and how assessments of its collection, representation, and communication vary. This discussion is organized around the concepts in figures 1 to 3. The concentric circles in figure 1 show information that is individual, private, intimate, and sensitive, with a unique core identity at the centre. Figure 2 describes kinds of information that may be gathered. Figure 3 builds on this; it is more analytic and identifies cross-cutting dimensions that can be used to unite seemingly diverse, or to separate seemingly similar, forms. This permits more systematic comparisons and some conclusions about how the nature of the information gathered by surveillance is likely to be viewed.

Given the complex, varied, and changing nature of the realities we seek to understand, I approach the task of classification tentatively and note some limitations.

The concepts are not mutually exclusive. A given type of information about the person can fit in more than one category – either because the concepts deal with different aspects such as the body, time, place, relationships, or behaviour, or because the information has mixed elements. For example, voice print, handwriting, and gait analysis are biometric and behavioural in contrast to a form such as DNA, which is entirely biometric.

My emphasis is on offering a framework that usefully captures the major kinds of personal information the technologies gather and create,

Figure 1. Types of information on the embodied

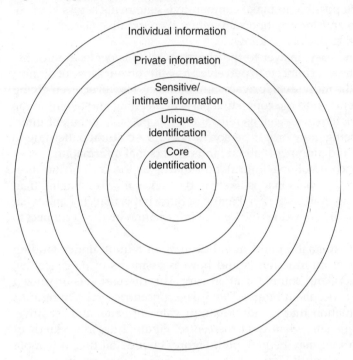

Individual information

Private information

Sensitive/
intimate information

Unique
identification

Core
identification

rather than on presenting a fully logical, operationally defined system, which can come later.

Information about Persons

What is personal information? One approach defines it in property terms as any information over which the individual has certain decisional rights. Thus facial image, copyrighted material, or the contents of one's medicine cabinet are personal – partly because they 'belong' to the individual. However, a control definition is limited in that once control has been given up, personal information is still present. New technologies further complicate matters, as they often blur the line between self and other, copy and original, and the multiple meanings of public and private. Current digital rights management (DRM) controversies, for example, are about whether property one 'owns' in the form of DVDs or software can be altered at will, or must not be changed or copied without permission (Cohen 2003).

Figure 2. Some types of descriptive information connectable to individuals

1. Individual identification (*the who question*)

Ancestry
Legal name
Alpha-Numeric
Biometric (natural, environmental)
Password
Aliases, nicknames
Performance

2. Shared identification (*the typification-profiling question*)

Gender
Race/ethnicity/religion
Age
Education
Occupation
Employment
Wealth
DNA (most)
General physical characteristics (blood type, height) and health status
Organizational memberships
Folk characterizations by reputation – liar, cheat, brave, strong, weak, addictive
  personality

3. Geographical/Locational (*the where, and beyond geography, how to reach question*)

A. Fixed
    Residence
    Telephone number (land line)
    Mail address
    Cable TV

B. Mobile
    E-mail address
    Cell phone
    Vehicle and personal locators
    Wireless computing
    Satellites

4. Temporal *[the when question]*

Date and time of activity

5. Networks and relationships (*the who else question*)

Family members, married or divorced
Others the individual interacts/communicates with, roommates, friends, associates,
Others co-present (contiguous) at a given location (including in cyber-space) or
  activity including neighbours

Figure 2. (*concluded*)

6. Objects (*the which one and whose is it question*)[1]

Vehicles
Weapons
Animals
Communications device
Contraband
Land, buildings, and businesses

7. Behavioural (*the what happened question*)

Communication
Fact of using a given means (computer, phone, cable TV, diary, notes, or library) to create, send, or receive information (mail covers, subscription lists, pen registers, e-mail headers, cell phone, GPS)
Content of that communication (eavesdropping, spyware, library use, book purchases)
Economic behaviour – buying (including consumption patterns and preferences), selling, bank, credit card transactions
Work monitoring
Employment history
Norm and conflict related behavior – bankruptcies, tax liens, small claims and civil judgments, criminal records, suits filed

8. Beliefs, attitudes, emotions (*the inner or backstage and presumed "real" person question*)

9. Measurement characterizations (past, present, predictions, potentials (*the kind of person, predict your future question*)

Credit ratings and limits
Insurance ratings
SAT and college acceptability scores
Civil service scores
Drug tests
Truth telling
Psychological tests and profiles
Occupational placement tests
Medical

10. Media references (yearbooks, newsletters, newspapers, TV, Internet)

1 Such objects complicate issues of identification. In the case of an inappropriately used device such as a car, gun, telephone, or computer it is necessary to determine which device was used, to whom it is registered, and finally who in fact used it. An anonymous dog bite is the same for the first two issues.

Thus in the discarding of a pill container or a magazine subscribed to personal information usually remains. In public settings others may generally record the image and sounds the individual gives off. These do not cease to reflect the person as a result, even if they are re-creations and not 'really' the person. Or consider a person's DNA obtained from dental floss discarded as garbage. A California court has held that an ersatz garbage collector (as part of a suit to prove paternity) could perform DNA analysis on it. Yet the DNA still distinctively reflected the discarder.

Another issue raising unresolved control questions is telephone numbers. This became clear in the controversies over Caller-Id. The act of *paying* for phone service (and paying even more for an unlisted number) would seem to imply control over the number. Yet as legislation and regulations generally imply, the phone number for land lines is rented and 'belongs' to the phone company. The question of formal ownership is distinct from the conditions of use and whether the number can be released if the phone company so chooses. However, its potential for reaching the individual and for probabilistic geo-demographic analysis brings a personal component to it.

As the above cases suggest, control is an important dimension with policy implications. Yet something beyond control or possession is involved in defining personal information. We need a broader conception.

Another approach is to view any datum attached to a corporeal individual (identified by distinguishing characteristics of varying degrees of specificity) as 'personal' because it corresponds to a person. This could include such things as being identified as a citizen of the United States, a watcher of the Superbowl, middle-aged, or owner of an SUV. But information about an individual is not necessarily equivalent to *personal* information in a more restricted sense.

Knowledge about the kind of car one drives when millions of people drive a similar car is a pale form of 'personal' information. It is more like impersonal information, although at a general level it serves to differentiate owners from non-owners and may convey symbolic meaning. Whether one owns a red convertible sports car or a black mega-pickup truck with flames painted on the front makes a statement, whether intended or not, about the owner.

When we refuse or resent having information about ourselves taken, it is often because that information is seen as 'personal' or 'none of your business.' Here something additional is present beyond many of the kinds of general information that can be associated with an individual.

Private personal information needs to be located within the larger category of individual information.

Even within the more limited category, all private information (defined as information that is not automatically known about the other but which is subject to the actor's discretion to reveal it or give permission for it to be revealed) is not the same. Some goes to the very centre of one's person while other information is peripheral or trivial (e.g., sexual preference versus city of birth). Considerations of the private and personal involve both a content and a procedures component. I emphasize the former.

Concentric Circles of Information

We can think of information about persons as involving concentric circles of individual, private, intimate, and sensitive information.

The outermost circle in figure 1 is that of *individual information,* which includes any data/category that can be attached to a person. The individual need not be personally known, nor known by name and location by those attaching the data. Individuals need not be aware of the data linked to their person.

Individual information varies from that which is relatively impersonal with minimal implications for an individual's uniqueness, such as being labelled as living in a flood zone or owning a four-door car, to that which is more personal, such as illness, sexual preference, religious beliefs, facial image, address, legal name, and ancestry. The latter information has clearer implications for selfhood and for distinctly reflecting the individual.

The information may be provided by, or taken directly from, the person through such things as remote health sensors or a black box documenting driving behaviour. Or it can be imposed onto persons by outsiders, as with the statistical risk categories of a composite nature used in extending credit.

The next circle refers to *private information* not automatically available. Such information is defined by discretionary norms regarding revelation, absent special circumstances to compel disclosure as with providing social security number for tax purposes, or a subpoena or warrant for a search. An unlisted phone or credit card number and non-obvious or non-visible biographical and biological details ('private parts') are examples of private information. We can refer to information about the person not known by others whose communication the individual can control as existentially private.

In contrast to such personal information of which the individual is aware is information with implications for life chances imposed from the outside of which the individual is often unaware. Much organizational categorization of individuals of the kind Foucault first called attention to is encompassing, routine and invisible to the subject, and artifactual. It is also a form of social sorting fundamental to current social organization (Lyon 2002; Bowker and Star 1999).

Labelling by judicial, mental health, or commercial organizations may involve imputed identities (e.g., a recidivism risk category) where the individual is unaware either of the existence of the information or its' content. Such imposed classifications are better seen as secret than as private. The information may be considered personal and even sensitive were it known by the individual.

Another distinction is whether, once known, an organization's information corresponds to how persons see themselves. This raises fascinating questions involving the politics of labelling and measurement validity. The disparity between technical labelling and self-definition may increase and become more contested as abstract measurements claiming to characterize the person and predict future behaviour based on comparisons to large data bases become more prominent.

Even when there is no disparity such labelling serves as a new source of identity, as with a high SAT scorer or a low cholesterol person. Organizational labelling has also become a marketeable commodity – among many other forms, note the selling of background and credit rating scores at some commercial establishments.

Sensitive/Intimate Information

The next innermost circle is that of *intimate* and/or *sensitive* information. The word intimate derives from the Latin *intimus*, meaning inmost. Used as a verb it means to state or make known, implying that the information is not routinely known. Several forms can be noted.

Some 'very personal' attitudes, conditions, and behaviours take their significance from the fact that they are a kind of currency of intimacy selectively revealed to those we trust and to whom we feel close. Such personal information is not usually willingly offered to outsiders, excluding the behaviour of exhibitionists and those seen to be lacking in manners. The point is not that the behaviour that might be observed is personal in the sense of necessarily being unique (e.g., sexual relations), but that control over access affirms respect for the person and sustains the value of intimacy and the relationship. Persons who prematurely reveal their hole cards or private parts are likely to do poorly at both cards and love.

We can also differentiate an intimate *relationship* from certain forms of information or behaviour that can be intimate independent of interaction with others. E.M. Forster captures this in noting that we 'radiate something curiously intimate when we believe ourselves to be alone.' This suggests a related form – protection from intrusions into solitude or apartness. Whether alone or with trusted others, this implies a sense of security, of not being vulnerable, of being able to let one's guard down, which may permit both feelings of safety and of being able to be 'one's self.'

When protected from others' observations by physical structures and manners, as in the case of bathroom activities, this apartness generally protects not against strategic disadvantage or stigmatization, but rather sustains respect for personhood.

The privacy tort remedy of intrusion attempts to deal with the subjective and emotional aspects of harm from incursions into solitude when personal borders and space are wrongly crossed. These are harder to define than a harm such as unauthorized commercial use of a person's data, reporting them in a false light, or reporting private facts. The former involve actions taken by the other *after* possession of information and usually some type of publication, even if no more than a sign in a shop window. In contrast with intrusion, it is the process itself which is objectionable.

Consider also moments of vulnerability and embarrassment observable in public, for example, the expression of sadness in the face of tragedy – as with a mother who has just lost a child in a car accident. Here manners and decency require disavowing, looking the other way, not staring, let alone taking and publishing a news photo of the individual's grief.

Some information is 'sensitive,' implying a different rationale on the actor's part for information control and the need for greater legal protections. This includes strategic information that could be useful for an opponent in a conflict situation or a victimizer, or a stigma that would devalue individuals in other's eyes, or subject them to discrimination.

Various U.S. laws recognize information about finances, health, and children as sensitive. The European Union's data protection directive requires special protection for 'sensitive data,' which it defines as involving information on race and ethnicity, political, philosophical and religious beliefs, health, and sexual life.

More broadly a central theme of Erving Goffman's (1959) work is that the individual in playing a role and in angling for advantage presents a self to the outside world that may be at odds with what the individual actually feels, believes, or 'is' in some objective sense. Through manners and laws, for most purposes, modern society acknowledges the legitimacy of there being a person behind the mask.

Unique and Core Identity

The two final circles at the centre involve a person with various identity pegs. These (whether considered jointly or individually) engender a *unique identity* ('only you' as the song says) in being attached to what Goffman refers to as an 'embodied' individual who is usually assumed to be alive, but need no longer be. Knowing unique identity answers the basic question raised by Sesame Street, 'who is it?' The question assumes the point of view of an outside observer trying to be honest, since individuals may prevaricate, have fluid identies, or, in rare cases, not know 'who' they are. It is from, and to, this identity that many other sources of potential information are derived or attached (radiating outward as well as being added) to the person.

The elements that make up the individual's uniqueness are more personal than those that do not and as the degree of distinctiveness increases so does the 'personalness' of the information.

Traditionally, unique identity tended to be synonymous with a *core identity* based on biological ancestry and family embedment. Excluding physically joined twins, each individual is unique in being the offspring of particular biological parents, with birth at a particular place and time. Parents and place of birth of course may be shared. Yet even for identical twins, if we add time of birth to the equation, the laws of physics and biology generate a unique *core identity* in the conjunction of parents, place, and time of birth. This identity may of course be muddied by unknown sperm and egg donors, abandonment, and adoption.

For most persons throughout most of history discovering identity was not an issue. In small-scale societies, where there was little geographical or social mobility and people were rooted in very local networks of family and kin, individuals tended to be personally known. Physical and cultural appearance and location answered the 'who is it?' question.

Names may have offered additional information about the person's relationships, occupation, or residence. They are still sometimes presumed to offer clues to ethnicity, nationality, and class origin, and first names usually reflect gender. Titles such as 'Mrs' or 'Dr' convey additional information. Names popular in one time period that subsequently go out of fashion may also offer unwitting clues to approximate age.

However, the literal information offered by a name is of little use when the observer, neither personally knowing nor knowing about the individual in question, needs to verify the link between the name offered and the person claiming it. With large-scale societies and the increased mobility associated with urbanization and industrialization core identity came to be determined by full name and reliance on proxy forms such as a birth certificate, passport, national identity card, and driver's licence (Caplan and Torpey 2001).

Yet given adoption of children, the ease of legally changing or using fraudulent names in the United States, and widely shared common names, name may not be an adequate indicator of core identity. Nor, given technologies for forgery and the theft of identification, is the mere possession of identity documents sufficient for determining this.

The conventional paper forms of identification have been supplemented by forms more inherent in the physical person, though even here we must remember that the measurement offers a representation of something inherent and this reflects choices rather than anything 'given' in nature. Such measurements and transformations are a form of simulacrum (Baudrillard 1996; Bogard 1996). According to the dictionary this can be a neutral 'image of something' as well as a darker 'shadowy likeness,' 'deceptive substitute,' or 'mere pretense.' Among the meanings of 'simulate' are 'imitate' and 'counterfeit.' The contrast offered by these definitions is of course what all the fuss is about. Just how far in distorting the richness of the empirical should a simulation or a symbol go before it is rejected as invalid, inauthentic, inefficient, or ineffective?

With the expansion of biometric technology a variety of indicators presumed to be unique (and harder to fake) are increasingly used. Beyond improved fingerprinting, we see identification efforts based on DNA, voice, retina, iris, wrist veins, hand geometry, facial appearance, scent, and even gait. The ease of involuntarily and even secretly gathering many of these may increase their appeal to control agents, who need no longer deal with the messy issues of informed consent.

Validity varies significantly here from very high for DNA and fingerprinting (if done properly) to relatively low for facial recognition. There are also many ways of thwarting surveillance efforts. Even when validity is not an issue biological indicators are not automatic reflections of core identity, although they may offer advantages such as being ever present and never forgotten, lost, or stolen. To be used for identification there must be a record of a previously identified person to attach the indicator to. These need not lead to literal identification, but rather whether or not the material presumed to reflect a unique person is the same as that in a data base to which it is compared (i.e., 'is this the same person?' *whomever* it is).

Police files are filled with DNA and fingerprint data that are not connected to a core legal identity. With data from multiple events, because of matching, police may know that the same person is responsible for crimes but not know who the person is. Some jurisdictions, such as New York, have 'John Doe' programs in which charges can be filed based on DNA profiles alone, even though there is as yet no name with which to link the information. This is intended to permit prosecution should such a link be established.

The question, 'who is it?' may be answered in a variety of other ways that need not trace back to a biologically defined ancestral core or legal name. For many contemporary settings what matters is determining the presence of attributes warranting a certain kind of treatment, continuity of identity (is this the *same* person), or being able to locate the individual, not who the person 'really' is as conventionally defined.

A central policy question is how much and what kind of identity information is necessary in various contexts, in particular,

1) whether identification of a unique person is appropriate and if it is,
2) what form it should take.

Thus far we have noted ancestral, legal name, and biometric forms of identity. Let us also consider *locational* identity, *pseudonyms*, and *anonyms*. Figure 2 lists ten broad types of descriptive information commonly connectable to individuals.

Location

You are where you live.

Claritas Corporation

Location refers to a person's 'address.' It answers a 'where,' more than a 'who' question. Address can be geographically fixed, as is the case with most residences and workplaces, land line phones, and post office boxes. A person of interest may or may not be at these places or be using a device. Yet such addresses are a kind of anchor point on a tether from which persons usually venture forth and return.

In contrast are geographically mobile (if always located *somewhere*) addresses, such as a cell phone number, e-mail address, or implanted GPS chip. These portable means need not reside or remain in a fixed place and stand in a different relationship to the person than a fixed geographical address, which is believed to offer greater accountabililty.

Two meanings of address are 'reachability,' involving an electronic or other communications address and the actual geographical location in latitude and longitude of the person in question at that time. These may but need not be linked, nor do they require core ID. Recent developments in the rationalization of U.S. postal addresses and the linking of census block data to GPS coordinates mean that the actual location of every address and every land line phone can now be known to within a limited number of metres.

Distance-mediated and remote forms of cyber-space interaction are ever more common, and the ability to reach and be reached is central for many activities. This ability to locate postal and conventional phone addresses is now matched by the ability to locate mobile cell phone users.

Communication location is increasingly important (beyond whatever additional knowledge it might contain, such as an address in Harlem or Beverly Hills, a phone number spelling a name, or as evidence indicating that a person was or was not at a given place). For many purposes being able to 'reach' someone becomes as or more important than knowing 'where' they literally are geographically. The 'locator' number for bureaucratic records is related, although it connects to the file, not to the person.

The 'where' question need not be linked with the 'who' question. The ability to communicate (especially remotely) may not require knowing who you are communicating with 'really' is, only that they be accessible and assessable. Knowing location may permit taking various forms of action such as communicating, blocking, granting or denying access, penalizing, rewarding, delivering, picking up, or apprehending.

The 'who' may also be known and the 'where' unknown. For example, the identity of fugitives is known but not how they can be reached. Even where both name and location are known an individual may be unreachable, as when blocking is present, or there is no extradition treaty.

Nor need the specific identity be known in order to take broad actions based on categorical/group inferred attributes. As the slogan 'you are where you live,' suggests the 'where' question for many purposes is a statistical proxy for the 'who' question. This has implications for trying to shape behaviour through appeals and advertisements, environmental engineering, setting prices (such as for insurance), and the location of goods and services. Claritas's location data offers customers 'segmentation, market research, customer logistics and site selection.'

Such typification based on statistical averages is increasingly a cornerstone of organizational decision making for merchants, employers, medical doctors, police, and corrections agents. From a standpoint of presumed rationality and efficiency it is understandable. But actions at the aggregate or group level, however logical, may clash with expectations of individualized treatment. Our sense of fairness involves assumptions about being treated in a way that is personal in reflecting the individual's particular circumstances. There is contradiction or irony here in that in order to have such fine-grained treatment vast amounts of personal information are required, creating the potential and temptation to treat individuals in general terms.

In reducing the several hundred thousand census blocks to a limited number of geodemographic types Claritas tells us that 'people with similar lifestyles tend to live near one another.' One of its 'products' – PRIZM – 'describes every U.S. neighborhood in terms of 62 distinct life style types, called clusters,' while its MicroVision 'defines 48 lifestyle types called segments' (www.Claritas.com). Such characterizations combine census, zip code, survey, and purchase data. You are what you consume. For many marketing purposes there is no need to know core identity if fixed or mobile location is known. The latter has received a very large boost with the appearance of the 911E system now required by federal law in response to the rapid spread of cell phones not tied to a fixed residence.

Location information has a very special status in that it can both identify and monitor or track movement over time. In addition it offers the potential to both *take from* and *impose upon* the borders of the person. As with monitoring of internet behaviour and subsequent advertisements, it can sequentially join two forms of personal border crossing. This contrasts with most other kinds of personal information collection (e.g., a photograph, a name, a medical record, or an overheard conversation), which only involve taking from the person. Location information in a sense offers two for one. The substantive information it provides can be compared to predictive models (or used to build them) that then serve to direct how the individual is responded to. In that regard it is like a supermarket saver card. But it also offers a means of action – knowing where the person is may permit 'reaching' them, either literally, as with 911 responders, or through targeted communications.

Pseudonyms and Anonyms

Apart from legal name and location, unique or at least somewhat distinctive identification may involve pseudonyms such as a nickname, alias, pen name, *nom de plume* or *nom de guerre*. Alpha-numeric indicators are functionally equivalent to a pseudonym. A numeric or alphabetic identification is often intended to refer to only one individual. While names can be held in common, letters and numbers are sufficient as unique identifiers, although they may also be keys to additional common information, such as where and when issued, and a rating. Unlike birth parents or DNA material, they are merely convenient and need have no intrinsic link to core identity.

As buffering devices, pseudonyms may offer a compromise solution in which protection is given to literal identity or location, while meeting needs for some degree of identification, often involving continuity of identity. They vary in the degree of pseudo-anonymity they offer. Whether intentionally or unintentionally, they may often be linked back to an individual known by name, location, or other details.

The number on a secret Swiss bank account permits transactions while keeping the owner's identity unknown to outsiders. An online service such as AOL likely knows the true (or at least claimed) identity of its customers, even as they are permitted to have five pseudonyms for use in chat rooms and on bulletin boards. The *nom de plume* of authors wishing to shield their actual identity will usually be known to their publishers. The true identity of protected witnesses, spies, and undercover operatives not using their real names is known by the sponsoring agency. Licence plate number also fits here. The identity of the owner is no longer publicly available, even though it is known by the state and can be made available under appropriate conditions.

There are also time-lagged forms such as census records, sealed court and arrest records, and other government documents that protect identity and content for a fixed period of time and are then made available. In other forms an organization may be explicit in offering confidentiality. Persons calling tip hotlines may be given an identifying number which they may later use to collect a reward. Those engaged in illegal activities, leakers, whistle blowers, and the stigmatized may be offered full anonymity. For example, persons encouraged to return contraband during a grace period are told, 'no questions asked.' In some areas, the name of those tested for AIDS is not requested and they are identified only by a number.

There are many settings where anonymity and the absence of a documentary record are implied, at least in the sense that individuals are not asked for personal information. Barter and cash purchases, many kinds of information request, and being in a public place are traditional examples. In other settings the core identity of the person is deemed irrelevant, even though some information and continuity of identity (apart from literally knowing it) is required. Many contexts require that the individual have some general characteristics warranting inclusion or exclusion, access or its denial, or a particular kind of treatment.

Eligibility certificates (or tokens) offer a way of showing that one is entitled to a given service without requiring that the user be otherwise identified. Theatre tickets and stamps on the hand at concerts, or cards purchased with cash for using telephones, computers, photocopy machines, mass transit, and EZ pass transmitters for toll roads are examples. With smart card technologies such forms are likely to become much more common.

While possibly offering some information, such as where and when obtained and where used, such certificates do not indicate who obtained or used them (although a hidden video camera may reveal that). Tokens offer a way of maintaining accountability – for example, proof of eligibility – while otherwise protecting personal information.

Beyond these certifications other forms of shared, but not uniquely personal, information can be seen in the possession of artifacts and knowledge or skill demonstration. Thus uniforms, badges, and group tattoos (and other visible symbols such as a scarlet letter or cross) are means of identity. These represent their possessor as a certain kind of person, whether involving eligibility or presumed social and moral character, with no necessary reference to anything more. The symbols can of course be highly differentiated with respect to categories of person and levels of eligibility. But the central point is their categorical rather than individual nature.

The possession of knowledge can also serve this function (e.g., knowing a secret hand signal, a PIN or account number, or a code, or the 'colour of the day' (used by police departments to permit officers in civilian clothes to let uniformed officers know they are also police). Demonstrating a skill, such as passing a swim test in order to use the deep end of a pool, can also be seen as a form of identity certification.

When no aspects of identity are available (being uncollected, altered, or severed) we have *true anonymity*. With respect to communication, a variety of forwarding services market anonymity. Consider a call forwarding service that emerged to thwart Caller-Id. The call, after being billed to a credit card, was routed through a 900 number and the originating phone number is destroyed. Calls from pay phones also offer anonymity not available from a home telephone. There are various mail forwarding services and anonymizing/anomizer web services that strip (and destroy) the original identifying header information from an e-mail and then forward it.

Among the dictionary meanings of anonymity are 'unknown name,' 'unknown authorship,' and 'without character, featureless, impersonal.' The traditional meanings of the term are somewhat undercut by contemporary behaviour and technologies. In writing of the 'surveillant assemblage' Haggerty and Ericson (2000) note 'the progressive "disappearance of disappearance"' as once discrete surveillance systems are joined. Genuine anonymity appears to be less common than in the past.

The line between pseudo-anonymity and anonymity is more difficult to draw today than it was for the nineteenth and much of the twentieth century. More common are various forms of *partial anonymity*.

'Anonymous' persons may send or leave a variety of clues about aspects of themselves (apart from name and location) as a result of the ability of technology to give meaning to the unseen, unrecognized, and seemingly meaningless. Consider a DNA sample taken from a sealed envelope or a straw used in a soft drink, heat sensors that 'see' through walls and in the dark, software for analysing handwriting and writing patterns, face and vehicle recognition technology, the capture of computer ID and sections of a web page visited, and composite profiles. Anonymity ain't what it used to be – whatever the intention of the actor wishing to remain unrecognized and featureless.

Patterns and Culture

Social leakage, patterned behaviour, and a shared culture are other reasons why it is relatively rare that an individual about whom at least something is known will be totally 'without character' or 'featureless.'

Some identification may be made by reference to distinctive appearance, behaviour, or location patterns of persons. Being unnamed is not necessarily the same as being unknown. Some information is always evident in face-to-face interaction because we are all ambulatory autobiographies continuously and unavoidably emitting data for other's senses and machines. The uncontrollable communication of some data is a condition of physical and social existence. This varies from information the individual is aware of offering, such as appearance or when using a cell phone, to that which he or she is likely to be unaware of, such as scent, gait, radiation and brain waves, hidden identification symbols on property such as cars and documents, and the subtle inferences available to various types of identification specialists, whether medical or juridical.

It may also involve leakage that the individual would like to control, but will often be unable to, such as facial expressions when lying (Eckman 1985, 2003) or faced with sudden surprise, as with unexpectedly seeing a police officer. New technologies have greatly expanded the potential to turn leaky data into information.

A distinction Erving Goffman makes between knowing a person in the sense of being acquainted with them, and knowing of them, applies here. The patterned conditions of urban life mean that we identify many persons we do not 'know' (that is, we do not know their names, nor do we know them personally). Instead, we know some form of their social signature – whether it is face, a voice heard over mass media or phone, a location they are associated with, or some distinctive element of style. In everyday encounters, for example, those made by virtue of regularly using public transportation, we may come to 'know' others in the sense of recognizing them.

Style issues fit here. Skilled graffiti writers may become well known by their 'tags' (signed nicknames) or their distinctive style, even as their real identity is unknown to most persons (Ferrell 1996). Persons making anonymous postings to a computer bulletin board may come to be 'known' by others because of the content, tone, or style of their communications.

Similarly, detectives attribute re-occurring crimes with a distinctive pattern involving time, place, victim, and means of violation to a given individual, even though they do not know the person's name (e.g., the Unabomber, the Son of Sam, the Red Light Bandit, Jack the Ripper).

With the development of systematic criminal personality profiling, detectives, steeped in the facts of previous crimes, make predictions about the characteristics and behaviour of unidentified suspects. Using general cultural knowledge, research data, data from other cases, and facts from the crime scene, along with intuition and the ability to imagine how the other thinks and feels ('verstehen'), they develop profiles that are intended to help understand and apprehend perpetrators (see, e.g., Douglas 1992; Ressler 1992). They claim to 'know' persons without literally knowing them with respect to core identity.

There are also pro-social examples such as anonymous donors whose gift is distinctive ('in memory of Rosebud') or who give in predictable ways that make them 'known' to charities. They differ from the anonymous donor who gives only once in an indistinct fashion.

Style is of course often linked with identity – a marketing goal sought by celebrities and advertisers relying on public recognition. This varies from seeing a painting and knowing it is a Van Gogh to the ability to identify popular singers. In such cases for those in the know, style and identity are merged. One kind of specialization is the ability to read clues others offer.

Composite Identity

Identification may involve social categorization. Many visible forms do not differentiate the individual from others sharing them (e.g., gender, age, skin colour, disability, linguistic patterns, and general appearance). Inferences about others concerning education, class, sexual orientation, religion, occupation, organizational memberships, employment, leisure activities, and friendship patterns are often volunteered or easily discovered.

Dress or simply being at certain places at particular times or associating with particular kinds of people can also be keys to presumed aspects of identity (e.g., being seen at a gay bar or with political dissidents). Folk wisdom recognizes this profiling logic in claiming 'birds of a feather flock together,' or 'you are known by the company you keep.'

Attributes which are common and readily available to others are not usually thought of as being private matters, although manners may require tactfully disattending, and anti-discrimination legislation formally ignoring, what is obvious.

The compartmentalization (isolation from each other) of various aspects of the person passively protect the individual's privacy and uniqueness. Any given general attribute may reveal little – such as being male or living in a rural area. However, when an increasing number of categories are combined, the individual may be uniquely (or almost) identified through a *composite identity*. Privacy may be invaded and a distinctive personal mosaic created with the merging of a number of seemingly non-personal general items (gender, ethnicity, age, education, occupation, census district). The greater the number of categories involved, the more specific the information and thus, other factors being equal, the closer to the unique individual and the more personal, or at least one major meaning of it.

David Shulman (1994) observes that the skill or art of the detective is to use readily accessible individual information (available partly because it tends to be neutral, non-controversial, or non-discrediting) to locate hidden and discrediting sensitive information.

Traditionally, personal information such as social security, telephone and licence plate number, or birth date were viewed as substantively neutral (in contrast to information regarding HIV status or a criminal record). Relatively little attention was given to keeping the above numbers private.

Yet with linked computer records (whether involving a single identifier or merging of multiple identifiers) they take on a more personal meaning. As locators they can be used to learn about and find or reach the person, going from public-neutral to private-discrediting information, generating a mosaic image of the person through linking previously unconnected information, and generating predictions based on comparing the person's data to that of similar others about whom much more is known.

Cultural Knowledge

Even possessing only one or two pieces of information may be revealing when there is general cultural knowledge. Here visibility refers not to what we literally see, but to what we know (or can discover) as participants in the culture. Sherlock Holmes's success, for example, lay partly in his ability to deduce applicable facts from his broad knowledge of culture and society. He often used his general knowledge to understand and locate a culprit.

There is a great line in the film *Annie Hall* in which Woody Allen, on learning that a young woman he meets is from Manhattan, presumes to tell her where she went to school and summer camp and about her taste in politics and art. She sarcastically replies, 'I love being reduced to a cultural stereotype.' She is reacting against Allen's categorization, not his surveillance per se – as might be the case if she discovered he was eavesdropping. The reaction against such labelling involves the assertion of individualism against the predictable commonalties presumed by culture and roles. In the tradition of Sartre 'authenticity' involves refusing to accept societal labels, particularly when they are stigmatizing. Freedom exists in rejection of the expectations and positions offered by the culture.

Broad knowledge of a culture, whether through socialization into it, or simply becoming familiar with it, may offer information about an individual about whom nothing specific is known. To know that an 'anonymous' individual is a citizen of the United States may already offer some information about language, thought patterns, and some general aspects of behaviour. With additional knowledge, such as age, may come familiarity with major news events and popular culture experienced by the individual's age cohort. To identify an individual as male is to presume other things about the person, such as a known physiology, a male voice, or lesser life expectancy. When correct, culture offers individual, but not personal or private information.

This ability to go from generalized knowledge to assumptions about unknown individuals is a cornerstone not only of detectives, but also of marketing research and of social research based on samples more broadly. In the latter cases, what is learned from a carefully chosen sample of persons with particular demographic characteristics (e.g., urban, college-educated, males between eighteen and twenty-five), who are paid to volunteer information about themselves also reveals information about others with those characteristics. Those in the sample are presumed to speak for, and be representative of, their group. There is an unrecognized value conflict here between the individual who, in consenting, also in a sense 'consents' for others who are not asked.

Form

The form and elements of the content of personal information are connected, at least initially (e.g., a video lens gathers the visual). But the tool and the resulting data can be subsequently disconnected in the presentation of results. Video or audio recordings are likely to be most powerful and convincing if directly communicated in the way they were recorded. However, their information can also be communicated in other ways – a written narrative of events or a transcript of conversations. Conversely, observations and written accounts can be offered as visual images – as with a sketch of a suspect or a video reconstruction of a traffic accident for court presentation.

Many new surveillance forms are characterized by some form of conversion from data collection to presentation. Thus physical DNA material or olifactory molecules are converted to numerical indicators which are then represented visually and via a statistical probability. With thermal imaging the amount of heat is shown in colour diagrams sometimes suggestive of objects. The polygraph converts physical responses to numbers and then to images in the form of charts. Satellite images convert varying degrees of light to computer codes which are then offered as photographs. Faces and the contours of body areas can be blocked or otherwise distorted when the goal is not identification, but searching for contraband.

When not presented with intent to deceive, the visual and audio, in conveying information more naturalistically, seem more invasive than a more abstract written narrative account or numerical and other representations relying on symbols. The visual in turn is more invasive than the audio. The Chinese expression 'a picture is worth a thousand words' captures this, although it must be tempered with some scepticism with respect to whether 'seeing is believing.' Direct visual and audio data implicitly involve a self-confessional mode which, initially at least, enhances validity relative to a narrative account of a third party, or even a written confession after the fact by the subject. Such data create the illusion of reality.

The other senses, such as taste, smell, and touch, play a minimal role, although they may be used to bring other means into play, as with justifying a visual search on the basis of smelling alcohol or drugs. The senses, however, may not be as independent as is usually assumed. Recent research in neurology suggests the senses are interconnected and interactive and cannot be as easily separated as common sense would have us believe (Sacks 2003).

Assessing Surveillance Data

Let us move from the descriptive categories of information in figure 2 to the dimensions in figure 3 which can be used to compare and contrast the former. This permits some conclusions with respect to how the kind and form of data relate to the prevalence and assessments of surveillance. The dimensions should be seen on a continuum, although only the extreme values are shown.

Figure 3. Some dimensions of individual information

1. *personal*			
yes		no (impersonal)	
2. *intimate*			
yes		no	
3. *sensitive*			
yes		no	
4. *unique identification*			
yes	(distinctive but shared)	no (anonymous) core	non-core
5. *locatable*			
yes		no	
6. *stigmatizing (reflection on character of subject)*			
yes		no	
7. *prestige enhancing*			
no		yes	
8. *reveals deception (on part of object)*			
yes		no	
9. *strategic disadvantage to subject*			
yes		no	
10. *amount and variety*			
extensive, multiple kinds of information		minimal, single kind	
11. *documentary (re-usable) record*			
yes [permanent?] record		no	
12. *attached to or part of person*			
yes		no	
13. *biological*			
yes		no	
14. *naturalistic (reflects "reality" in obvious way, face validity)*			
yes		no (artifactual)	
15. *information is predictive rather than reflecting empirically documentable past and present*			
yes		no	
16. *shelf life*			
enduring		transitory	
17. *alterable*			
yes		no	

The multidimensional nature of personal information, the extensive contextual and situational variation related to this, and the dynamic nature of contested social situations prevents reaching any simple conclusions about how the information gathered by surveillance will (empirically) and should (morally) be evaluated. Such complexity may serve us well when it introduces humility and qualification, but not if it immobilizes. Real analysts see the contingent as a challenge to offer contingent statements.

In that spirit I hypothesize that other factors being equal, attitudes towards surveillance will be more negative the more (both in terms of the greater the number and the greater the degree) the values on the left side of figure 3 are present. These factors combine in a variety of ways. They might also be ranked relative to each other – the seriousness of a perceived violation seems greatest for items 1–10. But such fine-tuning is a task best left for future hypothesizing.

Now let us simply note that there is an additive effect, and the greater the presence of the values on the left side of the table, the greater the perceived wrong in the collection of personal information. The worst possible cases involve a core identity, a locatable person, and information that is personal, private, intimate, sensitive, stigmatizing, strategically valuable, extensive, biological, naturalistic, and predictive and reveals deception, is attached to the person, and involves an enduring and unalterable documentary record.

Some of these factors apply across cultural contexts, such as location and identification, because of their implications for taking actions involving the person. The content of others, such as what is stigmatizing, is likely to show considerable variation across cultures. Consider for example Elias's (1978) analysis of the rise of manners and changing views of the public expression of bodily functions from the Middle Ages to the modern period.

The sense of indignation of course deepens when other elements are considered (e.g., invalid, inappropriately applied, or irrelevant means, lack of consent or even awareness on subject's part, an illegitimate goal). But my point here is to hold those constant and ask what conceptual mileage might lie in simply considering what it is that is surveilled.

It is one thing to list characteristics of results likely to be associated with attitudes towards surveillance and the perception of normative violations. Proof and explanation are quite a different matter. The assertions drawn from figure 3 are hypotheses to be empirically assessed. If this patterning of indignation (or conversely acceptance) of the stuff of surveillance is accurate what might account for it? Is there a common thread or threads traversing these? Several of the most salient can be noted.

For things not naturally known, norms tend to protect against inappropriately revealing information reflecting negatively on a person's moral status and strategic interests (e.g., his or her employment, insurance, safety, non-discrimination interests). The policy debate is about when it is legitimate to reveal and conceal information (e.g., criminal records after a sentence has been served, evidence of unpopular or risky but legal lifestyles, contraceptive decisions for teenagers, genetic data given to employers or insurers, credit card data provided to third parties). It is also about the extent to which the information put forth may be authenticated, often with the ironic additional crossing of personal borders.

Another distinct factor is the extent to which the information is unique, characterizing only one locatable person, or a small number of persons. Protection of information that would lead to unique and even core identity is more a means to the factors in the preceding paragraph than an end in itself. This is one version of the idea of safety in numbers, although the anonymity and lack of accountability may ironically lead to anti- or unsocial behaviour.

Also more in the means category, and set apart from immediate instrumental consequences, are norms that sustain respect for the person in protecting zones of intimacy, whether the insulated conversations and behaviour of friends, actions taken when alone, or the physical borders of the body. The protection of these informational borders symbolically and practically sustains individual autonomy and liberty. Backstage behaviour can also be a resource in which strategic actions are prepared.

This paper has analysed some aspects of the substance and nature of personal data that can be collected and suggests a framework for more systematic analysis. The kind of data gathered is one piece required for a broader ecology of surveillance. When the dimensions noted here are joined with other dimensions for characterizing surveillance – the structure of the situation and the nature of surveillance means and goals – we will acquire a fuller picture and a framework for systematic analysis and comparison at the micro and middle ranges. A formally, if not substantively, equivalent framework is also required as we look comparatively across time periods, societies, and cultures.

I am grateful to Richard Ericson, Kevin Haggerty, David Altheide, Albrecht Funk, Richard Leo, Glenn Muschert, Jeff Ross, and David Shulman for helpful comments. An earlier version was delivered at the meetings of the Society for the Study of Symbolic Interaction, Tempe 2003.

REFERENCES

Allen, A. 2003. *Accountability for Private Life*. Lanham, MD: Rowman and Littlefield.

Altheide, D. 1995. *An Ecology of Communication: Cultural Forms of Control*. Hawthorne: Aldine de Gruyter.

Baudrillard, J. 1996. *Collected Writings*. Edited by M. Poster. Stanford, CA: Stanford University Press.

Bogard, B. 1996. *The Simulation of Surveillance: Hyper Control in Telematic Societies*. New York: Cambridge University Press.

Bowker, G., and S. Star. 1999. *Sorting Things Out*. Cambridge: MIT Press.

Brin, D. 1998. *The Transparent Society*. Reading, MA: Perseus Books.

Caplan, J., and J. Torpey. 2001. *Documenting Individual Identity*. Princeton, NJ: Princeton University Press.

Cohen, J. 2003. 'DRM and Privacy.' *Berkeley Technology Law Journal*. 18:575–617.

Douglas, J. F. 1992. *Crime Classification Manual: A Standard System for Investigating and Classifying Violent Crimes*. Lanham, MD: Lexington Books.

Ekman, P. 1985. *Telling Lies*. New York: W.W. Norton.

– 2003. *Emotions Revealed*. New York: Times Books.

Elias, N. 1978. *The Civilizing Process*. Oxford: Blackwell.

Ericson, R., and K. Haggerty. 1997. *Policing the Risk Society*. Toronto: University of Toronto Press.

Etzioni, A. 1999. *The Limits of Privacy*. New York: Basic Books.

Ferrell, J. 1996. *Crimes of Style: Urban Graffiti*. Boston: Northeastern University Press.

Garfinkel, S. 2000. *Database Nation*. Sebastopol, CA: O'Reilly.

Goffman, E. 1959. *The Presentation of Self in Everday Life*. Harmondsworth: Penguin.

Lyon, D. 2001. *Surveillance Society: Monitoring Everyday Life*. Buckingham: Open University Press.

– 2002. *Surveillance as Social Sorting: Privacy, Risk and Automated Discrimination*. New York: Routledge.

Marguilis, S. 2003. 'Privacy as a Social Issue and Behavioral Concept.' *Journal of Social Issues* 59:1.

Marx, G. 2002. 'What's New about the New Surveillance? Classifying for Change and Continuity.' *Surveillance and Society* 1:1.

Ressler, R. 1992. *Whoever Fights Monsters.* New York: St Martins Press.

Rosen, J. 2000. *The Unwanted Gaze.* New York: Random House.

Sacks, O. 2003. 'The Mind's Eye.' *New Yorker.* 28 July.

Solove, D. 2002. 'Conceptualizing Privacy.' *California Law Review* 90:1087–1156.

Shulman, D. 1994. 'Dirty Data and Investigative Methods: Some Lessons from Private Detective Work.' *Journal of Contemporary Ethnography* 23:214–53.

5 Struggling with Surveillance: Resistance, Consciousness, and Identity

JOHN GILLIOM

- In late 2001, white vans began appearing at ten problematic inter-sections in the vicinity of Honolulu, Hawaii. The vans belonged to Affiliated Computer Services, which had been hired by the state to undertake a new for-profit traffic surveillance system. Using a combination of lasers, cameras, and embedded sensors, crews in the vans took automated photographs of the licence plates of cars that were either speeding or crossing the intersection against the light. Citations were then mailed to the owner of the vehicle and the firm's share of the fine was $29.75 per ticket. Things got off to a bad start as computer programming problems, legal disputes over administration, and extensions to the trial period created a wave of troublesome news coverage. On the first day of full operation, over nine hundred speeders were ticketed. At the same time, Honolulu's KITV NEWS reported that Eliminator licence place covers were seeing strong sales. These Canadian-made plastic covers sell for about $35 and block the viewing of a licence plate from the side angle so that the vancams, as they came to be called, cannot success-fully photograph them. A cheaper alternative was to smear mud on the plates. Other media outlets began to cover public opposition to the vans, local politicians and drive-time radio personalities began to get involved, more legal problems emerged, the American Civil Liberties Union mobilized, and after hearings in the state legislature made it clear that a thumbs down vote was in the offing, the gover-nor halted the program in early April of 2002, just four months after it began.
- In the fall of 2002, word spread about the new Total Information Awareness (TIA) project launched by the Bush administration and

run by John Poindexter (see chapter 3). As explained in EPICS's information on TIA, 'The project calls for the development of "revolutionary technology for ultra-large all-source information repositories," which would contain information from multiple sources to create a "virtual, centralized, grand database." This database would be populated by transaction data contained in current databases such as financial records, medical records, communication records, and travel records as well as new sources of information. Also fed into the database would be intelligence data' (http://www.epic.org/privacy/profiling/tia/). With access and control enhanced by data-mining technologies and biometric identification, the program would be a true showcase of surveillance technologies. The response was rapid and multifaceted. Even in the post-9/11 climate, newspaper editorials across the nation raised concerns about privacy and Big Brotherism. Privacy organizations mobilized. Unconventional politics broke out and drew significant media attention. After a San Francisco newspaper published Admiral Poindexter's home address and telephone numbers, he was inundated with calls. A web page called the Total Poindexter Awareness Project was set up and posted satellite photos of Poindexter's home, as well as information about his staff and his neighbours.

- Despite the efforts of CASPIAN – Consumers Against Supermarket Privacy Invasion and Numbering (see http://www.nocards.org/) – customer loyalty cards and the monitoring they engage in have spread rapidly. Giving retailers the power to track the purchasing habits of customers, identify and target their most profitable sectors, and offer special sale prices to preferred customers, the cards have quickly become an important element in retail marketing. And, reportedly, those outside the grocery industry, like direct marketers and insurance companies, are thirsting over the valuable databases that are being created. It has even been reported that 'federal agents reviewed the shopper-card transactions of hijacker Mohammed Atta's crew to create a profile of ethnic tastes and terrorist super-market-shopping preferences' (Baard 2002; see http://www.nocards.org/). Yet other supermarkets make it a point to advertise that their prices come with 'No Cards Required,' as some shoppers refuse to patronize stores with card programs, and others (the author included) provide false information to the card-issuing store.

Each of these struggles over surveillance gives us a glimpse into the enigmatic role of everyday resistance. Licence plates are obscured or intersections avoided; public officials are mocked and find their own high-technology tools turned against them; false names are given to supermarkets or alternative stores are chosen. No grand battles; no great protests in the streets; no sweeping promises of definitive success in besting the powers of surveillance. But this is, apparently, a wide-spread pattern of unconventional politics through which ordinary people can express and mobilize their opposition to surveillance policies while at the same time achieving short-term gains that are important in their daily lives. These 'weapons of the weak' can include everything from misrepresentation to avoidance, masking mockery, and subterfuge – they are the smaller, often unorganized actions that individuals use in their struggles with emergent systems of observation, management and control (see discussion in Marx 2003; Gilliom 2001; Scott 1985). This paper will argue that such politics should be taken seriously not just in their own right, but as a key vantage point through which to learn about both the consciousness of the surveilled subject and the nature of sur-veillance itself.

These brief stories about new surveillance initiatives and the oppo-sition which they often engender also help us to see the numerous questions that surround the interplay of surveillance, resistance, and subjectivity. For one, they show that in the huge and diverse landscape made up of the countless surveillance initiatives that have marked recent years, the popular response has been one of almost indescribable variety. In a dozen communities, automated traffic cameras go up with-out a fight, yet in another they are shut down. One new national defence surveillance program is attacked as dozens of others slip qui-etly into place. Most people don't seem to notice their supermarket shopping cards, while some are outraged and boycott stores that use them. Many people welcome the increasingly total network of video surveillance cameras, while others make dark comments about '1984' and 'Brazil.'

Further, those who engage in opposition – that is, the more public and organized efforts to block or modify a surveillance policy – or resistance – the often hidden everyday struggles to thwart or evade an established surveillance system – use widely divergent approaches. Litigation, petitions, letter writing, protests, public awareness cam-paigns, and other conventional modes of political action have been used by the opposition movement. More quietly – and more central to this essay – millions more become experts at the fine art of under-reporting, skirting the systems, evading detection, or masking forbidden action. Indeed, in a recent article, Gary Marx identifies eleven different categories of tactics that individuals use in their struggles against surveillance (2003). At the same time, vast majorities consciously or unconsciously acquiesce to the new systems of power.

Finally, these stories of opposition and resistance, when completely told, remind us how partial and temporary any 'victories' against surveillance appear to be. For every Honolulu, there are so many more cities where a system went up without a hitch. For the now-limited Total Information Awareness Project, we see CAPPS II – its sibling in the air passenger surveillance network – and a whole new generation of privately developed programs. And for each of these stories that we take note of there are so many more that we barely notice – new systems for exchanging and monitoring financial data, law enforcement records, insurance, or employment information. All told, a massive and increasingly interwoven network of surveillance technologies is surrounding and defining contemporary societies (Lyon 2001). Resistance must be understood as taking place *within* that context and not something which can prevent or undo it in any systematic way.[1]

This cacophony of diversity, variation, and transience in surveillance policies, popular reactions, political conflict, and the long-term evolution of surveillance systems suggest that one of the greatest challenges facing the emergent field of surveillance studies will be developing a grasp on the chaos. This essay, through a series of related discussions, explores my own effort to meet part of this burden by focusing empirical research on the experience and consciousness of the subject of surveillance.

It begins with a brief visit to my 2001 book, *Overseers of the Poor*, with the goal of pointing to three issues that emerge as central concerns for research on the politics of surveillance: resistance, consciousness, and the subjectivities of surveillance. Each of these areas is then discussed in the subsequent pages. *Overseers* argued for the importance of accounting for practices of everyday resistance that people use to defeat the

---

1 Recognizing this Ericson and Haggerty (2000: 609) have written that 'In the face of multiple connections across myriad technologies and practices, struggles against particular manifestations of surveillance, as important as they might be, are akin to efforts to keep the ocean's tide back with a broom – a frantic focus on a particular unpalatable technology or practices while the general tide of surveillance washes over us all.'

goals and commands of a surveillance system. We then turn to the issues inherent in research on opposition ideology – privacy and the post-privacy alternatives. After briefly revisiting some of the leading critiques and drawbacks of the right-to-privacy paradigm in surveillance politics, I discuss the merits and dangers of the more contextualized and necessarily fragmented forms of critique that were identified in *Overseers of the Poor*. Finally, I make a case for the need to focus research on the consciousness and identity of the *subjects* of surveillance in our efforts to understand this complex phenomenon.

Overseers of the Poor

In Ohio's Human Services bureaucracy there is a grand computer system known as CRIS-E (Client Registry Information System – Enhanced). CRIS-E is an ambitious software package that manages all case information about welfare clients in the state, storing and handling data pertaining to issues of identity, paternity, health concerns, employment and educational history, financial need, and any other of the myriad points of information collected by the welfare system. The system also manages the different fraud control programs, including number matches with federal data, welfare data from other states, and claimant data from agencies that serve the unemployed, injured workers, and the elderly. As it combines these different streams of data and information, CRIS-E issues automated warnings, termination threats, and demands for recertification visits.

Living with this system are thousands and thousands of young families, primarily mothers and their children, who struggle to subsist on the few hundred dollars a month in cash aid that the state gives out.[2] These families confront CRIS-E as a major force in their lives, a force that could cause them to lose their state aid or even go to jail if they ran afoul of the system. They also face ongoing questioning from their caseworkers and confront widely distributed posters and ads about the evil and danger of welfare fraud. This combination of status and experience, I argue, makes these families valuable experts on the nature and politics of surveillance.

The key goals of the research projects were to map the perception,

---

2 Since the time of my research, the programs have changed dramatically under the so-called welfare reform of the 1990s; CRIS-E, however, remains as a central player with an increasingly important role.

language, and action of a group of closely surveilled people. We went into the project sceptical about prevailing discourses such as the privacy and due process frameworks that so dominate the worlds of law, the mass media, and the organized opposition. Our hope was that we could tap into something close to the vernacular language and action of surveillance subjects to see how pervasive this dominant discourse is; what, if any, alternative framings might exist; and what, if any, oppositional struggles were taking place in a setting which, at first glance, appeared to be one of total domination and acquiescence. Since my hope was to get as close to an unstructured ethnographic visit as possible, the project was designed to avoid any connections with the human services bureaucracy, any use of structured questionnaires or surveys, and any gathering of information that could be used to identify the women to whom we spoke. Two local women, themselves recent welfare mothers, assumed the task of gathering the fifty taped interviews that took place in the homes of our research subjects. As much as possible, they 'snuck up' on the issues of surveillance, asking about their subjects' experiences of welfare, their caseworkers, how they were getting by, and only introducing questions about the computer system late in the conversation.

The stories these women told taught us much about life as the subject of surveillance. It is critical to note at the outset that this was no passive and unimportant system in the background of their lives. Because the monthly stipend from welfare in Ohio, and most states, is well below the amount that a family needs to survive on, these women are compelled to produce extra income. And because this extra income is the sort of rule infraction that the system and all of its data were designed to catch, they lived in regular danger of apprehension and sanction. Consequently, their practices of resistance brought enormous personal risk. Not only was there the possibility of straightforward apprehension through standard practices of fraud control and investigation, but a further context of risk was created by the fact that virtually everyone broke the rules *and* virtually everyone knew it. The women live in something like a state of perpetual potential blackmail as caseworkers, neighbours, former spouses, indeed, virtually anyone, can choose to report them, invoke dormant rules, or instigate investigation (Gilliom 2001: 87–9).

It is hard to understate how desperately these women need the help that they receive – living in an underdeveloped rural region with little industry and the highest poverty rates in the state, they experience a profound need which translates into extraordinary vulnerability to disruptions or terminations in their welfare support. As Linda told us: 'I had to apply for the welfare because my now ex-husband left with all of our money. He left me owing a lot of bills. Soon after he left, my father passed away. I was evicted from my home because [my husband] took all the money to go back to his home. We lived in an unheated, unfinished garage. We had no running water. No toilets. No bathrooms. We used a port-a-pot and we used to carry water. I cooked on a small wooden heating stove and we had cement floors. Snow sifted through the ceilings when it was snowing and I knew I had to do something to get out of there.' And, as Mary makes clear, this need has clear political implications: 'You have to go along with the system, like I said, in order to get a cheque ... You are just a poor person that absolutely needs welfare and, yeah, you lose your rights when you go on assistance.'

The combination of need, complicated and often unobeyable rules, and the constant threat of termination of sanction translates into what struck us as one of the most important dimensions of these women's responses to their particular surveillance regime – fear. As Eleanor told us,

> My caseworker one time lost some papers, some copies of some paycheque stubs and called me and told me that I was gonna get it. That if they came out from Columbus and pulled my file that I would be arrested for welfare fraud and that I would go to prison and I was pregnant for my daughter at the time. And see it was her mistake, where she had lost the copies that I had gave her and she tried to blame it on me. And I had to take them all back up there again because I had my receipt showing where I had brought them in and got them copied at the front desk. And I mean you know it is just scary. It's scary and if you are not worried about not being able to feed your children or have a home to sleep in you are worried about whether you are going to go to prison for welfare fraud. They make it hard any way you go.

The women also spoke of how the confusing and restrictive rules and regulations actually made it more difficult for them to provide for their children, and how they regularly had to violate rules in order to do basic things like pay the rent, gas the car, and buy things like shoes, diapers, and electricity.

As Moonstar explains:

The only way that you can make it, if you make it, is by working under the table ... Welfare don't give you enough money to barely make it and you have to do little things just to keep your head afloat ... I have no money in the end because I pay all the money. I have no money at the end. And the ADC is supposed to be for dependent children. How can I take care of my kids when I have got to pay everything in the household and not have no money to take care of my children? ... I have to go out and make a little extra money because I don't get enough to support my family, pay the bills, and be able to buy my kids shoes. If I have to go out and mow a yard for ten dollars that will get my kids extra shoes. Because my bills takes all of my money, every bit of it ...

In *Overseers*, I argue that the mothers' defiance of the rules and besting of the system through petty fraud, subterfuge, and other tactics (see also Marx 2003) manifests a pattern of 'everyday resistance' to the surveillance regime. As James Scott explained in *Weapons of the Weak*, everyday resistance encompasses 'the ordinary weapons of relatively powerless groups: foot dragging, dissimulation, false compliance, pilfering, feigned ignorance, slander, arson, sabotage, and so forth' (Scott 1985: 29). He continues, 'When a peasant hides part of his crop to avoid paying taxes, he is both filling his stomach and depriving the state of gain ... When such acts are rare and isolated, they are of little interest; but when they become a consistent pattern (even though uncoordinated, let alone organized) we are dealing with resistance. The intrinsic nature and, in one sense, the "beauty" of much peasant resistance is that it often confers immediate and concrete advantages, while at the same time denying resources to the appropriating classes, *and* that it requires little or no manifest organization' (296).

Along with these extraordinary politics went extraordinary ways of framing and expressing concerns over surveillance policy. Public struggles over surveillance in the United States are virtually compelled to take place in the legalistic vocabulary of the 'right to privacy' (with the occasional foray into the realm of 'Big Brother'). These are the dominant or hegemonic terms in which most analysis, conflict, and debate occurs. But there was little of it here. The few references to privacy that we heard were fleeting and often put in very personal terms, like 'they should mind their own business.' There was virtually no recourse to the tradition of rights or law as a vocabulary of protest or action. But the complaints and action were there.

As Coco put it, 'If you have kids, you will do anything for your kids. I mean, I do. So it's not really illegal.' Delilah and Cindy, one of the interviewers, discuss these issues further:

C: You said that sometimes you cut hair, and sometimes you help your brother wallpaper. And you know all these other people who have to do something to make ends meet when they are on ADC. How do you feel about that?

D: I think as long as someone is using what they are doing for their home or they are buying something that their kids need, I don't see anything wrong with it. If they are going out and they are doing it and they are boozing it up and they are using drugs, I think that's a shame ... [for] me it always went to my daughter. It always went into the house or into my car or my gas tank or maybe for something that Kelly needs that she would not have otherwise.

C: Do you think most of the people are like yourself?

D: I think they are doing it, most, the majority are doing it to better their home and their family.

Importantly, these women's explanations for their resistance were advanced with reference to their quite mundane daily needs to feed, clothe, and provide for their families. This frontline battle against a system of surveillance appears to be rooted in the everyday struggle to get by.

After an evaluation of the various economic and political gains and losses of their tactics of everyday resistance, *Overseers* argued that 'you are not likely to find a storefront office with a sign in the window reading 'Welfare Mothers Against Big Brother.' But what you will find is a widespread pattern of complaint, evasion, and resistance as welfare mothers struggle with the system that defines their condition and enforces the law through ongoing surveillance. It is a pattern of resistance that has clear results: desperately needed material benefits; the maintenance of a zone of autonomy in the face of the dependency of life on welfare; the sustenance of a shared identity of mothering; and the undermining of the surveillance mission itself' (112).

In thinking about how to apply the lessons of *Overseers* to the ongoing study of the politics of surveillance, three topics come to the forefront of a research agenda. First is the subject of everyday resistance and the many questions surrounding its nature, importance, and potential in anti-surveillance politics. Second is the topic of what we might call oppositional consciousness, or the ways in which those who experience surveillance think and speak about their emotions, decisions, and actions and how this relates to issues of status, context, and power. The third issue turns on issues of research method and epistemology and leads to a framework for studying the first two; here I discuss the proposition that continuing research on these questions should centre on understanding the complex questions of subjectivity and consciousness.

The Politics of Everyday Resistance

Examples of everyday resistance to surveillance policy go far beyond those identified in *Overseers of the Poor*. In 'A Tack in the Shoe: Neutralizing and Resisting the New Surveillance' Gary Marx argues that the individual subject of surveillance is 'often something more than a passive and compliant reed buffeted about by the imposing winds of the more powerful, or dependent only on protest organizations for ideas about resistance. Humans are wonderfully inventive at finding ways to beat control systems and to avoid observation' (2003: 372). Marx's exploration of the tactics of everyday resistance spans practices including means of detecting and avoiding surveillance, masking or obscuring identity, distorting data, blocking observation, breaking equipment, refusing to comply, achieving the assistance of frontline surveillance workers, and, as happened to Admiral Poindexter, turning surveillance against those who would survey. The widespread existence of everyday resistance is increasingly recognized, but there is no disputing that important questions remain about its role and importance in political struggle.

In the early 1990s, the prominent socio-legal scholar and advocate for the poor Joel Handler lashed out at intellectuals who glorified what he saw as petty and often self-destructive acts of everyday resistance to courts, bureaucrats, or other manifestations of legal power and domination (Handler 1992). Handler argued that progressive academics should keep the focus on organized movements, formal politics, and the sort of practical, collective actions that could produce substantive and widespread change in the political system. Students of everyday resistance, including myself, responded to Handler's challenge in several ways. One was to move away from some early romanticism and take seriously his call for a better reckoning of the actual costs, consequences, and pay-offs involved in practices of everyday resistance (McCann and March 1995; Gilliom 2001). But we also worked to demonstrate that, like it or not, patterns of everyday resistance are important and empirically undeniable elements of political struggle (see Ewick and Silbey 1998; Gilliom 2001). Further, it was argued, given structural changes in the nature of the state and its governance, everyday resistance may well become even more important in the future.

As I argued in *Overseers of the Poor*:

Ewick and Silbey (1998), make the important point that new means of power and administration produce or call forth new means of resistance and opposition. Working off Foucault's arguments in *Discipline and Punish*, they suggest that more public and organized forms of opposition fit older public displays of the state's power. 'By contrast, the technical, faceless, and individuated forms of contemporary power defy the possibilities of revolt or collective resistance. The spatial and temporal restructuring of the world in a disciplinary regime disables the very communities that were once the site of social disturbance' (188). The women studied here experience the welfare bureaucracy on their own – it either comes in their mailbox or they are called into an office or cubicle. They are concentrated in neither the poor house of old nor the modern urban housing project but spread out along the countryside, connected only by their invisible status and the equally invisible powers of the state's administrative machinery. There are few, if any, home visits, no badges, and no midnight raids. With none of these grand and visible displays of power over groups of people, it should hardly surprise us that forms of opposition and resistance are equally discreet *and* discrete.

This point is particularly important to students of surveillance because of the ways in which many surveillance programs quietly and particularly separate, distinguish, and analyse not just individuals, but myriad aspects of individual lives (Lyon 2001: 116; Haggerty and Ericson 2000: 609). The everyday realities exist in numerous locales (work, home, cyberspace, etc.), numerous dimensions (health, finance, consumerism, criminality, etc.), numerous forms (cameras, police officers, computer analyses, customer loyalty cards), and at numerous times (childhood and student life, past and present financial conduct and health, even predictions of futures). Many of the most important types of surveillance programs quietly monitor, record, and calculate, exerting power without a visible display of force. There is, then, no 'public square' or 'hanging tree,' no shared and visible location or intelligible moment in which this thing called *surveillance* stands tall for all to see and react to. If it is true that such public and visible interactions played an important part in shaping the sorts of movement politics and uprisings that have marked earlier periods, then it may be that the skirmishes and gestures of everyday resistance will become a more definitive politics in our time.

Privacy and Oppositional Consciousness

Quite different from the scattered, often invisible, and unvoiced actions of everyday resistance is the public, formal, and well-organized movement built around the legal right to privacy. In these times, the right to privacy is at once increasingly irrelevant and ever more crucial. It is more crucial because it serves as one of the more visible and trafficked anti-surveillance terms in the media, the courts, and the public consciousness. Media coverage of surveillance issues is almost universally framed around the idea of privacy and studies show that these sorts of formulae are crucial to getting topics covered at all (Bennett and Grant 1999). The right of privacy has also delivered the occasional court victory, even at the hands of the very conservative and pro-government federal courts in the United States. Infrared scanning of homes for evidence of marijuana growing operations, for instance, have been ruled a full search under the Fourth Amendment with the attendant requirements of probable cause and a judicially approved search warrant (*Kyllo v. U.S.*). Privacy has also served as a rallying cry for effective organizations such as Privacy Inc., the Electronic Privacy Information Center, the American Civil Liberties Union, the Electronic Frontier Foundation and the Privacy Rights Clearing House (see Marx 2003).

Despite these victories and important leverage points there are numerous causes for concern. While words are in some sense infinitely malleable, many of the most important new surveillance initiatives simply do not fit into a privacy framework. Take the example of the cameras that automatically detect and photograph speeders or red light runners. While cries of 'privacy violations' have been heard on this topic, they ring hollow. The recording of a displayed licence plate number, on a registered vehicle, on a public street, in the open air, during a publicly illegal and dangerous act simply does not raise a plausible privacy claim. Similar concerns might be raised about CCTV systems in public settings, facial recognition technologies, and other instances of quite public surveillance.

Just as privacy is not all that relevant to many surveillance policies, it may not be all that relevant to some subjects of surveillance. My research in *Overseers of the Poor* showed that welfare mothers subjected to intense computer surveillance of their financial conduct made relatively little use of the idea of privacy in their complaints. They complained instead about degradation and humiliation, testifying to the need to break bureaucratic rules to acquire the forbidden extra money needed to take care of their children, and revealing the profound fear created by the tensions between the need to break rules and the distinct possibility of apprehension.

Other complaints about the 'right to privacy' framework cover relatively familiar ground:

- Because it is fundamentally a *legal* concept it pushes debate and conflict towards the experts and authorities of legal institutions, potentially reducing popular participation and movement building (Rosenberg 1991). Further, conventional forms of legal protection requiring the initiation of action by citizens have been shown to be ineffective and discriminatory (Bumiller 1987, 1988).
- Because it is an abstraction that codifies and condenses the experiences of different peoples experiencing new and changing situations, its dominance in our vocabulary may prevent new or alternative ways of speaking about surveillance (Tushnet 1984).
- Due to its roots in aging philosophical systems of liberal individualism, privacy may no longer reflect the complexity and interdependence of our world and may not fit within emergent philosophical frameworks that place less emphasis on individualism.
- Privacy is, for the most part, failing to serve as the focal point of a successful broad-based movement against surveillance. Polls clearly show that privacy is part of the public vocabulary, but continuing consent and apathy suggest it is not sparking the public imagination.
- Privacy may not be all that relevant for the various types of group or category surveillance emerging, such as insurance company risk assessments, racial and ethnic profiling, or consumer tracking based on categories rather than identities.

In place of the conventional right to privacy arguments, welfare mothers mobilized a critique of surveillance that centred on the mundane and everyday need to take care of their children: the surveillance system made it more difficult to engage in the petty infractions required to produce needed income beyond the welfare stipend; thus it was considered morally wrong and actively circumvented and opposed.

This, I argued, demonstrated that there is life beyond (or beside) the right to privacy – grounded critiques that are meaningful within the context of people's lives and could, demonstrably, be part of an action plan for resisting surveillance. For some people, like the women in *Overseers*, the specific issues may turn on family, security, and need, while, for others, there may be other dimensions of economic struggle, and still others may find their concerns turning on issues of autonomy and dignity. Once we set aside the presumption that privacy is *the* issue that brings people into conflict with surveillance policies, we will see a broad and diverse array of thinking. Much of it, as was the case in *Overseers of the Poor*, will reflect the economic and social conditions that shape the lives of those who are surveilled. While I suspect that there may be strong patterns reflecting the differences marked by such things as race, class, and gender, at this point, we know very little about these under-explored facets of surveillance. The next section makes the argument for moving these concerns to the forefront of a research agenda.

The Subjectivities of Surveillance

The preceding two sections discussed questions about everyday resistance and oppositional consciousness. These discussions raise more fundamental questions about the varied ways in which subjects of surveillance experience, understand, and are affected by the processes of surveillance in their lives. Too often, the billions of us who are subject to surveillance are lumped into an undifferentiated mass. In some hands, we are all undifferentiated citizens with our universal right to privacy. In others, we might all be 'subjects of the gaze,' a part of the masses laid bare by the unflinching vision of the panopticon – or simply a blank part of the puzzle because so little research has been done on the subjects of surveillance. Whatever the cause, it seems clear that significant research energy should be turned to studying those who are watched.

Several characteristics made the people in *Overseers of the Poor* particular and unique: gender, class, geography, parenthood. The welfare status of these women meant that they were more tightly under the control of one comprehensive surveillance system than might be typical of more affluent and mobile people. They lived in something closer to the original idea of the panopticon than others who face not so much a singular and powerful omnipresence but rather the numerous checkpoints and glances encountered in airports, the credit industry, the corporate workplace, and urban CCTV systems, to name a few. The welfare surveillance system was singular, it was named, it sent them mail, it demanded documentation and recertification, and it could trigger the process that would terminate their benefits and result in criminal prosecution.

We all live with a variety of surveillance programs tracking the basic patterns of our behaviour. Indeed, it seems clear that those with greater affluence and participation in mainstream institutions may be subject to more comprehensive surveillance than the economically marginal women studied in *Overseers of the Poor* (see Marx 2003). But here rise questions of context, status, and vulnerability – many of us live with ongoing surveillance, but few of us live with that sort of massive, conspicuous and terrifying presence that CRIS-E manifests in these women's lives.

That these poor, rural, welfare mothers are a unique and hardly representative group is precisely the point that needs to be made about the broader politics of surveillance and our efforts to understand the subjectivity and consciousness of surveilled populations. Simply put, it matters very much who you are, what you do, what your resources are, and how truly threatening a surveillance program is. Some of these arguments run parallel to one of the oft-recited maxims from surveillance advocates and practitioners – 'if you haven't done anything wrong, you have nothing to fear.' While I disagree with this statement, it is the case that surveillance programs are *different* for those who are wanting or needing to deviate from the norms and those who are not. Urinalysis drug testing may be a proud moment for some, a minor nuisance for others, burdensome and humiliating for still more, but it is frightening and threatening to the millions who use marijuana. Traffic surveillance programs may be an assurance to some, a vague symbol of big government to others, and the source of numerous expensive fines for fast drivers. Customer Cards at grocery stores may be a widespread symbol of the surveillance society to academic experts, but for most, they rapidly became one more card in the wallet and a largely unnoticed vehicle for accessing sale prices. When the effects of a surveillance program are perceived as either positive or at least nonburdensome to the subjects of surveillance, the surveillance, I would argue, *is a different thing* than it is for the noncompliant. For the one, it is a beneficial or inconsequential gaze, for the other it is a system of detection, judgment, and, often, punishment aimed at limiting freedom and channelling behaviour. These differences are tied up in a bundle of variance having to do with different positions in social structure, economies, and hierarchies tied to race and gender; to differences in dependence or desire relating to programs or institutions involved; to differences in legal and political consciousness; and to differences in technological and political awareness, among others.

Since the nature and impact of a surveillance program is so inseparably tied to the perception, behaviour, and nature of the surveilled population, our very understanding of what surveillance is must be tied to these differences in subjectivity. The definition and meaning of surveillance must be variant and changing because the term only has meaning in the sense that it is relational. It is a watching over, which means that there must be both watcher and the watched. A tour of the field suggests that we have been particularly good at studying the watchers – the police, the CCTV operators, etc. – but not so good at the necessarily messier, less institutionalized, and exploratory but absolutely crucial job of studying the watched – the real people and real bodies who are the subjects of these systems (see also Lyon 2001: 124). Until we are able to generate sufficient research to make plausible sense of how differently situated people – welfare mothers, prisoners, students, middle-class professionals – speak of and respond to their various surveillance settings, I would argue that we are fundamentally unable to define the powers of surveillance or, indeed, to devise a meaningful account of what surveillance is.[3]

3 An important limit of the research in *Overseers of the Poor* is the focus on one particular institutional setting when the wider experience of surveillance is of multiplicity, overlap, and combination (Ericson and Haggerty, 2000: 610). A whole range of known and unknown monitorings of financial, medical, social, behavioural, professional, and other types of behaviours and identities frames the life of a typical citizen. Some may be unimportant until, if ever, something triggers them. Some may become quite visible for a time, then slip away. In some cases the invisibility may be a sign of success, in others it will signal failure. In some settings we may actively seek out the verification of a surveillance system, in others we seek to avoid them. This diversity over subjects, times, and systems is a central part of telling the story of surveillance, as is struggling to account for any sense in which systems synthesize or come together in meaningful ways. It will be especially important to turn our attention to at least some people for whom surveillance does not seem to matter very much. Through some combination of affluence, isolation, compliance, or unimaginably boring normalcy, there are some for whom surveillance would seem to be a glance without consequence. For the people studied in *Overseers of the Poor*, every facet of their surveillance mattered very much, but this is simply not the way that most of us experience it.

Conclusion: The Identities of Surveillance

The research reviewed shows that there are significant differences in the ways that people respond to surveillance (such as acquiescence, resistance, opposition), talk about surveillance (invoking issues of privacy, fear, need, or autonomy), and experience surveillance (being unaware of its existence or responding with apathy, resentment, or terror). The patterns of experience, language, and action that construct an individual's consciousness of surveillance, I have argued, are integral to understanding just what surveillance itself is, as such meanings are necessarily grounded in the settings and relationships from which they emerge. Surely, we could arrive at a generic technical definition of surveillance without these elements of subjectivity, but to persuasively explain this complex process of social and political communication, domination and struggle, requires us to move beyond universalizing abstractions.

In an outstanding new book called *Rights of Inclusion: Law and Identity in the Life Stories of Americans with Disabilities* (2003), legal sociologists David Engel and Frank Munger study the landmark *Americans With Disabilities Act* (ADA), which mandated institutional accommodation to disabled people, and argue that the law's meaning and impact can only be fully understood through close exploration of the ways in which it affects the consciousness, self-understanding, and actions of the subjects of law. The centrepiece of their work is the construction of life stories or narratives for about sixty disabled people. The narratives focus on how the existence of the disability shaped their identity and the ways in which, if any, the passage of the *ADA* reoriented their opportunities, behaviour, or self-understanding.

As we might expect, the extent to which the legislation affects legal identity varies from person to person, depending on life stories, the nature of the disability, their occupation, if any, and other factors of variance. For example, although the *ADA* created significant new avenues for suing noncompliant organizations, no one in the study actually litigated: only a couple used such power as a threat, several thought it may have tacitly been in the background of workplace changes, and others never knew about it. What the narratives teach us is that the meaning of the law depends on the identities of the subjects – on the everyday experiences and personal contexts that unfold in work, home, play, and solitude. As the authors wrote in an earlier article, '"Legal consciousness" emerges from the continual interplay of law, everyday life, and individual experience' (Engel and Munger 1996: 14; see also

Ewick and Silbey 1998: 34–7; McCann 1994: 283). In this sense *Rights of Inclusion* stands as an important challenge to a paradigm of legal studies that has often treated 'The Law' or 'Rights' as a singular and singularly knowable system of commands that created universal conditions for citizens. It argues for the idea that our understanding of the very fundamental question of *what the law is* must include this attention to legal consciousness and, therefore, the sort of ground-up, case-by-case research developed in their work.

Research on how people respond to surveillance has touched on things like resistance, litigation, opposition, compliance, and apathy, but made relatively little progress in understanding how these dimensions are linked to different contexts, programs, and people. Similarly, research on oppositional consciousness has only just begun to move beyond the assumptions and presumptions of the paradigm of privacy rights to explore the different and perhaps more promising ways through which the subjects of surveillance may speak. All of this, to me, comes back to a need to devote significant energy to research on the subjects of surveillance. Like Engel and Munger argue for the law in general, I believe that some of the most important dimensions to answering the very question of *what surveillance is* can only be addressed by exploring the ways that it plays out in the differing and complicated everyday lives of the subject.

REFERENCES

Baard, E. 2002. 'Buying Trouble.' *Village Voice*, 24–30 July http://www
   .villagevoice.com/issues/0230/baard.php.
Bennett, C.J., and R. Grant, eds.1999. *Visions of Privacy: Policy Choices for the
   Digital Age.* Toronto: University of Toronto Press.
Bumiller, K. 1987. 'Victims in the Shadow of the Law: A Critique of the Model
   of Legal Protection.' *Signs* 12:421–39.
– 1988. *The Civil Rights Society: The Social Construction of Victims.* Baltimore,
   MD: Johns Hopkins University Press.
– 1996. 'Rights, Remembrance, and the Reconciliation of Difference.' *Law and
   Society Review* 30 (1): 7–54.
Engel, D.M., and F.W. Munger. 2003. *Rights of Inclusion: Law and Identity in the
   Life Stories of Americans with Disabilities.* Chicago: University of Chicago
   Press.

Ewick, P., and S. Silbey. 1998. *The Common Place of Law*. Chicago: University of Chicago Press.

Foucault, M. 1979. *Discipline and Punish*. New York: Vintage.

– 1980. *Power/Knowledge*, ed. C. Gordon. New York: Pantheon.

Gilliom, J. 2001. *Overseers of the Poor: Surveillance, Resistance, and the Limits of Privacy*. Chicago: University of Chicago Press.

Haggerty, K.D., and R.V. Ericson. 2000. 'The Surveillant Assemblage.' *British Journal of Sociology* 51 (4): 605–22.

Handler, J.F. 1992. 'Postmodernism, Protest, and the New Social Movements.' *Law and Society Review* 26 (4): 697–732.

Lyon, D. 2001. *Surveillance Society: Monitoring Everyday Life*. Philadelphia: Open University Press.

Marx, G. 2003. 'A Tack in the Shoe: Neutralizing and Resisting the New Surveillance.' *Journal of Social Issues* (May) 59 (2): 363–80.

McCann, M. 1994. *Rights at Work: Pay Equity Reform and the Politics of Legal Mobilization*. Chicago: University of Chicago Press.

McCann, M., and T. March. 1995. 'Law and Everyday Forms of Resistance.' In *Studies in Law, Politics, and Society* 15:201–36.

Rosenberg, G.N. 1991. *The Hollow Hope: Can Courts Bring about Social Change?* Chicago: University of Chicago Press.

Scott, J.C. 1985. *Weapons of the Weak: Everyday Forms of Peasant Resistance*. New Haven: Yale University Press.

Tushnet, M. 1984. 'An Essay on Rights.' *Texas Law Review* 62:1363–1403.

PART TWO

Police and Military Surveillance

In Part Two Reg Whittaker, Jean-Paul Brodeur and Stéphane Leman-Langlois, Aaron Doyle, Christopher Dandeker, and Kevin D. Haggerty explore some of the ways that surveillance has become integral to police and military operations. The first three contributions focus on surveillance as a response to criminal and terrorist threats. The last two chapters examine how surveillance technologies are altering the organization of the military and the speed and execution of combat.

Reg Whitaker details the expansion of surveillance and attendant threats to privacy that have resulted from the official response to 9/11. The terrorist threat now facing the United States and other Western nations is qualitatively different from previous threats to national security. Nonetheless, some aspects of the response seem to be following the general historical pattern: arbitrary and intrusive powers are introduced and legal due process is cast as an impediment to security in a period of heightened social anxiety.

Prior to 9/11 the U.S. government already possessed formidable surveillance capacities. However, the full potential of these surveillance capacities was inhibited by various legal and political factors. Surveillance capability therefore remained largely decentralized and unintegrated. September 11 has accelerated centralization and integration. For example, the U.S. *Patriot Act* expands electronic monitoring powers and eliminates many of the previous checks and balances that had constrained surveillance systems. Restrictions on acquiring warrants have been reduced and official abilities to monitor electronic communications have been expanded. The treasury secretary has been granted new powers to regulate and probe the domestic and international activities of financial institutions. Moreover, spying powers previously restricted to foreign espionage investigations have been extended to include domestic investigations. The FBI is now permitted to use commercial data mining techniques, and has been encouraged to cooperate more closely with the CIA and military.

National boundaries are being transcended through legal and technological developments. At the same time, new technologies that can monitor ostensibly private acts and communications call into question traditional distinctions between the public and private. The line demarcating the state and private sector also becomes more ambiguous as state agents increasingly draw from information collected in the private sector. Finally, informational barriers within government are being reconfigured. For example, the historically distinct capabilities of counterintelligence, foreign espionage, and criminal investigation operations are now being integrated.

One of the most telling proposals to emerge after 9/11 was the 'Total Information Awareness' (TIA) initiative. This initiative was proposed as a cutting-edge technological system to identify and track terrorists. A massive data matching effort was designed to scrutinize all manner of information about citizens. TIA was to provide: 1) an analytical dimension concerned with extracting patterns that might predict terrorist behaviour; 2) a transactional dimension, focused on monitoring a potentially voluminous amount of transactional data; and 3) the identification of individuals whose profile triggered an alert.

In the face of political opposition, the TIA program went through a series of rapid transformations and was ultimately abandoned. Nevertheless, TIA exemplifies a police and surveillance logic that is now being embraced in other contexts. Questions remain, however, about the capabilities of such programs, specifically about their ability to analyse the information that they collect.

The corporate sector has a complex relationship with these developments, and governments are being pushed in different directions by private interests. On the one hand, we are witnessing the emergence of a security-industrial complex where private interests stand to win a bonanza of governmental contracts from the state-funded insecurity market. At the same time, corporations have sometimes resisted the costs and inconvenience of making substantial security improvements. The new security infrastructure offers state officials the promise of a remarkable expansion of power that they might find difficult to resist in times of insecurity. This power, however, comes at the cost of surrendering freedoms, and suggests a form of hubris: the U.S. government's technological reach might now exceed its political grasp.

Jean-Paul Brodeur and Stéphane Leman-Langlois are also concerned with developments in surveillance prompted by the terrorist attacks, specifically on what these changes might mean for domestic policing. They suggest that these developments cumulatively represent a move away from a model of 'low policing,' which is essentially law enforcement, and the fulfilment of a 'high policing' paradigm.

To explore the applicability of high policing models to contemporary developments they concentrate on the Communications Security Establishment (CSE) in Canada and the TIA program in the United States. Although these are only two programs in a much wider universe of surveillance initiatives, they are indicative of the much greater investment in signals intelligence that has occurred since the attacks.

The precursor to the CSE was formed by the Canadian government in 1941 with a mandate to decipher enemy electronic communications. Today the CSE is comprised of civilian experts and military personnel who have both an offensive and protective mandate. They monitor the communications of non-nationals and offer advice to governmental agencies about telecommunications security. After 9/11 new enabling legislation expanded the CSE's mandate, allowing it to provide technical and operational assistance to law enforcement. While it remains unclear exactly what this will mean in practice, it is very likely that the CSE will become involved in intercepting private domestic communications.

The TIA program in the United States articulates a new policing logic where data mining is understood to offer answers to many security questions. In essence, TIA is a form of computerized scrutiny of massive amounts of information through data-matching strategies. It is an example of the increased importance of deductive policing, where informational patterns are extracted from a sample of terrorist acts. If a monitored transaction fits a predictive pattern an individual is singled out. Much of the actual operation of TIA would have been outsourced to private firms, many of which are engaged in a form of convenience marketing, supplying the state with whatever informational technologies are currently available rather than those that have had demonstrated successes. Although TIA was officially abandoned in September 2003, this was largely a result of the limited political acumen of its proponents. The logic that underlies the initiative continues to gain momentum, with new programs using comparable techniques emerging in different governmental offices.

Brodeur and Leman-Langlois contemplate these developments in light of the four main attributers of high policing – a model of policing that originated in seventeenth-century France, which is characterized by political surveillance in the service of the status quo. *The* characteristic attribute of high policing is that it collects and hoards a great deal of information, something that is clearly exemplified in the current security environment.

Second, high policing conflates legislative, judicial, and executive powers. This dimension of high policing seems the most removed from recent developments. While the police occasionally offer curbside justice, there continues to be a formal distinction between legislative, judicial, and executive powers. Such demarcations are complicated by new legislative powers that allow the police to circumvent long-standing due process rights.

Third, high policing is concerned with preserving the existing political regime, not necessarily protecting civil society. As such, it relies on some notion of the 'enemies of the state.' Today, the fact that the interests of the citizenry are now assumed to coincide with those of the state suggests that the high policing opposition between state and public interests does not apply. In practice, however, things are more complicated. One complication is the blurring of the line between nationals and foreigners who reside in Western nations. Since highly publicized police abuses of the 1960s, the police have only been permitted to conduct political surveillance of foreign nationals. Amendments to the *Patriot Act* have changed this restriction, breaking down the barrier between political surveillance and law enforcement, and weakening the distinction between nationals and aliens.

Finally, high policing makes extensive use of informants, both human and technological. In the current situation there is evidence of the increased use of both types of informants. Some analysts are keen to develop more human sources, particularly after the demonstrated failures of human intelligence to anticipate 9/11. The prospect of doing so, however, is complicated by a number of factors. The fact that new enemies often have a different nationality, and require initiates to engage in extremely violent behaviour, makes infiltration very difficult and limits the prospects for relying on human intelligence as a policing strategy for the foreseeable future. Instead, there has been a turn to technological sources of information.

The embrace of new forms of electronic monitoring exacerbates possibilities for producing errors. Rather than concentrating exclusively on the surveillance dynamics of Orwell's Big Brother, Brodeur and Leman-Langlois suggest that we now also face the prospect of 'Big Bungler,' Big Brother driven mad by too much power and speed.

The greater availability of data and images produced by surveillance systems produces an attendant increase in the possibilities for novel uses of such information. Aaron Doyle draws attention to the trend towards broadcasting surveillance video footage on news and entertainment programs, accentuating the more emotional and symbolic dimensions of surveillance. Audiences have become increasingly accustomed to surveillance camera footage appearing in news broadcasts as well as entertainment shows.

Analysing how surveillance camera footage is broadcast allows us to re-think the prominent and contentious ways in which the media might influence society. To this end Doyle draws on the 'medium theory' of Joshua Meyrowitz, who has explored how the fixed properties of a medium can reshape social situations. This process can occur through the media's broad ideological consequences as well as more specific institutional effects. Doyle expands upon medium theory by emphasizing how larger power relations help dictate the nature of such recontextualization.

The proliferation of surveillance cameras has produced an abundance of available images for broadcast. The practice of broadcasting such images coheres with a culture of informing, and connects with trends in the TV industry to broadcast 'real' events. The police are also eager to broadcast such images, which they see as a means to help identify wanted individuals and promote police successes. Broadcasting surveillance footage, moreover, helps the police to maintain good relationships with journalists.

The willingness of the police to participate in such broadcasts suggests that this process tends to work to the benefit of powerful institutional actors. Doyle proposes four reasons for this tendency. First, the images that are displayed reproduce the police ideology. Second, the images cohere with media practices, especially a focus on recent events and a particular fixation on individual crimes, violent acts, and the deviance of the poor. The third factor concerns broader cultural dynamics, including how the images resonate with and reinforce a cultural conception of crime that is passionate, punitive, and sees crime as the product of individual failings. Finally, a series of factors related to the properties of TV as a medium help broadcasting surveillance to benefit institutional actors. These include the emotive nature of TV, and the fact that it tends to visualize the body and consequently has an affinity towards violence. Television is also epistemologically forceful, due in part to the belief that 'seeing is believing.' And because television shows the few to the many, it positions prominent individuals in ways

that make both them and the newsworthy events in which they participate seem larger than life.

In addition to having broad ideological consequences, broadcasting
surveillance footage can influence criminal justice institutions and processes more directly. It can assist police investigations at the same time
that it criminalizes certain events simply because they have been witnessed. These images provide a ritual of public shaming and can contribute to a form of trial by media. Broadcasting surveillance images
can also fuel the further spread of surveillance cameras.

Broadcast surveillance images are prominent components of the contemporary spectacle of crime. The rise of the mass media has shifted the
form and location of spectacular power, making it more public and
collective. As such, it differs in important ways from the spectacle of
punishment that surrounded the gallows. Today the media-centric spectacle of crime occurs much earlier in the criminal justice process and is
entwined with the broader symbolism and authority of the police.
Moreover, the public protest and dissent that came to characterize the
spectacle of the gallows is notably absent in the current spectacle of
crime.

Cumulatively, the trend towards broadcasting surveillance camera
footage represents an alternate current in the contemporary dynamics
of social control. It runs counter to the actuarial elements in social
control, which are defined by their low visibility and routine control of
categories of people at a distance. Instead, broadcasting surveillance
fits with a form of vengeance-oriented spectacle, or what David Garland has called the 'Criminology of the Other.' Where the database
exemplifies actuarialism, broadcast TV epitomizes its alternative through
its focus on the emotive, individualizing, random, physical, irrational,
public, and collective dimensions of crime. Its logic is precautionary,
concerned with dramatic but highly unlikely events.

It is not only criminal behaviour that the media have spectacularized
in recent years. Contemporary military conflicts have provided the
media with opportunities to broadcast unprecedented combat footage.
This development has been aided by a host of new communication and
visualization technologies. Christopher Dandeker examines the transformations that these new military surveillance and information systems are fostering, and addresses whether they are poised to prompt a
step-change in military power.

Historically, the military was at the vanguard of surveillance and
forms of organization and as such was seen as a model for business

enterprise. However, over the twentieth century businesses enterprise began to overtake the military in superiority of both surveillance and organizational development. There are now a number of parallels between developments in the world of business and the military. For example, the military now emphasize flexible organizational structures to allow for a swift response in uncertain environments in collaboration with other forces. The decline of immediate state-based threats parallels the lack of a stable business market. Likewise, reductions in the size of the military are comparable to corporate downsizing. The military need to resort to operations other than war in a global and multinational environment mimics the need for companies to respond to global markets. Military discussions about contracting out some functions also echo developments in the civilian business sector. Finally, both the military and business worlds are seeking to make the best use of new information technologies in all aspects of their organization.

The military have also turned to informational models and surveillance systems that allow simultaneous and synoptic monitoring. There are five connected developments in this regard:

1) The modularization of the all-volunteer force and an emphasis on joint operations;
2) An emphasis on panoptic and simultaneous surveillance to apply destructive power within fine calibrations;
3) A prioritization of informational superiority that is ushering in a form of combat where informational availability, speed, and combined networks is crucial;
4) 'Full dimension protection' entails efforts to overwhelm an opponent so as to avoid matched combat, resulting in a depression of the military decision cycle and an attendant escalation in the cycle of war;
5) All of the above prompt a fundamental change in military procurement and logistics as the defence industries must try and keep pace with the rapid pace of technological change.

The potential successes of the current transformation will hinge on how a series of open questions and tensions are resolved over the upcoming years. First, the military superiority of the United States produces a need for other states to be able to cooperate with them and to perceive that doing so is to their mutual benefit. While some nations will embrace a series of specialized tasks, others are apt to see this as a form of political subordination. Second, there is a danger that the current emphasis on maintaining deployable forces abroad will neglect defending the homeland. Third, rather than lift the fog of war, new technologies will enhance the tempo of war and create greater opportunities to make mistakes with more dramatic political consequences. Fourth, new technologies and operational roles produce a form of operational complexity that raises questions about the ability of an enemy to respond with asymmetrical weapons and tactics. Moreover, the level of mission complexity is itself increasing as a result of the rise of operations other than war and efforts to bring about regime change.

The greater complexity of military operations is at least partially related to the blurring of the division of labour between political elites and military officials. New political imperatives make it desirable for officials to monitor low-level military decision making while severely constraining the number of enemy and allied causalities. The contemporary media can now scrutinize aspects of military operations that were previously hidden. Technological developments which foster such scrutiny can themselves increase the pace of political events, as the technologies of military power and communication compress the levels of war. All of this serves to give small tactical units major strategic importance and raises questions about the role of the individual discretion of fighting soldiers operating in a fast-paced and visible environment.

A characteristic attribute of new developments in surveillance is how monitoring can prompt interacting processes of control and evasion. This effect is especially salient in determinations of the effective site of military decision making. There has been a long-term process of dispersion of military authority to lower levels of the command chain. As warfare becomes faster and more visible, leaders have less time to digest information, and their decisions are likewise scrutinized quickly. At the same time, however, the centralization of control encouraged by new communication and visualization technologies can also introduce the temptation to micro-manage warfare. This tends to compress the strategic, operational, and tactical levels of war and to accentuate a tension between the dispersion of control versus a desire to manage combat from afar. This tension can only be mitigated through measures to build trust between political and military officials.

Soldiers must become aware of the broader political frameworks in which they are operating, which leads Dandeker to accentuate the importance of the 'soldier-scholar.' The greater depth and breadth of media coverage of war means that soldiers must be more adept at handling the media. This imperative is even greater in complex reconstruction and peacekeeping operations.

The media's relationship with the military now entails interaction between panoptical and synoptic surveillance. While the media can bring both the many and the few into view, it tends to present highly selective images to a mass audience. Our ability to see war more immediately might come at a cost of reduced understanding, as such images tend to be presented with little attempt to situate them in a broader political and historical context.

Kevin Haggerty's analysis of surveillance and warfare flows directly from the insights and concerns introduced by Christopher Dandeker. In particular, Haggerty is interested in how new military surveillance devices introduce processes that can both accelerate and decelerate the speed of combat.

The context for these changes in the tempo of combat is the move to permanent warfare discernible in the United States since the end of the Second Word War. Some of the main dimensions of this development involve ongoing military gamesmanship, simulations, and the institutionalization of an extensive military industrial complex. Such developments are motivated by a commitment to invent and use high technologies to secure military advantage. New tools are routinely scrutinized for potential military applications, with surveillance and communications devices being particularly important in this regard.

Paul Virilio draws attention to the importance of speed as an integral attribute in establishing and reinforcing social hierarchies. Virilio's 'dromology' involves the historical study of speed in many different contexts, with a particular emphasis on how warfare has been affected by new vectors of speed. Informational and visualization technologies have accelerated the speed of combat and contributed to the automation of combat functions. However, his analysis of the place of technologized speed in warfare is only one side of the coin. Virilio ignores how the same technologies that increase speed can themselves paradoxically decelerate the speed of war. This deceleration can occur through information overload, centralization of command and control, and the necessity for prolonged military preparations.

One of the first places where information overload was expressly identified as a military problem was Vietnam. New reporting systems designed to enhance the speed and decisiveness of decision making ultimately produced so much information that they fostered a form of command paralysis. Computers were introduced to try and manage this problem but they, in turn, produced a greater specialization that itself created ever-more information. These systems also helped to increase the level of information required to make even comparatively simple decisions, slowing down the decision cycle even more.

New information and visualization technologies can adversely affect command and control by delaying and impeding decision making. Greater amounts of information tend to produce inordinately long decision cycles. Haggerty shares Dandeker's concern that new technologies of distanciation, which allow distant commanders to watch combat in real time, can tempt these commanders to intervene. At the same time, local field commanders become increasingly aware that they are being monitored, a situation that can introduce hesitancy and delays in a military context where decisive action is often preferred. Moreover, the valorization of information in this model of warfare can accentuate any inevitable gaps in knowledge, prompting ongoing searches for ever-more information, and thereby reinforcing a tentativeness in command and control.

Finally, the demonstrated speeds of informational warfare can themselves be slowed by marshalling requirements. While the U.S. military is now capable of arriving at any location on the planet in record time, they are not necessarily then capable of fighting an informational war. Such combat can require highly complicated and painstaking efforts to ensure that information systems are established, operative, and properly calibrated. This requirement can actually slow the progress of combat, as leaders delay engagements in order to ensure that the appropriate systems are in place and operating appropriately.

These three factors accentuate Virlio's failure to engage with the contradictions, limitations, and paradoxes of new military technologies. The same technologies that allow for accelerated combat can also operate as a brake on the speed of war.

While a series of questions remain about the likely successes of informational warfare, the demonstrated abilities of the U.S. military have already produced a series of strategic ripples in the international community. Some nations are seeking to develop their own information warfare capabilities, while others see their only option as developing nuclear weapons capabilities. It also seems to be entirely conceivable that some groups and nations, recognizing their inability to compete in the world of informational warfare, will be tempted to embrace more unconventional terrorist strategies.

6 A Faustian Bargain? America and the Dream of Total Information Awareness

REG WHITAKER

FAUST: By spying, your all-knowing wit is warmed?
MEPHISTOPHELES: Omniscient? No, not I; but well-informed.

The 9/11 terrorist attacks, and the declaration of a War on Terrorism, have raised the spectre of a global surveillance regime. The technological capacity to achieve a global surveillance regime has been known for some time, and the final two decades of the twentieth century witnessed remarkable developments in surveillance technologies. The daily lives of people in the most industrially developed nations are now tracked and recorded. Electronic eyes scan the globe, from closed circuit cameras on the ground to satellites gathering sophisticated imagery from space. Voice communication is scooped out of the sky by electronic listening posts. Sophisticated search engines troll through e-mail traffic and internet use. Global positioning systems based on satellites can yield the precise location of targeted individuals anywhere on earth. Unique biometric identifiers such as palm and finger prints, iris patterns, facial and gait characteristics, and DNA sequencing are increasingly being recorded and stored in data banks and demanded to gain access to services or pass security screenings. Virtually every daily economic transaction adds to an electronic trail that can potentially construct a unique social profile of an individual. Detailed medical records and significant genetic information can profile individuals in a remarkably intimate manner.

What is particularly alarming to privacy advocates is that the new information technologies, based on the universal language of digitization, permit the seamless transfer and matching of information gathered by different agents for different purposes. Databases can 'talk' to each other, and, in so doing, create the capacity for decentralized 'dataveillance,' a surveillance society in which the 'files' exist in no central location, and are under no central control, but which in their totality may exercise far more intrusive capacity to gaze into the private space of individuals than the Big Brother surveillance state of the past. Yet these same technologies offer numerous benefits that must be balanced against the threats they pose to privacy: to take one instance alone, the potential benefits to health of detailed medical and genetic databases are immense. The same developments that alarm some excite others. Indeed, most typically, they alarm and excite the same people.

The political and administrative problem of privacy protection in the era of dataveillance has generally been posed as how best to constrain and control this technical potential so as to protect a reasonable degree of personal privacy, while at the same time retaining the economic and social benefits promised by the new technologies.

Prior to 9/11, there seemed good reason to believe that there were structural impediments to total surveillance. The new information technologies, although often originating in the defence sector, owe their rapid global diffusion primarily to private sector research, development, and marketing. Their commercial potential had been best exploited by the corporate sector, with governments by and large reaping the benefits as technological spin-offs. But a paradox appeared in the 1990s. The toolbox of surveillance was increasing exponentially while the totalitarian surveillance state which had been so threatening a feature of the twentieth century, whether in fascist or communist form, was in dissolution, and even the liberal capitalist state that had borrowed, in less virulent form, many of the surveillance capacities of the authoritarian state appeared to be in retreat before the forces of the market.

If states collect and use information on their citizens primarily as a means of social and economic control, corporations collect personal data primarily for marketing purposes. There is a long tradition, especially in the United States, of resisting government threats to personal privacy. More recently, different concerns have been raised about threats to privacy arising from the vast, unregulated, and rapidly accumulating corporate data banks. These concerns can be best described as a shift from the Big Brother surveillance state to the Little Brothers surveillance society (Whitaker 1999). Legal responses have varied. Europe and Canada have adopted public regulatory regimes that stipulate that personal data collected by private entities for a specific commercial purpose may not be sold or traded to third parties without the express consent of the individuals from whom the information was drawn. The United States, with its anti-government traditions, has so far tended to rely more on corporate self-regulation, enforced largely by private litigation. In both cases, the emphasis is on firewalls separating public from private data collections, and separating different private data collections from each other. In the absence of effective barriers, data matching and linkage quickly threaten personal privacy. Yet despite the ominous growth trajectory of panoptic technologies, the decentred and dispersed quality of the information gathered by the electronic eyes and ears diminished their totalizing potential. Nobody, public or private, seemed to have the will or capacity to put it all together – until 9/11.

Even with the attention paid to the private sector, it was already apparent well before 9/11 that states, and the United States pre-eminently among them, did have some impressive, but underused, surveillance capacity. Two examples illustrate the potential for a renascent Big Brother state. Since the late 1940s the English-speaking countries, under the leadership of the U.S. National Security Agency (NSA), have maintained an extensive electronic eavesdropping partnership known as the UKUSA alliance. During the Cold War, listening posts across the globe closely monitored communications within the Soviet Bloc. Today the UKUSA countries tap into the Intelsat communications satellite system that relays most of the world's phone calls, faxes, telexes, Internet, and e-mail communications around the world. A system called ECHELON links all the computers among the UKUSA agencies using a set of keywords in a dictionary contributed by all the agencies; flagged messages are automatically routed to the country or countries that entered the particular keyword flag. In the United States, the FBI had, before 9/11, begun deploying CARNIVORE, a super search engine which, when installed on Internet service providers, is capable of trolling through e-mail traffic and flagging communications of interest to the agency based on the identities of senders and receivers, keyword recognition, and so forth.

A second example of state surveillance capacity is the machinery for tracking money-laundering trails centred in the U.S. Treasury Department in the Financial Crimes Enforcement Network (FinCEN). FinCEN 'links the law enforcement, financial and regulatory communities together for the common purpose of preventing, detecting and prosecuting money laundering and other financial crimes' (United States

Department of the Treasury). It relies on worldwide monitoring of large financial transactions, and reportedly has at its disposal sophisticated artificial intelligence software capable of detecting anomalous or suspicious patterns in the vast daily volume of transactions, flagging transactions that might require closer attention or criminal investigation. Technology permits the instantaneous flow of capital across national borders by computer key stroke, which has assisted in the rapid development of transnational financial networks. Governments, the bankers and financiers assert, cannot interfere effectively in this globalizing process, and should stay out. Yet these networks are also threatened by illicit financial flows, funding criminal or terrorist enterprises, and here the private sector needs, and demands, the intervention of governments. The same technologies that foster licit financial flows also enable national states working together, under U.S. leadership, to monitor and investigate illicit flows.

Prior to 9/11, these powerful state surveillance systems faced certain limitations, both legal and political. Each of the participating UKUSA agencies was constrained by domestic law or practice from listening in on its own citizens. The awesome potential of the ECHELON system drew the critical attention of the European Parliament, especially of its French members, who complained about a global 'Anglophone spy network' that might be used against European economic interests (Davies 1997). FinCEN's potential was limited by the reluctance of many countries to cooperate fully with new disclosure laws, and of many transnational financial corporations to open their books and their clients' financial data to Uncle Sam's prying eyes.

The need to control the dark side of globalization, whether organized criminal or terrorist networks, had given greater urgency to the development and deployment of international policing and surveillance. Yet structural limitations on global surveillance power were also apparent. If the diffusion of new information technologies extends the reach of surveillance, discontinuities in the distribution of technology set limits to the prying gaze. While the developed world is increasingly luminous and transparent, much of the less developed world remains dark and opaque, simply because it is less 'plugged in' to sophisticated communications technologies and networks. Recent military interventions in the Third World may, in one way, be seen as violent attempts to expand surveillance coverage into obscure and thus threatening corners. The invasion of Afghanistan in 2001 required a huge mobilization of intelligence resources to apply to a country that, by dint of sheer underdevelopment, was resistant to the tools of contemporary surveillance, and largely remains so, even after forcible regime change. The invasion of Iraq in 2003 encountered an additional surveillance problem in a poor country: a regime dedicated to hiding itself from prying eyes. The gross errors made by the external surveillance regime of the United States and United Kingdom with regard to Iraq's presumed, but never discovered, weapons of mass destruction are an embarrassing indication of the continuing blind spots in the global surveillance regime.

The problems of the global surveillance regime are also the problems that inevitably fuelled calls for that same regime to be stepped up and enhanced. The Little Brother surveillance society had demonstrated its technological potential. The Big Brother surveillance state, after a period of outsourcing and privatization, was waiting to be recalled to active duty and once again take charge. September 11 triggered that call.

How 9/11 Differs from Past Global Crises

Like the two most recent historical antecedents of this war, the Second World War and the Cold War, the War on Terrorism has ramifications for domestic politics, especially for civil liberties and minority communities. The forcible relocation of the Japanese populations from the west coast of North America in 1942 and the excesses of McCarthyism in the early Cold War offer notorious examples of how the search for security can generate injustices. More worrisome is the potential long-term damage to the fabric of civil liberties that may persist long after the emergency passes.

The historical cycle in which violent threats generate the expansion of arbitrary and intrusive powers of government is being repeated, most noticeably in the United States. Once again, the constitutional protection of rights is being dismissed, sometimes from the highest offices in the land, as an inconvenient impediment to safety.[1] And yet again a panicked public is encouraged to trust in action over deliberation, results over due process.

---

1 Astonishingly, in two cases of suspected terrorists who hold American citizenship (José Padilla and Yaser Esam Hamdi), the Justice Department has repeatedly asserted a claim that their constitutional protections can be set aside, and that the government's decisions must not be subject to review by the judicial branch. At the time of writing the Justice Department had at least partially avoided adverse court rulings in these cases.

In certain ways, the present crisis bears even more dangerous potential than earlier wartime emergencies. The terrorist threat is qualitatively different from the threats posed in previous emergencies. As Washington struggles to find the most effective response to a new, and, in many ways, unprecedented threat, the very novelty and uncertainty of the situation tempts government to reach for new powers, while heedlessly discarding old forms and conventions. Phrases such as 'thinking outside the box' and 'connecting the dots' may be appropriate for policy makers moving in unfamiliar and uncharted territory, but they can be dangerous guides for encroaching upon rights.

There *is* much that is novel about the organizational structure and operating methods of the terrorists. Al Qa'eda is a contemporary product of globalization: flexible, adaptable, diversified, transnational, decentred, a network of networks. Like transnational organized criminal networks, al Qa'eda operates very much as a paradigm 'new economy' corporation. As part of the 'dark side' of globalization, terrorist networks assiduously cover their global tracks, evading the scrutiny of law enforcement and security agencies rooted in national jurisdictions. As borderless enterprises, they utilize the most up-to-date technologies of communication that facilitate the instantaneous transfer of ideas, capital, and financial resources across national borders and continents. The intelligence and security failure of 9/11 clearly signalled the need for agencies like the CIA and the FBI to modify, if not reinvent, new and more appropriate intelligence-gathering mechanisms. Just as clearly, emphasis on new information technologies and more intrusive and extensive surveillance was inevitable.

Disappearing Boundaries between Public and Private

In response to the borderless terrorist threat, government has tried to dissolve, or at least to weaken, a series of boundaries that had previously been erected to demarcate spheres that were believed best kept as distinct from one another as possible. Foremost among these are national jurisdictions, demarcating national sovereignties. The Bush administration has actively and aggressively moved in numerous ways since 9/11 to extend its surveillance and its coercive reach across national boundaries, to the extent that many of its allies have become increasingly resentful.

Within the United States, there has been a concerted drive to break down various firewalls that have been maintained in the past to protect the private sector from government control. The tracking of those directly responsible for 9/11 – the biggest single criminal investigation in history – pointed to private sector databases (involving, for instance, credit card, telephone, and air miles data) for essential clues in reconstructing the trail of the terrorists. Money laundering investigations, involving reporting and automatic surveillance of a very wide range of private financial transactions, had already begun in relation to criminal activities, but since 9/11 have been greatly stepped up to track terrorist financing.

The appropriation of private sector data has proved very useful in the investigation of terrorist trails. Yet however effective from a criminal investigation standpoint, post hoc inquiry is the least interesting use of all-source data collection for counter terrorism. To technical enthusiasts, the more enticing prospect is *risk profiling*, concerned not with the past, but with the future; less with who have already engaged in terrorist acts, but with who *might be* terrorists. Profiling is about risk calculation. The more information available, so the tempting and seductive argument goes, the greater the likelihood there is of constructing accurate high-risk categories and thus actionable profiles of potential terrorists. Breaking down the firewalls separating private databases from public scrutiny is in this sense a necessary first step in terrorism prevention.

More subtly, a series of internal firewalls that once separated data banks held within government itself have also fallen under attack. The most notable of these is the distinction between counterintelligence investigations of foreign espionage operations in America and criminal law enforcement, with lower standards with regard to legal safeguards for the targets of surveillance in the former. In general, faced with borderless threats, government has sought to gain access to a seamless web of information on citizens and non-citizens, drawing on all potential sources.

This thrust has plausible arguments in its favour, rooted in the specific nature of the terrorist threat, with a ready constituency among a fearful public. It is also important to point to the inherent dangers. There are very good reasons why the various firewalls have been erected around data collected for different purposes by different agents, and why restraints have been imposed historically on government access to personal information. The right to privacy, long viewed as a fundamental element of a free society, is increasingly in question in the era of new information technologies.

U.S. Anti-terrorist Surveillance Powers

Introduced within a week of the 9/11 attacks and rushed through Congress under great pressure, the *USA Patriot Act* was signed into law on 26 October 2001.[2] Its provisions significantly expand the electronic surveillance powers of federal law enforcement authorities, often without providing appropriate checks and balances to protect civil liberties. Ideas that had previously been put forward by the executive and subjected to strenuous criticism were whisked through the legislative process after 9/11.

With regard to communication surveillance, the Act relaxes restrictions around required warrants and court orders. With regard to methods of identifying senders and receivers of telephone communications, section 216 extends court orders to cover e-mail messages and Internet use, and to cover the entire United States as opposed to the former limit to the judicial district in which the court has jurisdiction. Critics point out that, even though the capture of message content is specifically prohibited, e-mail header information, which may include subject headings, or the addresses of specific web sites visited, may be much more revealing than the simple telephone numbers previously captured.

This provision may provide sanction to the FBI's CARNIVORE program, even though once installed by an Internet service provider, CARNIVORE may monitor all the communications of all subscribers, not just those targeted by a court order. Service providers are immunized from legal liability for surrendering their customers' privacy when they cooperate in 'good faith' with government, even before court orders have been obtained, and they are promised reasonable compensation for the costs of assisting authorities. Although the Act imposes no positive obligation on service providers to modify their systems to accommodate law enforcement needs, anecdotal information suggests that after 9/11, Internet service providers have come under increasing pressure to assist, even in the absence of specific court orders (McCullagh 2001). The extent to which providers have installed CARNIVORE is unknown: public disclosure of such information is prohibited.

The *Patriot Act* lowers the barriers between criminal investigations and foreign intelligence information. Facilitating closer cooperation between criminal investigators and foreign intelligence collectors is prob-

2 P.L. 107-56, 115 Stat. 272 (2001), Uniting and Strengthening America by Providing Appropriate Tools Required to Intercept and Obstruct Terrorism.

ably not a controversial intention in itself, but the extension to criminal investigations of the much laxer standards of protection required under the *Foreign Intelligence Surveillance Act* (*FISA*) is debatable. *FISA* was originally a response to concerns about domestic spying that surfaced during the 1970s. While the need for special surveillance powers and secret evidence in relation to foreign espionage activities in the United States was widely accepted, it was generally believed that such methods were unacceptable when employed against American citizens, especially those engaged in First Amendment activities, as with the Vietnam War protests. Under *FISA*, government operates on a lower threshold for gaining authorization for intrusive surveillance against targets suspected of espionage and foreign intelligence operations in the United States than is the case for criminal investigations. For instance, under *FISA*, investigators gain access to records from car rental agencies, motel accommodations, and storage facilities, as well as easier recourse to physical searches and wiretaps. Special *FISA* courts meet in camera, with non-disclosure of evidence deemed to be of a sensitive intelligence nature.

Under the *Patriot Act*, *FISA* is amended in certain crucial and significant ways. Originally, a *FISA* surveillance order required certification that '*the* purpose for the surveillance is to obtain foreign intelligence information.' If evidence of a criminal offence was uncovered from *FISA* surveillance, prosecution could not be based on *FISA* surveillance, but would have to be based on a surveillance order under Title III of the *Omnibus Crime Control and Safe Streets Act of 1968*, a narrower and more restrictive authorization for wiretap orders. After 9/11, the Justice Department sought to amend *FISA* to read simply that '*a* purpose' was to obtain foreign intelligence. Congress baulked at this very low threshold, but instead provided in section 218 that foreign intelligence gathering be a '*significant* purpose' to trigger a *FISA* surveillance or search order. Less threatening to privacy than the administration's proposal, this compromise nevertheless departs significantly from the clear distinction originally drawn between surveillance for criminal law enforcement and surveillance for foreign intelligence. The Act encourages cooperation between law enforcement and intelligence investigators, but in blurring the line between the two, invites the Justice Department to opt for the lower threshold. Finding a 'significant' foreign intelligence component to terrorist investigations will probably not tax the resources of law enforcement officials unduly.

Not content with lowering the threshold, the *Patriot Act* widens *FISA*'s powers. It permits 'roving surveillance,' that is, orders that are not tied to a particular place or particular means of communication. While it is not unreasonable to point out that those seeking to evade surveillance will constantly shift their means and location of communication, there are obvious dangers that roving surveillance will capture communications among persons not named in the order. The Act extends *FISA* to cover e-mail as well as telephone communication. It extends the duration of surveillance and physical search orders, in some case providing extensions of up to a year. It expands the scope of access to certain kinds of business records to include seizure of any 'tangible items,' regardless of the nature of the business in possession of the records.

Questions about how government would use expanded access to *FISA* orders were raised in 2002 when the *FISA* court issued a stunning rebuke to the Justice Department (Shenon 2002a; Eggen and Schmidt 2002), identifying more than seventy-five cases in which it says it was misled by the FBI in documents in which the bureau attempted to justify its need for wiretaps and other electronic surveillance. The FBI and the Justice Department were said to have violated the law by allowing information gathered from intelligence eavesdrops to be used freely in bringing criminal charges, without court review, and criminal investigators were improperly directing the use of counterintelligence wiretaps. In August 2002 the Justice Department appealed this opinion to the little-known Foreign Intelligence Surveillance Court of Review, arguing that the *Patriot Act* now permitted *FISA* to be used to obtain evidence for a prosecution if the government also has a significant non-law enforcement foreign intelligence purpose (Shenon 2002b).

On 18 November, the Review Court, in its first ever decision, concluded that *FISA*, as amended by the *Patriot Act*, supports the executive branch's position, and that the restrictions imposed by the *FISA* court are not required by *FISA* or the Constitution.[3] 'Proclaiming a major victory in the war on terrorism, Attorney General Ashcroft said the decision "revolutionizes our ability to investigate and prosecute terrorists" because it permits criminal investigators and intelligence agents to work together and to share information' (Savage and Weinstein 2002). The *New York Times* called the decision 'A Green Light to Spy' (*New York Times* 2002). As if in confirmation, the attorney general announced

3 United States Foreign Intelligence Surveillance Court Of Review: In re: Sealed Case No. 02-001 Consolidated with 02-002, On Motions for Review of Orders of the United States Foreign Intelligence Surveillance Court (Nos. 02-662 and 02-968). Argued 9 September 2002, decided 18 November 2002.

plans to intensify secret surveillance, including the designation of special intelligence prosecutors in every federal court district, and the creation of a new FBI unit to seek intelligence warrants (Eggen 2002a). In March 2003, Ashcroft revealed that he had personally authorized secret electronic surveillance and physical searches without immediate court oversight in 170 'emergency' cases since the 9/11 attacks – more than triple the emergency searches authorized by other attorneys general over the past twenty years (E. Schmitt 2003).

The final section of the *Patriot Act* that raises issues about privacy protection is the expanded authority of the treasury secretary to regulate and probe the activities of financial institutions, especially their relations with foreign entities, in pursuit of money laundering. New money laundering crimes are defined in the Act, mainly to extend the existing legislation to include a wider range of terrorist and cybercrime activities, and to extend U.S. jurisdiction to prosecute money laundering offences abroad. The most controversial aspect of the expanded surveillance powers in the Act is the concerted attempt to extend the cooperation of financial institutions, securities dealers and brokers, commodity merchants, and so forth with law enforcement agencies with regard to suspected terrorist associated money laundering. In effect, the Act seeks to enlist financial institutions as active participants in the government's surveillance program, yet another example of the tendency to break down barriers between the public and private sectors. Failure to actively cooperate could result in severe penalties, including seizure of assets.

Early in 2003, a draft version of a new and expanded *Patriot Act* was leaked to a public interest advocacy group. The *Domestic Security Enhancement Act* of 2003 would, among other things, prohibit disclosure of information regarding people detained as terrorist suspects; create a DNA database of 'suspected terrorists'; place the onus on suspects to prove why they should be released on bail; and allow the deportation of U.S. citizens who become members of or help terrorist groups. The attorney general has refused to comment on the status of the draft before formal introduction to Congress, but the drift seems clear.[4]

Nor have the appetites of the executive and Congress for intrusive surveillance powers been exhausted by the language of the *Patriot Act*.

---

4 The Center for Public Integrity, Justice Dept. Drafts Sweeping Expansion of Anti-Terrorism Act: http://www.publicintegrity.org/dtaweb/report.asp?ReportID= 502&L1=10&L2=10&L3=0&L4=0&L5=0.

The *Homeland Security Act*, signed by the President in November 2002, contains additional powers for electronic surveillance. It permits Internet service providers voluntarily to provide government agents with access to the contents of their customers' private communications without those persons' consent based on a 'good faith' belief that an emergency justifies the release of that information. Twenty-year prison terms are provided for computer hackers who recklessly cause or attempt to cause serious bodily injury – a threat based on the idea that hackers could maliciously break into and disrupt the computer systems of critical infrastructures like nuclear power plants or air traffic control systems. It facilitates the sharing of sensitive intelligence information and the content of electronic intercepts with state and local authorities. There are some limited privacy protections in the Act (Electronic Privacy Information Centre 2002).

Surveillance in Practice

The attorney general has released new investigative guidelines for the Justice Department reflecting its mission to 'neutralize terrorists before they are able to strike,' and to shift emphasis from criminal investigation to crime 'prevention' (perhaps 'pre-emption' might be more appropriate than the traditional 'crime prevention'). Among these guidelines is authorization given to the FBI to conduct 'online research' for counterterrorism purposes, 'even when not linked to an individual criminal investigation.' Moreover, the FBI is authorized to use 'commercial data mining services to detect and prevent terrorist attacks, independent of particular criminal investigations.' The FBI is further enabled to establish its own databases drawn from multiple sources, and to operate 'counterterrorism information systems' – although the guidelines prohibit the FBI from 'using this authority to keep files on citizens based on their constitutionally protected activities' (Electronic Privacy Information Centre 2002). Of course, if such activities are suspected of being carried out in association with terrorist activities, this protection is presumably waived.

One of the surveillance spin-offs from the *Patriot Act* that has come to partial notice is its application to libraries and bookstores. *FISA* court orders, now extended to e-mail and online communication, apparently include records of URLs visited in web surfing, or the detailed information such as keywords recorded in search engines consulted online. Persons seeking anonymity in Internet use may use library online facilities. Libraries have consequently been visited by the FBI seeking records of usage and identification of users, citing 'roving' warrants. The FBI may also have sought to have their surveillance software, like CARNI-VORE, installed on library systems. Nor do these intrusions stop at online monitoring. Section 215 of the *Patriot Act* requires any person or business to produce any books, records, documents, or 'tangible items.' Libraries and bookstores may be required to provide records of books borrowed, or purchased, by persons under suspicion. The incidence of such requests, and the degree of compliance, is not known: Catch-22 is that the FBI prohibits disclosure regarding such orders (Kasindorf 2002). This has led one commentator to suggest that Attorney General Ashcroft may have been reading Kafka in his local library (Hentoff 2002). This is uncomfortably reminiscent of the notorious FBI 'Library Awareness Program' that from the 1960s to the late 1980s attempted to bully librarians into snooping into the reading habits of patrons deemed left-wing or anti-American by the FBI (Foerstel 1991). Could we now be seeing attempts to track and identify persons reading about Islam or terrorism?

This raises another serious problem in evaluating the impact of the new powers: the administration is resistant to providing Congress access to information on the implementation of the Act. Congress, which was pressured into passing the *Patriot Act* without due deliberation, did add sunset clauses to many key sections, under which authority is terminated by 31 December 2005, unless extended by Congress. If sunset provisions are to have any real force, Congress must have access to sufficient information to make informed judgments on whether the extension of new powers is justified. Yet when a bipartisan request was made by Congress in July 2002 that a wide range of questions concerning the implementation of the *Patriot Act* be answered by the executive, the government failed to respond to most of their questions. Even threats of subpoenas have not succeeded in dislodging much more information. In early November 2002 the President signed a bill authorizing Justice Department appropriations, including statutory authorities relating to federal law enforcement activities. The bill contains a number of clauses inserted by Congress imposing substantial obligations on the executive for reporting information on a wide array of matters related to law enforcement. In signing, the President attached a 'Signing Statement' that indicated his interpretation of the reporting obligations, asserting his right to withhold information 'the disclosure of which could impair foreign relations, the national security, the deliberative processes of the Executive, or the performance of the Executive's constitutional duties,' as defined in each case by the President.[5]

There are plausible arguments in favour of secrecy when issues of security against terrorist attack are at stake. But most of the critics of government stonewalling do not challenge the reasonableness of keeping genuinely sensitive and potentially damaging information out of the public realm. Their expectations are simply for a level of information that will permit democratic accountability and allow the public to make informed judgments on government's performance in the War on Terrorism. The excessive secrecy and the active resistance to requests for information, even from the legislative branch, follow a familiar pattern of governments with authoritarian tendencies: as they expand their intrusive gaze into the interstices of private life, they seek to render their own actions opaque to the citizenry. This was the precisely the role of the Inspector in Bentham's panopticon, the paradigm for all modern surveillance. Yet in the contemporary world of decentred, multidirectional surveillance, attempts to prevent the watched from watching their watchers are doomed to limited success in practice. Technologies of surveillance are so widely diffused, and so cheaply deployed, that the state's attempts to shroud its own actions in secrecy inevitably fall under criticism, and even ridicule. Since 9/11, the balance in public access to government information has undoubtedly shifted towards non-disclosure, but government actions in a relatively open liberal democracy like the United States are too exposed to publicity to escape indefinitely the scrutiny of critics. Some of the most ambitious U.S. government surveillance projects, as we shall see, have been exposed to such withering public derision that they have had to be retracted or severely curtailed.

Towards 'Total Information Awareness'?

The first surveillance project to overstep the invisible line between public tolerance and outrage was the Justice Department's ill-starred TIPS program. TIPS (Terrorism Information and Prevention System) envisioned mobilizing a volunteer army of informants for the 'stated purpose of creating a national information sharing system for specific industry groups to report suspicious, publicly observable activity that

5 Statement by the President, 'President signs Justice Appropriation Authorization Act,' 4 November 2002.

could be related to terrorism.'[6] Certain occupations that provided access to private homes, such as utilities personnel, letter carriers, and cable and telephone repair people, were identified as particularly useful sources, but a national TIPS hotline was also established for concerned private citizens to report suspicions about neighbours or local happenings to a central FBI data bank. Reaction was immediate and vociferous. The U.S. postal service formally stated that its personnel were directed not to cooperate with the program, and numerous expressions of resistance from other quarters were recorded in the media. Ashcroft's idea was likened to the infamous Stasi secret police in the former East Germany. Other critics pointed out that the FBI did not need the task of sifting through neighbourhood gossip while trying to track global terrorist threats. Exposed to a mixture of indignation and ridicule, TIPS was first scaled back, and then withdrawn altogether, vanishing from government websites as if it had never existed (Eggen 2002b).

TIPS was an old-fashioned, low-tech, prosaic surveillance project that was understandable enough to be widely recognized as unacceptable in a free society. Another project, one that defines itself by its cutting-edge technology and which was conceived in expansive, global terms, has also been met with fierce resistance. This initially went under the Orwellian title of 'Total Information Awareness' (TIA).

Some background is required to situate TIA. The American government's surveillance plans are not confined to domestic law enforcement. Foreign intelligence agencies, despite the setback administered by their failure to anticipate and prevent 9/11, possess surveillance capacities, previously directed at targets abroad, that are tempting tools for an aggressive government to employ at home as well. After all, the logic goes, terrorists operate across borders, so why should the United States deny itself the awesome powers of foreign surveillance for homeland security? Since the administration has made every effort to break down the barriers between criminal investigation and intelligence collection,[7] it is a simple step further to reduce the barriers set up against

6 'Operation Tips Fact Sheet' from the Justice Department website, summer 2002. For reasons that will be made clear, all TIPS information later disappeared from this and other U.S. government websites. The only remaining records are those now archived by independent observers, and media accounts at the time.

7 See most recently: Attorney General, 'New Guidelines to Share Information Between Federal Law Enforcement and the U.S. Intelligence Community,' 23 September 2002. WWW.USDOJ.GOVTDD

intelligence agencies spying on Americans. Close cooperation between the FBI and the CIA has been encouraged at the top level since 9/11. The CIA is reportedly 'expanding its domestic presence, placing agents with nearly all of the FBI's 56 terrorism task forces in U.S. cities, a step that law enforcement and intelligence officials say will help overcome some of the communications obstacles between the two agencies that existed before the Sept. 11, 2001, attacks' (Priest 2002). The NSA, which shares some of the blame for the 9/11 intelligence failure, has access to the vast surveillance capacities of the UKUSA network and its ECH-ELON system, but is barred by law from snooping on Americans. Should tracking terrorists abroad come to a halt at the U.S. border, when it is America that is squarely in the terrorist sights? The NSA director has mused on this question in an appearance before the Senate.[8]

There were of course very good reasons why agencies like the CIA and the NSA were barred from domestic spying, and these reasons have not vanished since 9/11. There may be compelling and plausible arguments in favour of greater cooperation, but a domestic role for foreign intelligence collection agencies must at the very least be accompanied by rigorous checks and balances and strong accountability mechanisms – unfortunately the very kinds of safeguards that the Bush administration has been assiduously limiting since 9/11.

The true visionary for a post-9/11 global surveillance state is neither John Ashcroft nor the directors of any of the existing security and intelligence agencies. He is Vice-Admiral John M. Poindexter, of Iran-Contra affair notoriety (see chapter 3). Poindexter has been rehabilitated by the Bush administration and made director of the Information Awareness Office in the Defence Advanced Research Projects Agency (DARPA). DARPA has a distinguished lineage in the information age. It was from this office that the Internet sprang. In 1969 DARPA set up a pioneer computer network among defence scientists called ARPANET, which, after a series of transformations and freed of its original sponsor, eventually evolved into the Internet. DARPA has a mandate to develop and apply new surveillance technologies post-9/11.

The concept of 'Total Information Awareness' was initially revealed to the world on a website featuring a logo of a large eye atop a pyramid

---

8 Statement For The Record By Lieutenant General Michael V. Hayden, USAF, Director, National Security Agency, before the Joint Inquiry Of The Senate Select Committee On Intelligence And The House Permanent Select Committee On Intelligence, 17 October 2002: http://intelligence.senate.gov/0210hrg/021017/hayden.pdf.

scanning the globe, under the slogan '*scientia est potentia*' (knowledge is power). Poindexter's office was putting together a series of research teams working on advanced surveillance technologies of various kinds, from bio-recognition technologies to sophisticated translation systems to bio-surveillance providing early warning of attacks of biological agents. He envisaged a cutting-edge system for detecting, classifying, identifying, and tracking terrorists. He wished to 'punch holes in the stovepipes' that separate different data collections and develop a seamless global system that mines data from all possible sources, public and private, American and foreign. One of the 'significant new data sources that needs to be mined to discover and track terrorists is the *transaction space*.' Terrorists move easily under cover, but they leave an 'information signature. We must be able to pick this signal out of the noise.'[9]

The *Washington Post* (2002) editorialized that 'anyone who deliberately set out to invent a government program with the specific aim of terrifying the Orwell-reading public could hardly have improved on the Information Awareness Office.' The archconservative *New York Times* columnist William Safire, usually a strong Republican supporter, denounced this 'supersnoop's dream.' The person of Poindexter himself, with his dubious past, tended to darken public perceptions of the program. But the very idea of such a global intelligence system – tracking the 'transaction space' within the United States as well as around the globe – set off alarm bells from all quarters and all sides of the political spectrum about Big Brother and the end of privacy. Critics pointed out that privacy laws and regulations in the United States (not to speak of Europe and Canada, where extraterritorial invasions of privacy would be certain to rouse opposition) could not be blithely and unconstitutionally set aside by Washington simply by invoking national security.[10] The administration began backtracking and the TIA website began, bit by bit, to vanish, beginning with the sinister logo and the motto, followed by the laundered biography of Admiral Poindexter as well as the biographies of his senior associates and graphics depict-

---

9 Dr John Poindexter, Director, Information Awareness Office of DARPA, 'Overview Of The Information Awareness Office,' Remarks as prepared for delivery by, at DARPATech 2002 Conference, Anaheim, Calif., 2 August 2002, http://www.fas.org/irp/agency/dod/poindexter.html.

10 Background in U.S. law may be found in a report for Congress by the Congressional Research Service of the Library of Congress: Gina Marie Stevens, 'Privacy: Total Information Awareness programs and related information access, collection, and protection laws,' updated 14 February 2003.

ing the sweeping global ambitions of the program. Eventually the site was modestly retitled 'Information Awareness Office' and offered numerous assurances that nothing contemplated by the TIA program will overstep privacy protections.[11]

Bipartisan agreement in Congress supported an amendment to hold up funding for TIA until the administration explained it in detail to Congress, including its impact on civil liberties, and to bar any deployment of the technology against U.S. citizens without prior Congressional approval (Clymer 2003; Safire 2003). The American Civil Liberties Union produced a devastating brief subjecting the claims of TIA to profile high-risk terrorists to statistical analysis: even with an extraordinarily high rate of 97 per cent accuracy, the ACLU pointed out, 9 million innocent Americans would be 'caught' (ACLU 2003b). Renamed the Terrorism Information Awareness program, TIA deposited a hefty report with Congress which detailed information on the specifications of various TIA research projects along with some considerations on how the negative impacts on civil liberties could be minimized. Despite the President's admonition against denying him 'an important potential tool in the war on terrorism,' Congress was not mollified. The Senate voted to halt all funding, while the House preferred somewhat less stringent constraints. Just as the two houses were set to resolve their differences in a joint committee, DARPA hit the headlines with yet another bizarre construction from Admiral Poindexter's fertile imagination: a terrorism futures market in which select investors would bet on terrorist actions, the idea apparently being to harness the power of the profit motive to predict the terrorist future. Already indignant Congressional politicians of both parties went ballistic. Republican Senator John Warner of Virginia, chair of the armed services committee, declared: 'That is two strikes now. Do you have to throw a third strike?' Deputy Defense Secretary Paul Wolfowitz declared himself shocked, and announced that the program had been terminated. The *New York Times* declared that Poindexter had to go (Hulse 2003). Within days, his resignation was announced, Defense Secretary Donald Rumsfeld indicating that 'the admiral had become too much of a political lightning rod and that it was time for him to go' (E. Schmitt 2003).

Whatever the fate of TIA post-Poindexter, there are any number of TIA-like programs springing up. DARPA has another project, separate from the TIA office, called LifeLog, which seeks to amass every conceiv-

---

11 http://www.darpa.mil/iao/news.htm.

able bit of information that can be gathered from every source (including audiovisual sensors and biomedical monitors) about an individual's life, and download it all into a vast, searchable database. Someone with access to such a database could supposedly recall an individual's memories from many years ago through a search-engine interface. Just why DARPA would want such an expensive high-tech version of Marcel Proust's *petite madeleine*, the taste of which set the writer on his seven-volume *Remembrance of Things Past*, is a bit of a mystery, but it seems to suggest a Promethean desire to know all. As one critic put it, it is 'TIA cubed' (Shactman 2003).

Congressional roadblocks have not discouraged the U.S. government from floating TIA-like schemes under other names and other auspices. A new Terrorist Threat Integration Center (TTIC) began work in May 2003, combining the resources of the CIA and the FBI to analyse foreign and domestic intelligence collected throughout the government to better 'connect the dots' and prevent future terrorist attacks. The center, according to the White House, 'will have unfettered access to all terrorist threat intelligence information, from raw reports to finished analytic assessments, available to the U.S. government' (Anderson 2003). The Transportation Security Agency recently announced that the Computer Assisted Passenger Pre-screening System II (CAPPS II) will soon begin testing at several airports around the United States. CAPPS II is a system for conducting background checks on all airline passengers and categorizing them according to the level of risk they pose. The ACLU warns that CAPPS II is based on the same concept as TIA, which proposed 'massive fishing expeditions through some of our most personally sensitive data' (ACLU 2003a). On 10 July 2003, the Senate voted to withhold funding for CAPPS II until the Transportation Security Administration (TSA) provides more information about procedural and technological safeguards.[12]

Even the discredited TIPS snoop net is being recreated at the local level. 'Watching America with Pride, not Prejudice' is the motto of the New Jersey-based Community Anti-Terrorism Training Institute, with the chilly acronym of CAT Eyes, an 'anti-terrorist citizen informant program being adopted by local police departments throughout the East Coast and parts of the Midwest' (Takei 2003). The program's founder ambitiously envisions an eventual 100 million informers, a

---

12 'Senate Requires Reporting For CAPPS II; Extends TIA Moratorium', *EPIC Alert*, 10:15, 22 July 2003.

ratio of watchers to watched of about one to two, as compared with the East German *Stasi* ratio of one to eight.

Almost sure to provoke reactions when revealed in the harsh glare of publicity, TIA-type programs flourish in the absence of critical attention, and in their cumulative effect habituate the citizenry to the normality of being regularly watched. Such programs, moreover, fit rather easily into the technological and economic realities of the early twenty-first century. As one newspaper report put it, 'it is increasingly possible to amass Big Brother-like surveillance powers through Little Brother means. The basic components include everyday digital technologies like e-mail, online shopping and travel booking, A.T.M. systems, cellphone networks, electronic toll-collection systems and credit-card payment terminals' (Markoff and Schwartz 2002). TIA-type programs simply look towards developing software that can put all this diffuse private and public data together in ways that will flag suspicious behaviour patterns for attention.

Yet serious questions arise about the justification for TIA-type programs, even in terms of their own stated objectives. There is a consensus among informed experts that the 'intelligence failure' of 9/11 is attributable less to a collection deficit than to an *analytical* deficit, along with a bureaucratic incapacity to manage what amounts to an information overload. There were many bits and pieces of information concerning the threat from al Qa'eda, and even the imminence of a major attack on American soil. The key failing of the intelligence community was its inability to put the pieces together and make sense of the bigger picture. It has been widely observed that as a general rule, the collection capacity of intelligence agencies has outstripped their analytical and managerial capacities. TIA-type schemes actually threaten to worsen this imbalance, swamping overworked analysts with too much information, almost all of it irrelevant but requiring processing. The idea that out of a deluge of detailed information from banking, credit, debit, air miles, and other databases actionable profiles of potential terrorists will somehow emerge is more a matter of faith than of science. As such it is not entirely unlike the limitless faith in computers and American know-how that fuels the scientifically dubious scheme for a fail-safe anti-missile shield.

It is possible to become over-alarmed about the Orwellian prospect, especially if one accepts too readily the techno-hype behind it. The intentions may be alarming, but the means of delivering TIA are as yet more suspect than enthusiasts like Admiral Poindexter believe. There are questions surrounding many of the technologies that are being promoted by the private sector to the U.S. government as quick fixes for terrorism. When the Defense Department tested face-matching technologies, their results were less impressive than the figures claimed by the companies peddling them.

There are also some important technological limitations on Big Brother's surveillance capacities. Chief among these is the universal accessibility of encryption systems that defeat the decryption capabilities of the NSA and all other intelligence agencies, American or foreign. Clumsy and unsuccessful attempts were made during the Clinton administration to impose a 'Clipper Chip' that would offer government a trapdoor entry into encrypted messages, and to embargo the export of encryption software outside North America. Programs such as the FBI's Magic Lantern – a virus sent to capture encryption keys from a remote targeted computer – look to technical fixes of this problem, but government snoopers are by and large resigned to living with an inability to read intercepted e-mail messages. The U.K. government, prior to 9/11, went so far as to legislate criminal sanctions for persons who refuse to disclose their encryption keys to police or security officials who demand them, but the U.S. government has declined to follow this example. The reason for U.S. reticence is not hard to find: opposition to the British legislation came not only from civil libertarians, but more influentially, from e-commerce interests irritated at government intrusion into security systems, the integrity of which is essential to e-commerce transactions.

This points to an inherent contradiction in the relations between public and private actors as barriers between state and corporate surveillance systems are lowered. There are two very different, and sometimes antagonistic, concepts of 'security' at work. To the public sector, security comes from accessing and controlling the 'transaction space.' To the private sector, security means guaranteeing the integrity of transactions with clients, whether consumers or other businesses. In the face of the terrorist threat as embodied in 9/11, there is good reason for business to buy into government surveillance programs to make their own operations more secure. Indeed, there is a lot of money to be made by private companies in equipping government surveillance operations. Yet the corporations are not always on the same page as government, or indeed on the same page as each other. These confusions lead to a peculiar, stuttering dynamic in the development of government surveillance programs.

Nowhere are these confusions more apparent than in the operations of the office of special adviser to the President on cyber security, Richard Clarke. A former adviser on terrorism to President Clinton, Mr Clarke unusually survived to serve his Republican successor, until his recent resignation. On 18 September 2002, the White House officially released its draft 'National Strategy to Secure Cyberspace,' which had emerged from Clarke's office.[13] It is an anodyne document, full of good intentions and good advice to the private sector about how to maintain security against hackers and cyber-terrorists, with no teeth to enforce the advice. The economic damage to national security that might be exacted by a concerted cyber attack on the U.S. private sector and critical infrastructures has been the subject of many warnings. Yet many in the private sector do not wish to assume the costs associated with adopting stricter security systems, or to shake consumer confidence by pointing to the vulnerability of their existing security systems. Hence the gutting of Mr Clarke's report, and his precipitous departure.

Actually, the contradictions are even more acute. The vulnerabilities of private sector security are hyped and indeed over-hyped by the private security industry that has a vested interest in advancing sales of its software and hardware systems. The very notion of cyberterrorism has been characterized as overblown and alarmist (Green 2002; Lewis 2002). Dire predictions of an 'electronic Pearl Harbor' almost certainly run ahead of the actual potential for damage. Many critical infrastructures, such as air traffic control and power systems, communicate within 'intranets' not connected to the Internet and are 'air-gapped' to provide protection from malicious hackers cruising the net looking for targets. Government is thus being pushed in different directions by different private interests at the same time as it extends its surveillance of the private sector.

A Security-Industrial Complex

To assess the impact of 9/11 and the future direction of the surveillance state it is necessary to look at the relations of government with the private sector, or more precisely, with various private sectors. The Bush administration has sought to dramatically extend its surveillance reach as a counter-terrorist strategy. In doing so, it has constructed close links with certain corporate interests in the high-tech, dot.com sector. Home-

13 *Draft Strategy to Secure Cyberspace:* http://www.whitehouse.gov/pcipb/.

land security is quickly shaping up as the biggest government contract bonanza since the end of the Cold War. If President Eisenhower warned in his farewell address in 1961 of a military-industrial complex, the War on Terrorism has generated an emergent security-industrial complex (Koerner 2002). Homeland Security Secretary Tom Ridge has been an enthusiastic advocate of public-private partnership: 'We look to American creativity to help solve our problems and to help make a profit in the process' (Mitchell 2001). Tens of billions of dollars are on the table and are being snapped up by companies that, like the Cold War defence industries, have in the first instance a single customer, Uncle Sam. This has been a godsend for a sector recovering from the huge hit of the dot.com collapse prior to 9/11. Their profitability rests on a continued market in government for new surveillance and security technologies, and on the hope for commercial spin-offs as business and society look for technological security fixes. In both cases, the security-industrial complex has a stake in joining with government in pumping up the threat level, just as the defence interests and the government pumped up anxiety over the Soviet threat in the past – including 'missile gaps' that never existed.

Yet it is also reported that 'while the Bush administration has waged its campaign to strengthen homeland security since the Sept. 11 terrorist attacks, many of the nation's largest and most influential businesses have quietly but persistently resisted new rules that would require them to make long-term security improvements' (Pianin and Miller 2002). Corporations in the banking, retail sales, chemical manufacturing, and nuclear power industries are baulking at government efforts to impose tougher and more expensive security standards. The private sector wants security but is traditionally suspicious of government regulation and intrusion. Ironically, the Republican Party tends to share this suspicion, even as a Republican administration imposes unparalleled government intervention in the name of national security. Although many qualms about big government may have been swallowed in the wake of 9/11, there are ideological contradictions within the governing party that are bound to intensify with time.

This is the complex context in which the Bush administration's efforts to amass new and intrusive surveillance powers must be assessed. In the end, it may well be that privacy protection laws and resistance by both civil libertarians and sections of the private sector, as well as over-reliance on technology, will ensure that the administration's reach exceeds its grasp.

A Faustian Bargain?

Despite obstacles and setbacks, the concept of TIA is by no means discredited within U.S. government circles. Data mining, or data-veillance, under government direction in the name of fighting terrorism is an idea whose time has clearly arrived. Implicit in this is the concept of a global surveillance regime. Since the terrorist threat knows no boundaries and ensuring 'homeland security' is the object of counter-terrorism, exempting Americans from the panoptic gaze makes little sense. Nor can walls erected between private and public data sources be functionally justified, especially when the private sector rushes to offer its troves of personal data to government in the name of patrio-tism and public spirit, which has been happening frequently since 9/11 (Dror 2003). In this security optic privacy concerns take on a certain nuisance quality, and constitutional constraints become impediments to effective action.

The full-blown, original manifestation of TIA, before its emascula-tion, suggests much more than a mere bureaucratic or technocratic response to the challenge of terrorism. *Knowledge is Power* is a motto fraught with significance. The idea that knowledge is power was a driving force of the twentieth century, fuelling the huge state invest-ments in science and technology and the mobilization of science in wars and cold wars, as well as the hyper-development of intelligence gather-ing to steal the knowledge of other states. The history of nuclear weap-onry is emblematic of such processes. In the twenty-first century, the information technology revolution has proceeded much more impor-tantly in the private than in the public sector. But the Promethean possibilities of harnessing the vast information resources diffused throughout the decentred, multiple surveillance systems around the globe are, to those on the commanding heights of an American state unchallenged as the globe's only remaining superpower, too tantalizing to resist.

Nor is it any accident that the TIA concept arose within the U.S. Department of Defense. The dominant military philosophy that drives American military power and interventions in the twenty-first century derives from the 'revolution in military affairs' that has produced 'net-work centric warfare.' Information and control over information is key to this concept, in which intelligence guides the organization of the 'battlespace,' and the delivery of devastating and sophisticated weap-onry with a degree of precision never seen before (see chapters 9 and 10 in this volume). American command and control of space and space-based communications and intelligence is crucial. The enemy's communications infrastructure will be targeted in the first wave of attack, rendering the adversary 'blind' and 'deaf,' while the attacker 'sees' and 'hears' everywhere. Clausewitz's famous 'fog of war' and his dictum that all military action is undertaken in a 'resistant medium' are now considered passé, obsolete wisdom from an earlier age, before penetrative surveillance and global positioning systems could be employed to see through the fog. The quick collapse of the Taliban regime in the face of American intervention was a preview of this new kind of warfare, and the equally rapid collapse of the Saddam Hussein regime in the face of the 'Shock and Awe' invasion of Iraq constituted a successful test for the new warfare system against what appeared at least to be a relatively well-equipped army. The evangelists of the new doctrine, firmly in command of the Defense Department and the Bush White House, are supremely confident that American military might (which equals half the military force of the rest of the world's states combined), combined with unparalleled American intelligence capacity, can control any 'battlespace' the United States chooses to engage, with minimal, acceptable levels of casualties, and that U.S. economic resources will be equal to the challenge of financing this global military hegemony (Walters 2001).

It is in this context that we must view the pronounced turn of the Bush administration towards unilateralism in foreign policy – abrogation of the Anti-Ballistic Missile Treaty; renunciation of the chemical and biological weapons verification protocol; non-compliance with the Kyoto Accord; insistence on American exemption from the International Criminal Court; the invasion of Iraq without UN sanction– and the Bush Doctrine claiming the right to initiate preventive war by intervening against any state that poses a real or potential threat to U.S. security, as determined by the United States alone. The present administration believes that building multilateral alliances and coalitions in support of U.S. objectives is desirable, where possible, and so long as these do not interfere with American goals and timetables, but ultimately unnecessary, and dispensable. Hence the breakdown in the Western Alliance over Iraq, and even the possible marginalization into irrelevance of the United Nations, is viewed with apparent equanimity as a small price to pay in exchange for the opportunity to redraw the geopolitical map of the Middle East on American terms.

This is the setting in which the concept of TIA has to be understood.

Particular programs, and particular personnel, may suffer setbacks and rebukes from Congress, commentators, and the public, but there is an underlying logic and thrust that is highly unlikely to be displaced by civil libertarian scruples. TIA fits comfortably within the dominant strategic doctrine of Bush's America. Some would call this doctrine optimistic, self-confident, ambitious, idealistic. Critics call it aggressive, imperialistic, power mad, and a grave danger to the peace of the world. Examining the prospectus offered by the enthusiasts of TIA, its language, its imagery, and its moral, if not moralistic, tone, strongly suggest that there are more than technocratic practicalities involved. It is as if America has been offered a Faustian bargain by the promise of technology.

Look down from the heights, Mephistopheles tells Faust, and see the world at your feet, rendered transparent to your gaze, all secrets uncovered. Your enemies, real and potential, will speak to you henceforth in the voice of Psalm 139:

> Such knowledge is too wonderful for me;
> It is high, I cannot attain unto it ...
> If I say, Surely the darkness shall cover me;
> Even the night shall be light about me.
> Yea, the darkness hides not from thee;
> But the night shines as the day:
> The darkness and the light are both alike to thee.

All this is yours, promises the voice of technological power. There will, of course, be the small matter of a quid pro quo: at the end of the day, you will have to abandon your constitutional protection of civil liberties and personal privacy, to the extent that these get in the way. This is regrettable, but Faust reflects that 'the other side weighs little on my mind' compared to the new world rising before his eyes – 'The rest concerns me not: Let come what will.'[14]

What Faust forgets is an earlier exchange with Mephistopheles, when he has interrogated the latter about his 'all-knowing wit.' The Prince of Darkness demurs, slightly: 'Omniscient? No, not I; but well-informed.' Undoubtedly, the vast surveillance powers available to the United States can make that country's government better informed, but to reach for

14 Quotations from Goethe, *Faust*, translated by Philip Wayne (Harmondsworth: Penguin, 1949).

omniscience, even as a distant but attainable goal, is to fly in the face of the limitations of the technology itself, as well as the capacity of analysts, however intelligent, objective, and well-intentioned (and these are not, by and large, the qualities that spring to mind in describing those currently in charge of the U.S. administration), to give meaning to the relentless, bottomless, ocean of 'facts' they can now conjure up. Total Information Awareness is a concept that suggests, more than anything else, hubris, the same quality that so much of the world today sees in American foreign and defence policy in general. The fiasco of the fruitless search for weapons of mass destruction in Iraq – the *casus belli* for the invasion – and the consequent decline in the credibility of the American and British governments, as well as their intelligence agencies, is a sign of the vulnerability of states whose intelligence reach exceeds their grasp.

It may also be the wrong answer to the right question. If neo-liberal globalization has recreated the old Hobbesian problem of insecurity and disorder, this time on a world scale, it may well be necessary to look for Hobbesian solutions. In seventeenth-century England the effects of the rising market on society had suggested to Hobbes that, without government, life would resemble a 'war of all against all.' The micro-rationality of individuals pursuing their self-interest resulted in the macro-irrationality of a world in which life was 'solitary, poor, nasty, brutish, and short.' His answer derived from enlightened self-interest: each person would transfer their power to the Sovereign, the Leviathan state, which would in turn guarantee the security of all.

Today, the dark side of globalization threatens the security of the world of Leviathan states that has, since Hobbes's day, internationalized the market. When the terrorists struck at the World Trade Center, symbol of global commerce, they simultaneously challenged the global order and challenged the American state, which failed in its fundamental Hobbesian task of protecting its own citizens. The United States resolved to put an end to the threat of disorder posed by the terrorists, and, in this resolve, were supported by most of the world's states and peoples. Certainly, American leadership and initiative is both necessary and welcomed by all who share the insecurity posed by the terrorist threat. The hubris of America is shown not in this resolve for leadership, but in America's pretension in assuming for itself the role of the new global Leviathan.

The terrorist networks, like the wider forces of the dark side of globalization, pose borderless threats, operating in the new global space of flows, leaving the territorially bounded legal and policing jurisdictions of national states largely impotent to control them. Global governance requires enforcement, but only national states possess effective enforcement powers. A twenty-first-century Hobbesian answer to global disorder is the Multilateral Leviathan – a broad alliance or network of states cooperatively coordinating their enforcement powers to contain and limit borderless threats that make each of them insecure, according to multilateral agreements and treaties, with international sanction. American leadership would have been at the heart of any such Multilateral Leviathan, but coalition building on this scale requires tact, diplomacy, the arts of compromise and negotiation, and a willingness to listen and learn from others – all qualities in notoriously short supply, if not entirely absent, from the Bush White House. As the President declared in the wake of 9/11: 'You are either with us, or you are on the side of the terrorists.' *With us*, it now appears to much of the world, means *under us*. The impressive global coalition of support assembled for the Afghanistan intervention has shrunk desperately for Iraq. The Americans speak of a 'coalition of the willing,' but what much of the rest of the world sees is a coalition of the bribed, the bullied, and the bilked. Of course, international relations has always been a field for great power domination, but the Cold War era did witness effective governance by the United States through an alliance architecture based on a decent respect for the opinions and interests of partners. In the contemporary era the actual practice of coalition compromise has withered, and even the forms have begun to be discarded.

Total Information Awareness, either as a program or a concept, represents overreach. It is a symbol of a greater hubris that threatens in the longer run to bring down America's plan to assume the role of a global, but unilateral, Leviathan.

REFERENCES

ACLU. 2003a. *CAPPS II Data-Mining System Will Invade Privacy and Create Government Blacklist of Americans*. 27 Feb.
– 2003b. *Total Information Compliance: The TIA's burden Under the Wyden Amendment: A Preemptive Analysis of the Government's Proposed Super Surveillance Program*. 19 May.
Anderson, C. 2003. 'Bush Announces New Counterterrorism Center.' *Associated Press*. 14 Feb.

Clymer, A. 2003. 'Conferees in Congress Bar Using a Pentagon Project on Americans.' *New York Times*. 12 Feb.

Davies, S. 1997. 'Spies Like US.' *Daily Telegraph*. 16 Dec.

Dror, Y. 2003. 'Big Brother Is Watching You – And Documenting.' *Ha'aretz* (English Language Edition). 20 Feb.

Eggen, D. 2002a. 'Broad U.S. Wiretap Powers Upheld.' *Washington Post*. 19 Nov.

– 2002b. 'Proposal to Enlist Citizen Spies Was Doomed from the Start.' *Washington Post*. 24 Nov.

Eggen, D., and S. Schmidt. 2002. 'Secret Court Rebuffs Ashcroft: Justice Department Chided on Misinformation.' *Washington Post*. 23 Aug.

Electronic Privacy Information Centre (EPIC). 2002. *Alert*. 19 Nov.: 9:23.

Foerstel, H. 1991. *Surveillance in the Stacks: The FBI's Library Awareness Program*. Westport, CT: Greenwood.

Green, J. 2002. 'The Myth of Cyber Terrorism.' *Washington Monthly*. Nov.

Hentoff, N. 2002. 'Has the Attorney General Been Reading Franz Kafka?' *Village Voice*. 9 Feb.

Hulse, C. 2003. 'Swiftly, Plan for Terrorism Futures Market Slips into Dustbin.' In Editorial 'Poindexter's Follies.' *New York Times*. 30 July.

Kasindorf, M. 2002. 'FBI's Reading List Worries Librarians.' *USA Today*. 17 Dec.

Koerner, B. 2002. 'The Security Traders.' *Mother Jones*. Sept./Oct.

Lewis, J.A. 2002. *Assessing the Risks of Cyber Terrorism, Cyber War and Other Cyber Threats*. Washington, DC: Center for Strategic and International Studies.

Markoff, J., and J. Schwartz. 2002. 'Many Tools of Big Brother Are Up and Running.' *New York Times*. 23 Dec.

McCullagh, D. 2001. 'Anti-Attack Feds Push Carnivore.' *Wired News*. 12 Sept.

Mitchell, A. 2001. 'Industry See Opportunity in U.S. Quest for Security.' *New York Times*. 25 Nov.

New York Times Editorial. 2002. 'A Green Light to Spy.' 19 Nov.

Pianin, E., and B. Miller. 2002. 'Businesses Draw Line on Security: Firms Resist New Rules for Warding Off Terror.' *Washington Post*. 5 Sept.

Priest, D. 2002. 'CIA Is Expanding Domestic Operations: More Offices, More Agents with FBI.' *Washington Post*. 23 Oct.

Safire, W. 2002. 'You Are Suspect.' *New York Times*. 14 Nov.

– 2003. 'Privacy Invasion Curtailed.' *New York Times*. 13 Feb.

Savage, D., and H. Weinstein. 2002. 'Court Widens Wiretapping in Terror Cases.' *Los Angeles Times*. 19 Nov.

Schmitt, E. 2003. 'Poindexter Will Be Quitting over Terrorism Betting Plan.' *New York Times*. 1 Aug.

Schmitt, R.B. 2003. 'U.S. Expands Clandestine Surveillance Operations.' *Los Angeles Times*. 5 Mar.

Shactman, N. 2003. 'A Spy Machine of DARPA's Dreams.' *Wired News*. 20 May.

Shenon, P. 2002a. 'Secret Court Says F.B.I. Aides Mislead Judges in 75 Cases.' *New York Times*. 23 Aug.

– 2002b. 'Justice Department Denounces Secret Court in Wiretaps.' *New York Times*. 28 Sept.

Takei, C. 2003. 'Building a Nation of Snoops.' *Boston Globe*. 14 May.

United States Department of the Treasury. 2000–5. *Financial Crimes Enforcement Network, Strategic Plan*.

Walters, G.J. 2001. *Human Rights in an Information Age: A Philosophical Analysis*. Toronto: University of Toronto.

Washington Post Editorial. 2002. 'Total Information Awareness.' 16 Nov.

Whitaker, R. 1999. *The End of Privacy: How Total Surveillance Is Becoming a Reality*. New York: New Press.

7 Surveillance Fiction or Higher Policing?

JEAN-PAUL BRODEUR AND
STÉPHANE LEMAN-LANGLOIS

In past and recent work, one of us tried to articulate a distinction between high and low policing (Brodeur 1983, 1992, 2000 and 2003). This distinction was initially formulated in France in the seventeenth century and elaborated upon by Napoleon's minister for policing, Joseph Fouché. In simple terms, low policing consists of law enforcement and high policing of political surveillance (L'Heuillet 2001). More precisely, high policing was defined by four features: (1) it was absorbent policing, hoarding all-encompassing intelligence on socio-political trends, while making parsimonious use of this information in the actual prosecution of individuals, neutralized only when deemed strictly necessary; (2) it conflated legislative, judiciary, and executive or administrative powers, the police magistrate enjoying all three; (3) its goal was the preservation of the political regime ('the state') and not the protection of civil society; (4) to this end, it made extensive use of informants that infiltrated all walks of society. These features were traditionally combined within a police paradigm where the protection of the political *status quo* was the primary goal of policing and where, furthermore, the interests of the regime were not seen to be coterminous with the interests of civil society.

Although the most potent symbol of political surveillance is now Orwell's Big Brother, high policing was not initially a totalitarian paradigm, even if it was not constrained by rules of accountability (Brodeur 1983). Indeed, as Brodeur initially tried to argue, high policing cancelled out the notion of police deviance. This point was neatly encapsulated by a member of the French Assemblée nationale, who recently delivered a report on the accountability of the French security services. He advocated that they be granted a large amount of unfettered discretion according to the principle that in the field of national security 'the rights of the State supersede the rule of law' (France, 2003: 3).

In the 1983 paper on high and low policing, it was argued that the coming years would witness the rise of high policing, and Brodeur later attempted to show, in a paper entitled 'Cops and Spooks' (Brodeur 2000), that high and low policing agencies shared an increasing common ground, albeit with various degrees of discomfort. Since 9/11, this conclusion cannot be doubted for obvious reasons reaching far beyond Brodeur's limited gifts of foresight, or anybody's for that matter.

This paper is divided in three parts. First, we provide a selective account of major developments in political surveillance that occurred in Canada and the United States, after 9/11. Second, we review the features of high policing, as described in 1983, asking to what extent they still apply in the present context. In a concluding section, we discuss a new set of features of political surveillance.

Developments in Political Surveillance

In such a short essay, we cannot give an account, however brief, of all the developments in political surveillance after 9/11. In Canada, Bill C-36, the main legislation enacted after 9/11, has over 180 pages, and it was followed by other legislation. The *USA PATRIOT Act* of 2001 has 10 titles and 182 sections. To this must be added the numerous provisions of the *Cyber Security Enhancement Act* passed on 19 November 2002 by the U.S. Senate as an Amendment to the *Homeland Security Act*, which brought together 22 agencies and some 170,000 personnel.

Such massive complexity is in direct contradiction with Habermas's argument that the cornerstone of a democratic society is the possibility of public debate. Not only were these laws adopted without any informed public debate, but it is questionable whether the legislators themselves were fully aware of what they were voting on. A recent feature in the *New York Times* has shown that when the Senate passed the US$80 billion bill to pay for the impending war in Iraq, it did not notice that it was also voting for dozens of other pork-barrel projects, among which were US$10 million to a research station at the South Pole that had had a hard winter, and US$3.3 million to fix a leaky dam in Vermont – not to mention allowing the Border Patrol to accept donations of body armour for dogs (Firestone 2003). A sprinkle of increases in the surveillance power of various agencies is as undetectable, in such huge bills of legislation, as pork-barrel.

In this wealth of developments, we shall focus on the Communications Security Establishment (CSE) in Canada and the Total Information Awareness (TIA) project in the United States, which appear to us to be among the most significant. The latter project was later christened *Terrorism* Information Awareness, because its reach into privacy was disturbing for the U.S. public, the civil libertarians, and even the political class.

Canada and the Communications Security Establishment

Although it was created under its present name in 1975, the CSE's history began earlier in 1941, when a secretive unit called the Canadian Examination Unit (XU) was created for the purpose of interception and decipherment of the enemy's electronic communications. The crucial development occurred in 1947, with the signing of the UKUSA treaty that brought together the United States, the United Kingdom, Australia, New Zealand, and Canada in an alliance devoted to the interception of radio communications, known today under the acronyms of COMINT (communications intelligence) and SIGINT (signal intelligence), as opposed to HUMINT (human intelligence), which is collected by persons rather than by machines. In Canada, the Canadian Security Intelligence Service (CSIS) is the main HUMINT agency. In the UKUSA treaty the United States was designated as the first party to the agreement and the other four nations as second parties (Bamford 1982: 399).

Like its American counterpart the National Security Agency, the CSE was not created by legislation but by a decree of the executive (Order-in-Council PC 1975-95). The agency, which is part of the Department of National Defence, has two components: the first is comprised of civilian experts who perform various tasks ranging from cryptology to intelligence analysis and covert operations (Frost and Gratton 1994); the second is manned by the military personnel who operate the Canadian Forces Supplementary Radio System (CFSRS). Originally, the CFSRS was to monitor communications in the Soviet Arctic, for example on Elsmere Island, using technology provided by the U.S. government.

The CSE is highly secretive. Its mandate was spelled out in full only in 1997, in the first report of the CSE commissioner, the civilian watchdog of the agency (CSE Commissioner 1997: 5–6). Initially, the CSE had a twin mandate. As Canada's cryptology agency, it collects and analyses foreign radio, radar, and other electronic emissions and through the provision of SIGINT it contributes to the government's foreign intelligence program. The CSE also manages Canada's Information Technology Program (ITS), which provides technical advice, guidance, and various other services in support of government telecommunications security. The CSE thus operates in both an offensive and a protective capacity. It always emphasized that its offensive capacity was exclusively directed at foreign communications and that Canadians were not targeted. Although earlier put in doubt (Sallot 1984), this claim has so far been vindicated by the CSE commissioner since he began to report in 1997.

The original pattern of targeting may now change. Following the 9/11 attacks in the United States, Canada enacted Bill C-36, which increased in various ways the surveillance powers of the police. Moreover, embedded in C-36 was a little-noticed Trojan horse: Bill C-36 actually provided the CSE with its enabling legislation, surreptitiously introduced as a sub-part of Part V of the *National Defence Act*, which bears on military justice and has no relationship at all with SIGINT. In all of his reports, the CSE commissioner had recommended that such legislation be the object of a wide public debate. This is precisely what did not happen. Instead, the enabling legislation was adopted under pressure, with few if any amendments. Yet such a debate was crucial, especially since the enabling legislation added a new and controversial component to the CSE mandate. On top of its foreign intelligence and ITS duties, the CSE is now also 'to provide technical and operational assistance to federal law enforcement and security agencies in the performance of their lawful duties' (s. 273.64(1)c). The federal agencies referred to are the Royal Canadian Mounted Police (RCMP), Canada Customs and Revenue, and CSIS, among others.

The CSE's involvement in policing may not be problematic in certain respects, as when it would provide help to other agencies in deciphering lawfully intercepted communications. However, it may itself assist federal law enforcement and security agencies in the interception of private communications. According to Canadian law, such interception requires judicial authorization that is granted on the basis of an affidavit explaining the reasons for the interception, its instruments, and its target(s). One of us (Brodeur) explicitly asked the personnel of the CSE commissioner's office whether such an affidavit would mention the

CSE's assistance in performing an interception. The answer was that this was indeed a good question ('you are right on the money') but, for reasons of national security, no answer could be given. There are other grey areas surrounding the CSE's implication in policing and law enforcement.

This situation is all the more troubling because of the massive investment of Western governments in SIGINT. In the United States, the budget of the NSA is much bigger than the CIA's. Because CSE personnel are recruited among civilians and military personnel, it is difficult to determine the true figures in relation to personnel and budget. In a comparative study of the CSE and of CSIS's respective manpower and budgets in the years preceding 2001, one of us found that the CSE was spared the drastic cuts that affected CSIS and that overall, its budget was probably higher than that of CSIS (Brodeur 2003: 231). This tentative finding was confirmed after 9/11. The Canadian government injected an additional $47 million in the protection of its national security shortly after the attack on the Twin Towers and the Pentagon. The CSE was granted the greatest portion of this money, $37 million as compared to $10 million for CSIS.

In a speech given on 30 January 2003, on the occasion of the release of his annual report for 2002–3, the privacy commissioner of Canada issued a dire and pressing warning that the federal government was on its way to destroying essential rights to privacy, with important consequences for the freedoms enjoyed by Canadians (Privacy Commissioner of Canada 2003). His call should be heard, but perhaps in a somewhat weaker, yet more crucial, sense than he meant it. It may be too late to protect privacy from state and non-state intrusion: all that is left would be to strengthen the means to guarantee that the data collected on us are valid, and that they are not riddled with gross mistakes as is generally found when personal data are assessed in quality control exercises (Laudon 1986).

In April 2003, the Canadian government announced that a much-criticized data bank on air travellers would be de-nominalized, a protection of privacy expedient that is also used in the United States (Buzzetti 2003a). What this means is that data are stored on patterns of activities and on transactions, while the name of the persons engaging in these activities is provisionally put in brackets. The name of a particular individual is sought only when a threatening pattern of behaviour has been discovered. We shall return to this topic at a later stage.

The United States and the Total Information Awareness Project

Following 9/11, the U.S. Congress conducted a joint House and Senate inquiry into the failure of the intelligence community to prevent the attacks against the World Trade Center and the Pentagon (U.S. Congress 2002). The joint committee's report is long on general findings and recommendations but short on details. However, the Vice-Chairman of the Senate Select Committee on Intelligence, Senator Richard C. Shelby, issued his own report – which is tantamount to shock treatment in reality therapy – on the efficiency of the U.S. national security apparatus in counterterrorism (Shelby 2002). Senator Shelby was not the only one to blow the whistle, as a scathing memo written by FBI agent Coleen Rowley has shown.

Senator Shelby documents two series of shortcomings which, even when we take into account the benefit of hindsight, appear grievous.

In the first case, two individuals, Khalid Al-Mihdhar and Nawaf Al-Hazmi, were spotted by the Malaysian security service attending an Al Qaeda meeting in Kuala Lumpur, where the terrorist operative who had organized the attacks against U.S.S. Cole, in Yemen, was also present. This information was passed on to the CIA. The United States has a surveillance program, TIPOFF, according to which the names of suspicious individuals trying to enter the United States, or to obtain a visa, are given to immigration and visa-issuing authorities, who then block the entry of the suspects into the country. For reasons that are unclear, the CIA failed to use TIPOFF until it was too late. Both individuals entered and left the United States on several occasions and finally settled under their own names in San Diego, where they took flying lessons. The CIA also refused to share with the FBI its knowledge of the presence of these two terrorists on American soil. Both were on the plane that crashed into the Pentagon.

The second case involved Zacharias Moussaoui. When he was arrested while taking flying lessons in the United States, the legal separation between criminal prosecution and foreign intelligence prevented the FBI agents investigating him to get a warrant to access his computer (where the name of Mohammad Atta, the chief organizer of 9/11, was later to be found). Advised that they had to connect him to a known terrorist organization, they tried in vain to link him to Chechen terrorists. When they finally obtained a warrant, 9/11 had already occurred.

These two cases are of particular interest for the clues they provide regarding the two key metaphors guiding the reform of political surveillance in the United States. The first metaphor is 'connecting the dots.' In these cases, the dots were not connected for two reasons: (1) the lack of horizontal integration, which prevented the various services from sharing their information, and (2) the dismal quality of intelligence analysis. The second metaphor is 'the wall' separating foreign security intelligence from its use in domestic law enforcement, thus preventing any kind of vertical integration between intelligence and field operations. This divorce is forcefully decried by Senator Shelby: 'Intelligence analysts would doubtless make poor policemen, and it has become very clear that policemen make poor intelligence analysts' (Shelby, 2002: 62).

Shelby's judgment does not seem to leave much ground for appeal: 'The Bureau's failures leading up to September 11 thus suggest the possibility that *no* internal FBI reorganization will prove able to effect real reform' (Shelby 2002: 70, emphasis in text). It is in this context that computerized surveillance, such as it is embodied in the Total Information Awareness program, seems to offer an alternative to human fallibility: 'I mention TIA here at some length because it represents, in my view, precisely the kind of innovative, "out of the box" thinking ... which Americans have a right to *expect* from their Intelligence Community in the wake of a devastating surprise that left 3,000 of their countrymen dead. It is unfortunate that thinking of this sort is most obvious in the Defense Department rather than among the Intelligence Community leaders ...' (Shelby 2002: 43, emphasis in text).

Before we describe TIA in further detail, we should make clear that funding for the program has now been cancelled by the Senate (by the *Department of Defense Appropriations Act* of 2004, s. 8120). Yet it remains a powerful example of a general trend in policing and surveillance. And as we will discuss later, the disappearance of TIA by no means implies the disappearance of the tactics and technologies it was developing. Note for instance that the newly created Department of Homeland Security (DHS) has its own 'advanced projects' department (the Homeland Security Advanced Projects Agency, HSARPA), which it will provide with US$500 million a year to develop technologies similar to those of the TIA program.

TIA was initially developed within the Pentagon by the Information Awareness Office (IAO), one of two data analysis projects developed by the Defense Advanced Research Projects Agency (DARPA). The other program is the Information Exploitation Office (IXO), which focuses on real-time battlefield operations. At its inception the IAO was headed by

Vice-Admiral John Poindexter, who was acquitted on appeal for his alleged role in the Irangate scandal. At the core of the TIA project is a computerized Big Brother that is, paradoxically, said to reconcile all-encompassing surveillance with privacy rights. This reconciliation would rest on a crucial difference between surveillance as exercised by humans (HUMINT) and surveillance as exercised by machines, that is, computers, which we shall call COMPUTINT. When a police officer or an intelligence agent listens to intercepted communications, he or she knows that only a small part of the information thus collected can be used in actual proceedings against the suspect. However, these listeners cannot erase from their memory all that they learn on the intimate life of the suspect, which can be potentially used for blackmail purposes, even if it was not initially relevant to the investigation at hand. This invasion of intimate life is felt to be particularly obnoxious by those submitted to surveillance (Plenel 1997). Computers, it is claimed, can be programmed to retrieve from a surveillance project only what is strictly relevant to its lawful purpose and not to keep on file what is unrelated to that purpose. In contradistinction from HUMINT, COMPUTINT can be made strictly selective. To a significant degree, the IAO's version of privacy is an offspring of this conception of the properties of computers. Needless to say, this claim rests upon a great deal of confidence in the computers' programmers.

TIA had three components. The first component was analytical. Ninety per cent of all terrorist incidents that have occurred were submitted to detailed analysis in order to extract, from their minute description, *patterns* that may be predictive of terrorist behaviour. To give a simple example, such a pattern may be (1) paying in cash (2) for plane tickets (3) purchased in a small or medium size airport (4) leaving the date of the return trip open. The second component of TIA, which we shall call *transactional*, consisted in the computerized monitoring of daily transactions occurring in the United States. This monitoring would be quite extensive as it would involve mining financial, educational, travel, medical, veterinary, country entry, place/event entry, transportation, housing, critical resources, government, and communications data (note that this information was removed from the IAO website even before the death of the project). The third and last component was to perform the actual *identification* of individuals who triggered an alert by engaging in a recognized behavioural pattern. The computer would then go through authentication biometric data and the operators of the system could try to intercept the individual(s) for questioning before they have a chance for further action. Even if pervasive monitoring and data mining was undertaken by the second component of the system, it was argued that privacy would not be intruded upon, as the computer would trigger an alert only if a person's behaviour matched a previously established threatening pattern. Thus, the data in the transactional component would have been, to all practical purposes, denominalized, as was the case with the previously mentioned Canadian data bank on travellers.

There are some additional features of TIA on which we want to briefly comment. First, most of it was heavily 'outsourced,' the realization of the project resting in the hands of private firms. Raytheon, Syntek Technologies (Mr Poindexter was its vice-chairman), Booz Allen, Hamilton Inc. (whose vice-president, Mr Mike McConnell, previously chaired the NSA), Hicks and Associates, Microsoft, Intel, and Veridian Corporation were important participants. New firms kept appearing at each visit of the now-defunct IAO website. Second, the marketing of high technology is not governed solely by a principle of *adequacy* (whether the purchased technology will aptly do the job), but also by a principle of *availability*. Although there is generally a relative balance between these two principles, with a slight tendency towards adequacy, the question of availability becomes prominent in crisis situations where *one must do with* what is to be found on the market, because of urgency. There is no better example of this than the use of automated translation software. TIA was to make use of automatic translation software – Translingual Information Detection, Extraction and Summarization (TIDES) and BABYLON, a version of TIDES that can be used on handheld computers – because the US intelligence community was faulted for its lack of foreign language skills. The problem with automatic translation is that it doesn't work beyond giving a vague impression of the content of a message. For example, the Canadian government stopped subsidizing the Université de Montréal, which devoted diehard efforts to automatic translation research, when it failed to develop software that would translate weather forecast bulletins, which use the simplest of language, between French and English. Still, the use of automated translation technology was expanded in the last version of TIA (May 2003) by the integration of GALE (Global Autonomous Language Exploitation), a new software program for mining data in verbal and written communications in foreign languages. Note that translation software projects all survived the demise of the IAO/TIA and simply became the responsibility of other entities. There are many other examples of computer technology integrated in the system with little quality control. Such wrong-headed can-do obstinacy may impede efficiency rather than enhance it when one is facing real-life terrorists.

TIA died while still in development. It was first envisaged to be ready by 2005–6, *before* the U.S. Congress decided, in the spring of 2003, to limit funding and to authorize only a limited application to foreign nationals. In September 2003 the Senate gave it the coup de grace and axed all and any TIA projects having anything to do with data mining. Yet, despite the early death of TIA, the concept it embodies – electronic profiling of individuals against a background of behaviour-predictive transactional patterns – is gaining ground. In March 2003, the U.S. Congress, while cutting funds to the TIA, granted authority to the federal Transportation Security Administration for developing a system for screening passengers which is a clone of TIA applied to air travellers. The Computer Assisted Passenger Prescreening System (CAPPS) is already under fire from privacy advocates.

The TIA spirit remains alive and well. In many quarters, including the already mentioned Transportation Security Administration, which is cutting down on the number of armed air marshals who were supposed to be assigned to flights in the United States, data mining and analysis is seen as the answer to most security problems, in spite of ongoing public relations nightmares. Those nightmares previously prompted the transformation of *Total* Information Awareness into *Terrorism* Information Awareness, and *Bio-Surveillance*, one of the TIA subprojects, to *BioAlirt* (Bio-event Advanced Leading Indicator Recognition Technology). It also caused the original IAO logo, which included an all-seeing eye and the motto *scientia est potentia*, to vanish from the publications of the office. These minor adjustments were not sufficient to offset the last scandal of the IAO in the summer of 2003: the FutureMap project, which invited investors to trade in terrorism futures on the Internet in order to exploit the claimed predictive powers of markets. This final straw caused John Poindexter to resign from office and Senate to cancel all funding.

It should be evident that the demise of the IAO and its TIA was due to a monumental lack of political acumen – or basic cultural awareness – on the part of those responsible, and not to a loss of confidence in the technology. That confidence is stronger than ever and even DARPA remains involved with projects such as LifeLog and 'Combat Zones that See' (CTS), which could erase the boundary between battlefield awareness and policing, on an extremely large scale. Some of the TIA subprograms may have been taken over by the Army Intelligence and Security Command (INSCOM), which was an active testing and development partner for TIA. The CIA has recently commissioned Systems Research and Development to develop a new data-mining tool named Anonymous Entity Resolution. Moreover, private enterprises and universities which built the TIA technologies are unlikely to simply scrap them and the research already done on the system. Finally, and once again in startling ignorance of popular culture, a program dubbed MATRIX – like the blockbusting series of movies depicting a totalitarian, machine-controlled dystopian fantasy – is implementing the TIA principle at state level with funding from the Department of Justice. (MATRIX is a contrived acronym for Multistate Anti-TeRrorism Information eXchange.) Note that MATRIX is operated by private (profit and non-profit) entities, and that its central server is located on the private premises of Seisint, inc., a computer technology firm founded by a known drug smuggler.

High and Low Policing Revisited

We shall now review the four features that were originally attributed to high policing in order to see if they fit the analysis of political surveillance undertaken in the first part of this paper. This review will also provide us with an occasion to probe deeper into the meaning of these traits.

Absorbent Policing

Absorbent policing is a mixture of profligacy and of parsimony. On the prodigal side, it is characterized by the accumulation of intelligence that purports to be all-encompassing, as if the whole informational content of civil society was sucked into the state's data banks. Mustering the resources of information technology, initiatives like TIA would appear to be the true accomplishment of this part of the program of high policing. Furthermore, the root metaphor of 'connecting the dots' underlines the structural nature of security intelligence: intelligence is not a mere pile of data but a network of cross-references.

On the stingy side, high policing was not only retentive in the sense that it hoarded information, it was also restrained in its use of this information in *public* prosecutions of individuals. It was *quiet* policing that apparently distanced surveillance from punishment. There were historical reasons for this restraint, as the targets of high policing were often members of the aristocracy who threatened the sovereign's power and who were powerful enough in their own right to make the government cautious in dealing with them. It was also progressively discovered that quiet policing was cost effective, because it instilled dread into the whole populace through the stealthy character of its operation. The efficiency of the panopticon lies not in watching all but in having all chillingly believe that they are exposed to constant surveillance.

This second feature of high policing is more controversial in the present context, where the alleged 'wall' between security intelligence and law enforcement is perceived as a major impediment to efficient protection. There does not seem to be a consensus on how to solve this problem. A first strategy is to break through the wall. It is clearly exemplified in Canada by legally directing CSE, the most remote of our intelligence services, to assist federal law enforcement agencies. In the United States, it is expected that police forces will have prime access to most data-mining tools, as they do now with MATRIX. According to this strategy, high and low policing will merge, the capacity of law enforcement agencies for high and 'intelligence-led' (as this last concept is now being floated in the United Kingdom) policing being enhanced.

On the other hand, it seems that some critics, like Senator Shelby (2002: 62), have nearly given up on the police as intelligence workers, viewing their shortcomings in the sharing and analysis of the wealth of information available to them as insuperable. In this second script, not only would the divorce between high and low policing be finalized, it may be that the leading agencies for high policing would fall under the military rather than the intelligence community. This development could have nefarious consequences. As skirmishes between the CIA and the Pentagon have shown, the military may be even more autistic than the intelligence community in its appraisal of situations (Risen 2003).

The Accumulation of Powers

The high police magistrate was not simply a powerful officer of the executive (the eyes of the sovereign). He also presided over a police tribunal that could adjudicate justice and order punishment, and he enjoyed wide regulatory powers that made law for most professions and occupations. This feature of high policing is, it would seem, the most remote from the present reality, because of the formal distinction between legislative, judiciary, and executive powers which has been embodied in our laws since at least the time of Montesquieu.

Yet in practice it is more or less recognized, since Jerome Skolnick's (1966) classic essay, that police provide 'justice without trial' and make law on the streets. This observation was made in reference to all policing and had no specific relationship to high policing. But today high policing is being reshaped by a massive call for the prevention of terrorism, and in this context, one should speak of pre-emption rather than prevention, since what is involved are proactive efforts to minimize risk as much as is brutally possible within a partisan interpretation of the precautionary principle. In this respect, police may not yet enjoy legislative powers, but they are certainly endowed, by executive fiat, with the power to circumvent the rules of due process and to name confidentially those who should be arraigned. Preventive detention of persons considered to be a threat to U.S. security bears witness to this fact, although its extent is not publicly disclosed. All incarceration amounts to punishment, regardless of whether it is preventive or repressive – preventive incarceration is actually considered as more punitive by the courts, because of its indeterminate character – and should be inflicted upon an individual only after a judicial hearing. This applies even more to the death penalty, which is the ultimate sanction. Yet both incarceration and, exceptionally, the death penalty are applied within the scope of high policing, as they are explicitly authorized by the executive, the first as preventive detention and the second as political assassination. With the imposition of incarceration and a fortiori the death penalty, high policing is thus becoming a substitute for proceedings normally conducted by the judiciary. Needless to say, the police exercise all their policing powers in respect to pre-emption. It would then seem that under a 'state of pre-emption,' high policing actually exerts powers which are normally separated.

Preservation of the Regime versus the Protection of Society

High policing is entirely devoted to the preservation of a political regime. It rests upon an identification of the internal opponent with the foreign enemy, against whom the state has to be protected. The partial recycling of national security agencies into 'economic security' is now the latest incarnation of this trait. Historically, 'enemy of the state' was an extendable category that included leaders from the aristocracy claiming a right to the throne and their partisans; all were perfunctorily said to be supported by a foreign power. It also included religious opponents, who were likewise suspected of being agents of countries that had officially established the religion that was banned in their own country (e.g., French Protestants were seen as potential agents of England, Holland, and Denmark). Furthermore, it included political and social agitators who could drive mobs to riot, for example, in the case of famine, and notorious criminals who enjoyed some popularity in their defiance of the state's rule. These were not believed to have been recruited by a foreign power, but their actions were seen as either destabilizing or an intolerable blemish on the majesty of the sovereign.

A similar state of mind, which tended to assimilate all internal dissidence to the great Communist threat, was pervasive during the Cold War and during the U.S. war in Vietnam; it culminated with President Nixon being obsessed with 'enemies' during his presidency. The abuses of high policing were finally investigated by the Church and Pike commissions, who published influential reports (respectively U.S. Congress 1976 and U.S. Congress, 1977). However, it would seem that the 9/11 attacks resulted in bridging whatever gap may have remained in the post–Cold War days between the state and U.S. citizens, and in generating a new consensus of fear. In respect to its present engagement in counter-terrorism, high policing cannot be considered as affording protection to the regime, as opposed to the citizens of the United States or other countries. Instead, high policing is now apparently devoted to the protection of the citizens, whose interests are assumed to coincide with those of the state. It would then seem that the hallmark feature of high policing – taking the side of the state against civil society – does not apply as well in the aftermath of 9/11.

There is an undeniable element of truth in this insight, but upon closer examination the situation shows far more complexity. We shall now argue that the former blurring of the line between violent terrorism and political dissidence is being reinvented as a progressive erosion of the boundary separating the nationals of a country from foreigners on its territory. The consequences of this undermining of the basis of citizenship are portentous in respect to surveillance.

Following the Church and Pike inquiries, measures were taken to curb police covert operations, such as the FBI's notorious COINTELPRO, particularly when they targeted U.S. citizens. Similar measures were taken in Canada in the early 1980s, following the Keable (Ministère des Communications, Québec 1981) and the McDonald (Minister of Supply and Services, Ottawa 1981) inquiries: when CSIS was created in 1984, its agents were strictly defined as intelligence officers and not granted policing (peace officer) powers, in order to bar them from engaging in 'disruptive tactics.'

The most wide-ranging reform, however, was to limit political surveillance to foreigners. As we have seen, the Canadian CSE is forbidden from intercepting private communications if either the originator or the recipient of the communication is a Canadian citizen or a legal resident of Canada. In the United States, the *Foreign Intelligence Surveillance Act* (*FISA*) was enacted in 1978. It was the main statute governing political surveillance in that country but its content was in this respect ambiguous, as shown in Senator Shelby's report, of which we shall quote two excerpts.

Much of the blame for the dysfunctional nature of pre-September 11 law enforcement agencies/Intelligence Community coordination can be traced to a series of misconceptions and mythologies that grew up in connection with the implementation of *domestic* intelligence surveillance (and physical searches) under the Foreign Intelligence surveillance act. (*FISA*) (Shelby 2002: 46, emphasis added)

...

The Bureau disseminated extraordinarily few intelligence reports before September 11, 2001, even with respect to what is arguably its *most unique and powerful domestic intelligence tool*: collection under the Foreign Intelligence Surveillance Act. (*FISA*) (Shelby 2002: 66–7, emphasis added)

What is rather striking in these quotations is the assertion that the main statutory instrument for *domestic* political surveillance derives from an Act that granted authority for the purpose of collecting of *foreign* intelligence information. This foreign intelligence orientation was strengthened by U.S. case law stating that the 'primary' purpose of the requested surveillance or search be the collection of *foreign* intelligence (Shelby 2002: 53). To all practical purposes, *FISA* made it difficult to target U.S. citizens, who were protected by *FISA* 'minimization rules' for handling information on U.S. citizens or lawful permanent residents. Political surveillance was apparently directed against foreign nationals. Even in this last case the law limited surveillance to the collection of intelligence, divorced from the bringing of criminal charges against a *FISA* target of surveillance (Shelby 2002: 52–3, see note 105). The ultimate result of such legislation, be it in the United States or Canada, and of its narrow interpretation, was to harden the dichotomy between nationals of a country and foreigners on its territory. Such a result can have dire consequences in times like ours, when the distinction between natives and aliens is blurred by strong migratory pressures that make it difficult to decide unambiguously who is a stranger in the land and who is not. In March 2003, in the midst of the war against Iraq, the FBI sought to interview between 3,000 and 11,000 Iraqi-born people, focusing on Iraqi immigrants rather than Iraqi-Americans. However, it seems that in the actual interviews, this distinction was not always followed (Hakim and Madigan 2003).

The *USA PATRIOT Act* of 2001 removed the limitations that jurisprudence had set to political surveillance under *FISA*. For years this law had provided that the primary purpose of political surveillance undertaken under its authority had to be intelligence collection. Section 218 of the *USA PATRIOT Act* stated that: 'sections 104(a) (7) (B) and section 303(a) (7) (B) (50 U.S.C. 1804(a) (7) (B) and 1823 (a) (7) (B)) of the Foreign Intelligence Act of 1978 are each amended by striking "the purpose" and inserting "a significant purpose."' The upshot of that amendment, which was later confirmed by the newly created *FISA* Court of Review, is that the amended *FISA* statute permits surveillance and physical searches even for undertakings that are primarily intended to result in the criminal prosecution of individuals, provided that a 'significant' intelligence purpose remains. This reinterpretation of *FISA* as granting authority both to surveillance and prosecution was a first and momentous step in undermining the divide between nationals and foreigners.

The so-called wall between intelligence and law enforcement actually ran parallel to the separation of national citizens from alien residents and protected the former while facilitating the targeting of the latter. This is strictly in line with high policing, as it was originally exercised. More significantly, perhaps, the breaking down of the barriers between political surveillance and law enforcement weakens the distinction between natives and strangers. It thus aggravates the situation for both. The round-ups of foreign suspects residing in the United States and preventively detained *incommunicado* is an ominous sign that aliens may become relatively free game in national territories. As for the nationals, there is a risk of their becoming strangers in their own country, if the coordination between law enforcement and intelligence agencies becomes an ordination to a high policing ministry. All of this is particularly ominous given that there is now a movement among U.S.

Republicans to abolish the five years sunset clause of the most intrusive dispositions of the *USA PATRIOT Act* and make them permanent (Lichtblau 2003a).

The questions that were put to the secretary of the newly created Department of Homeland Security when he testified in March 2003 before the Senate Appropriations Committee show that the Bush administration is making headway in this direction. Mr Ridge was asked the following question by Senator Arlen Specter (R-PA): 'My next question is what steps can you take when the FBI uses the wrong standard for probable cause under the Foreign Intelligence Surveillance Act? A couple of weeks ago, Director Mueller was here, and we explored that they were using the wrong standard, *more probable than not*, as opposed to *suspicion under the totality of the circumstances*, and they weren't getting the warrants they should have been getting. With you being responsible for homeland security, what can you do to see that the FBI uses the right standard' (U.S. Senate, Appropriations Committee 2003: 8, emphasis added). Secretary Ridge evaded the question, avoiding having to take sides between the director of the FBI and the influential senator. He referred to the 'respectful disagreement' between Director Mueller and Senator Specter and dropped the matter. Senator Specter interjected ominously that his disagreement with Director Mueller was 'not respectful.' The crucial point about this questioning is that it is wholly directed towards U.S. *internal* security. Not only is the person being questioned the director of homeland security, but the whole interrogation relates to the FBI, which is *not* an organization mainly targeting foreigners (although it does). The typical high policing conflation of surveillance and law enforcement can not only result in blurring the distinction between citizens of a country and foreigners, it could also trigger a regression to the times when internal dissidence ('subversion') was confused with violent action against the state. Unless it is rigorously monitored, should 'suspicion under the totality of circumstances' become a legal standard it may lead us back to the abuses of the past.

The Use of Informants

In police parlance there are two kinds of 'sources': human sources (undercover police and various kinds of informers) and technical sources (wiretaps, CCTV, and all kinds of electronic surveillance devices). The two examples that we discussed – the enlarged terms of reference of the

CSE and TIA – point to a substitution of technical sources for human sources in high policing. Although this needs qualification, the substitution is largely true and motivated by the present circumstances.

First, the time is past when one's enemy resembled oneself (Italian Red Brigades in Italy, the German Red Army Faction in Germany, and the IRA in Ulster). Today the most dangerous organizations targeted by high policing agencies no longer originate from the countries they attack (or in which they may reside for periods of time) and they operate on a transnational basis (Buzzetti 2003b). Consequently, they tend to have a very different ethno-racial make-up than the personnel of these security agencies, from whom they differ by language, religion, mores, and physical traits. This affords them no small degree of protection against infiltration. Furthermore, these organizations are very violent. Seeking further protection against infiltration, they ask a price for joining them – the commission of a heinous crime, generally murder – that they know no undercover agent, and no informant under the control of a handling officer, will be allowed to pay. Contrary to police fiction and journalistic delusions, undercover police and the agents they control are not allowed to commit murder or to put lives in danger in order to be accepted into a criminal organization or to maintain their cover therein, at least in democratic countries that respect the rule of law.

There are several ways out of this predicament. The first one is to recruit as informants persons who are already members of these organizations and who have previously paid their ticket of entry, thus moving from a strategy of infiltration to a strategy of 'exfiltration': offering various rewards to an insider for information and for his testimony in court. Generally speaking, the police cannot ignore for long the crimes of these informants when they have been committed on the home territory, and informants eventually end up in a witness protection program, testifying against their former accomplices in return for a reduced sentence for themselves. Not only, then, is their time as secret informants limited, their trustworthiness is doubtful and is increasingly being questioned by the courts, as they tend to bring justice into disrepute. A second solution is to rely on informants controlled by a friendly foreign service that does not mind running murderers as informants. The problem with this answer is that it doubles the risk of manipulation for the service attempting to run an informant, who will never be met, through a third party of unscrupulous colleagues. A last resort is to recruit one's own moles abroad, as the Soviet Union did with outstanding success in the United Kingdom and, more recently, in the United States, with Aldrich Ames. However, the possibility that such moles may be double agents is always present, as the FBI learned in the Katrina Leung case (Lichtblau 2003b). In organizations motivated by religion, which now appear to be the most lethal, this strategy seems to come up against a wall (although we may learn differently in time). Two years ago, US$25 million was offered for information leading to the neutralization of Osama Bin Laden, with no success to date (of course Bin Laden may now be dead, which would explain the dearth of information).

In conclusion, although it is repeatedly claimed that infiltration is the most efficient high policing tactic (Shelby 2002: 76–9), and although there have been clamours for a return to the use of human sources, many problems will have to be solved to make it viable and efficient in a transnational context characterized by sectarianism and ruthlessness. From what we know about the history of murderous regimes, they are not easy to infiltrate. The same would apply to murderous organizations. Nevertheless, governments have far from given up on infiltration and are taking measures to solve some of the problems that we have just discussed. In new legislation passed against organized crime in the beginning of 2003 (Bill C-24), Canada has granted permission to under-cover police *and their agents* to commit serious crimes in order to infiltrate criminal organizations and preserve their cover. The law does not list permitted and forbidden crimes in a legal 'schedule,' since that would defeat its purpose. While crimes such as murder, aggravated assault, and other grievous violent acts are not covered by this provision, police may be authorized to commit simple assault (an extendable offence) and deal in drugs in the context of sting and counter-sting operations. These authorizations confirm the notion that high policing cannot be law abiding in all respects and that it cancels out to a significant extent the notion of police deviance.

The neutralization of police deviance is even stronger in technological high policing. This is illustrated in several ways. As previously stressed, the CSE is prohibited from intercepting the private communications of Canadians. Yet since it scoops up all electronic emanations flying through the atmosphere, it is bound to mostly intercept Canadian communications. There is, we are told, software that is supposed to filter out such communications. But, if it were at times to fail, who (what) could be made accountable? A second example: in Canada and the United States one needs a judicial warrant to intercept telephone communications transmitted through land lines, which offer a reasonable presumption of privacy (hence the need for a warrant to intrude upon this privacy). However, when land lines are overloaded, a part of telephone communications is diverted for transmission purposes to telephone towers that transmit them without any wires. Does one still need a warrant to intercept a private communication that is suddenly switched onto a wireless telephone tower? We have no idea. Legislation always lags far behind the development of technology. Techno high policing is thus to a large extent outside the reach of any law, and the notion of its being deviant is completely short-circuited.

Finally, techno high policing is as potent for generating surveillance paranoia as was the notion of being surrounded by police informants. The belief that one is under surveillance at a distance generates a feeling of helplessness that is even more chilling than the fear of personal betrayal. One can at least try to uncover a snitch and remove him or her from one's environment. Trying to get rid of surveillance technology, however, is as self-defeating as attempting to do away with the environment itself.

Higher Policing?

We could conclude that the features of high policing still apply to surveillance as it is now exercised, albeit with some qualifications. The most significant of these qualifications concerns the now dominant role of technology, which could not be taken into account for historical reasons by the architects of high policing. However, the basic trait of high policing – its relative intractability to the rule of law or, it may be suggested, its alegality – is enhanced by our dependence upon technology and, more generally speaking, by the present evolution of political surveillance.

However justified, this conclusion is not on all points satisfactory. First of all, high policing is, as we have seen, essentially dependent upon the collection, analysis, and dissemination of intelligence. So fundamental is this link that high policing should be described as intelligence-leading rather than intelligence-led policing. In theory, it fills the blanks, moving from one piece of the puzzle to the total picture. In this respect, we have to take stock of the criticism of the capacity of law enforcement organizations and of their personnel to perform this function. Hence we should make a clear distinction between high policing as an ideal type, in the sense of Weber, and high policing as a very incomplete incarnation of this paradigm.

Second, whether invented in France at the end of the seventeenth century as high policing or reinvented in England at the beginning of the nineteenth as low policing, policing always stood as an *alternative to the military*. It is premature to say that this is no longer the case. Nevertheless, it must be recognized that the militarization of policing, in its diverse aspects, is now a real issue. This issue has taken a particular urgency because of the emergence of mass terrorism, which fits, in their literal sense, neither the legal category of crime nor the political category of war as an aggression perpetrated by an enemy state – although both concepts are metaphorically abused in referring to terrorism. For instance, the U.S. administration has taken to referring to both the police and the military as 'war fighters,' this metaphor being clearly tipped in favour of the military. Note that both of the agencies that we have described – the CSE and the IAO – answer to departments of defence (at their inception, at any rate).

These developments are significant enough to warrant a new characterization of political surveillance, one which does not so much contradict the high policing paradigm as it completes it.

Deductive and Inductive Surveillance

One of the most important differences between technological surveillance (SIGINT) and surveillance through informants (HUMINT) is that the former is exercised from a distance, whereas the latter implies closeness (infiltration that relies on proximity to the target). This ambivalence between the tactics of proximity that are the defining feature of community policing and the watching from a distance that is a growing trend of surveillance is characteristic of late modern policing.

Watching individuals from a close position is an *inductive process* that moves from the particular subject under observation to his or her inclusion into a category of like individuals or to a general conclusion (these persons are enemies of the state and so forth). An initiative like TIA, which relies on the computerized monitoring of a great mass of transactions, is innovative in relation to the traditional inductive conception of surveillance. As we saw, the main component of TIA relied on the extraction of hypothetically predictive patterns from a very large sample of terrorist acts: if possible the sample would include all known terrorist acts. These patterns are general, consisting of prototypical sequences of events revealing a hostile intention. Although they are not science, they perform a role that is nevertheless similar to scientific generalizations. Transactions are monitored by computer, without any particular attention being at first given to the individual(s) or groups of individuals engaged in them: the computer is simply attempting to apply general predictive patterns to transactions initially divorced from whomever is performing them (de-nominalized, as we said). However, if a series of transactions fits a predictive pattern, an alert is given and the system moves into its identification mode, which then directs field officers to intervene. The whole process can be formalized as a tentative syllogism:

Universal premise:	All who engage in pattern A are dangerous
Particular statement:	At least one X is engaging into the performance of this pattern

Alerting Inference:	At least one X (this or these particular X) is dangerous
Identification mode:	Who is X?
Effective identification:	X = A (a) ...... A (n)

Normative premise:	All dangerous persons must be stopped

Pragmatic conclusion:	A (a) ...... A (n) is (are) wanted/arrested for questioning

Deductive surveillance is not limited to political surveillance. It is part of a growing trend in policing, alongside the generalization of profiling. Profiling poses a dilemma where both alternatives are problematic. If the profile is too general, it generates an intolerable amount of false positives; if it is sufficiently precise to exclude false positives, it will also miss a number of true positives and the probability that it is spurious will increase.

Deductive surveillance is also closely in line with one feature of high policing, that is, the accumulation of powers that should remain distinct, as pseudo-scientific laws, predictive patterns, and profiles supersede legislation and provide norms for pre-emptive repression.

Militarization and Privatization

The feature of war that is most obscenely displayed nowadays is that techno-proficient countries no longer rely on military personnel in combat. The firepower of Western countries, most notably, the United States, is so overwhelming for their enemy that the majority of Western casualties in a war against a non-Western power are caused by 'friendly fire,' as Canadians learned in Afghanistan and U.K. troops learned in Iraq. It is more hazardous to wage war with the United States than against its enemies.

As President Eisenhower aptly said, the military are bound to the industrial complex, which develops its technology. Researchers have also emphasized the strong links between the military, technology, and private industry (Haggerty and Ericson 2001; Kraska 2002; Leman-Langlois 2003). *To the extent that policing may be under the sway of the military, it is also dependent upon private enterprise.* In this respect it is worthy to note that Dr Charles E. McQueary, a former executive of General Dynamics and Bell Laboratories, is the newly confirmed undersecretary for Science and Technology in the Department of Homeland Security. Since the creation of this department the Homeland Security Research Corporation was established in San José, California; so was the Homeland Security Industries Association, a trade group that has signed more than a hundred companies as members since it was incorporated in July 2003.

As we previously argued, the purchase of technology is not so much governed by what is adequate for the task at hand as it is by what is available on the market. In this respect the big armaments firm, Raytheon, has announced Project Yankee: 'a company wide effort to determine how its military expertise and products might be converted into counter-terrorism' (Shenon 2003). Our conviction is that change is not brought about by human policy as much as its imposed by the development of technology (think of the police cruiser or the personal computer). It is thus to be foreseen that the impact of technology originally conceived for the military will strongly affect domestic policing, for good, bad, or worse.

Enemies and Law Breakers

As always, changes in the stuff of policing will be reflected in our concepts of policing. Up until now, a clear distinction has been made between internal and external security. Internal security meant protection against criminal aggression directed towards one's person and one's property, such aggression being motivated by hatred, lust, or greed, but rarely by politics. It was first incumbent upon the public police and, increasingly, upon private security to provide such protection. External security referred to protection against aggression by another country, the army being the main instrument of this protection. It now seems that all of these distinctions can be questioned in various ways. The definition of war as violence taking place between different countries is not rich enough to accommodate asymmetric conflicts, where one state is the target of an aggression that is politically and/or religiously motivated, by an international organization based in several countries but belonging to none. The 9/11 attacks are the most dramatic example of asymmetric conflicts. Asymmetric conflicts are hybrids that challenge the utility of traditional classificatory schemes. Due to their complete disregard for the rules governing hostile relationships between states (e.g., ultimatums, declarations of war, etc.), they can be construed as crime and effectively they are. Because of the mass destruction that they bring, they can also be considered as acts of war, and again effectively they are. Being both crimes and warlike aggressions, events like 9/11 fall simultaneously within the province of internal and external security; as such they are also simultaneously under the purview of the police and the military. The very name of the U.S. Homeland Security Department is an obsolete umbrella, trying to shade agencies reaching far beyond the U.S. homeland, as even the most cursory examination of its make-up reveals.

We have belaboured the 9/11 example because of its vividness in our minds. However, as de Lint and Virta (2003) have argued, politics are permeating a great deal of crime and of policing. Indeed, it could be claimed that all organized crime, when it reaches beyond gangs that quickly dissolve, has political overtones (Brodeur 2000). The upshot of these remarks is that political surveillance is now a misnomer and we should simply talk plainly of surveillance, as Foucault did.

Surveillance Fiction

Some years ago Allan Greenspan, the chairman of the U.S. Federal Reserve, coined the expression 'irrational exuberance' to stigmatize the speculation fever that produced the stock market bubble that erupted shortly after his comment. This phrase also applies to the present context. Technology, it is brashly claimed, will remedy all the past failures of intelligence. This effervescent surveillance bubble may well burst, like the previous one, and create a score of victims found not among the surveillants but rather among their 'collateral' targets. The problem is that under national security secrecy legislation we will not know about it, and many people will be hurt without any external means of controlling the damage.

The now infamous FutureMap project (Futures Markets Applied to Prediction), which precipitated the end of the IAO, is a prime illustration of the surveillance bubble. This program was developed to predict international events based on how an 'events futures' market would behave. A more self-defeating project could not be devised. Predicting how stocks are going to evolve has always been a losing game, as nearly all heads of mutual funds who have not yet been fired can testify.

One would hope that such dud technology will not be used to plot the next moves of al Qa'eda, yet it probably will, in one guise or another, no matter how facetious. There is now such a craving for intelligence that it has produced an irrational sellers' market. One consequence of this predicament is the maximization of the possibilities of making mistakes. There is something that no software can do by itself: input data. The current illusion is that police, who are said to be poor analysts, will prove better as data collectors. They will not, if they are not better trained. In the present context marked by a high level of immigration, merely spelling a name consistently is a considerable challenge that is not met. The more encompassing and sophisticated our data collecting and processing devices, the higher the possibilities for making errors. Just recently, the U.S. Justice Department identified some three thousand criminal cases that could have been affected by flawed procedures and skewed testimony by FBI laboratory technicians up until 1997 (Associated Press 2003). The bigger the science, the bigger the shadow of mishaps that it projects. Scientists have developed a large safety net against such miscarriages, but the police have yet to do so.

In his original 1983 paper Brodeur claimed that the net cast by high policing mainly caught distraught fish that swam headlong unto them. Shortly following 9/11, the venerable French composer and orchestra director Pierre Boulez, who once headed the New York Philharmonic, was detained for several hours by the Swiss under suspicion of being a terrorist. It happened that under the pandemonium prevailing in France in May 1968, Pierre Boulez, then a young artist, proclaimed that opera houses should be 'burned down' (setting fire to something was trendy at the time). This was not forgotten by the police and Boulez was branded as a terrorist in a buried file, which resurfaced in a small border crossing booth, after 9/11. Being a celebrity, Monsieur Boulez was quickly released with the profuse apologies of the Swiss authorities who had invited him to their music festival. An unknown Arab student previously given to despondent rhetoric, as many students were at that time, might not have fared as well.

What we should be worried about is not the Big Brother, but the Big Bungler. Not only is he alive, well, and uncontrolled, he may thrive in the years to come, particularly now that Saddam Hussein has been made to shed his public skin as the Head of the State of Iraq. Big Bungler is in no way a fumbling Mr Bean substitute for Big Brother, about whom we need not worry. On the contrary, Big Bungler is Big Brother driven mad by too much power and too much speed. He offers little protection against the Meanies and might hurt all the wrong people.

REFERENCES

Associated Press. 2003. 'Errors at F.B.I. May Be Issue in 3,000 Cases.' *New York Times*. 17 Mar.
Bamford, J. 1982. *The Puzzle Palace: Inside the National Security Agency, America's Most Secret Intelligence Organization*. New York: Penguin Books.
Brodeur, J. 1983. 'High Policing and Low Policing: Remarks about the Policing of Political Activities.' *Social Problems* 30 (5): 507–20.
– 1992. 'Undercover Policing in Canada: Wanting What Is Wrong.' *Crime, Law and Social Change* 18:105–36.
– 2000. 'Cops and Spooks.' *Police Practice and Research* 1 (3): 299–321.
– 2003. 'The Globalization of Security and Intelligence Agencies: A Report on the Canadian Intelligence Community.' In *Democracy, Law and Security: Internal Services in Contemporary Europe*, ed. J. Brodeur, P. Gill, and D. Töllborg, 210–61. Aldershot: Ashgate.
Buzzetti, H. 2003a. 'Surveillance internationale. Ottawa rend plus acceptable sa banque de données sur les voyageurs.' *Le Devoir*. 10 April.
– 2003b. 'Rapport de la Vérificatrice Générale du Canada. Le Canada a perdu la trace de 36,000 immigrants illégaux.' *Le Devoir*. April 9
Communications Security Establishment Commissioner. 1997. *Annual Report, 1996-97*. Ottawa: Minister of Public Works and Government Services Canada.
DARPA. 2003a. *Fiscal Year (FY) 2004/FY 2005 Viennial Budget Estimates*.
– 2003b. *Report to Congress Regarding the Terrorism Information Awarenss Program. In Response to Consolidated Appropriation Resolution, 2003, Pub. L. No. 108-7, Division M*, § 111(b). http://wwwdarpa.mil/body/tia/TIA%20DI.pdf

de Lint, W., and S. Virta. 2003. 'Security From Politics: Low and High Policing Revisited.' Paper presented at the In Search of Security Conference, Montreal, Law Commission of Canada, Feb.

Firestone, D. 2003. 'Senate Rolls a Pork Barrel into War Bill.' *New York Times*. 19 Apr.

France. 2003. *Rapport à l'Assemblée nationale, renseignement; Rapporteur spécial: M. Bernard CARAYON*, Document #256. http://www.assemblee-nat.fr/12/budget/plf2003/b0256-36.asp

Frost, M., and M. Gratton. 1994. *Spyworld: Inside the Canadian and American Intelligence Establishments*. Toronto: Doubleday.

Haggerty, K., and R. Ericson. 2001. 'The Military Technostructures of Policing.' In P.B. Kraska, ed., *Militarizing the American Criminal Justice System*, 43–64. Boston: Northeastern University Press.

Hakim, D., and N. Madigan. 2003. 'Immigrants Questioned by F.B.I.' *New York Times*. 22 Mar.

International Herald Tribune Editorial. 2003. 'New Airport Profiling.' Mar.

Kraska, P.B. 2002. *Militarizing the American Criminal Justice System*. Boston: Northeastern University Press.

L'Hevillet, H. 2001. *Basse politique, haute police: Une approache philosophique de la police*. Paris: Éditions Fayord.

Laudon, K.C. 1986. *Dossier Society: Value Choices in the Design of National Information Systems*. New York: Columbia University Press.

Leman-Langlois, S. 2003. 'The Myopic Panopticon: The Social Consequences of Policing through the Lens.' *Policing and Society* 13 (1): 43–58.

Lichtblau, E. 2003a. 'Republicans Want Terror Law Made Permanent.' *New York Times*. 9 Apr.

– 2003b. 'F.B.I. Was Told Years Ago of Possible Double Agent.' *New York Times*. 12 Apr.

Ministère des Communications, Quebec. 1981. *Rapport de la Commission d'enquête sur des opérations policières en territoire québécois* (rapport Keable).

Minister of Supply and Services, Ottawa. 1981. *Freedom and Security under the Law: Second Report, Commission of Inquiry Concerning Certain Activities of the Royal Canadian Mounted Police (The McDonald Report)*.

Plenel, E. 1997. *Les mots volés*. Paris: Stock.

Privacy Commissioner of Canada. 2003. *Annual Report to Parliament, 2001–2002*, Ottawa. http://www.privcom.gc.ca

Risen, J. 2003. 'C.I.A. Aides Feel Pressure in Preparing Iraqi Reports.' *New York Times*. 23 Mar.

Sallot, J. 1984. 'Secret Agency Keeps Data on Individual 'Security Risks.' *Globe and Mail*. 21 Nov.

Shelby, R.C. 2002. *September 11 and the Imperative of Reform in the U.S. Intelligence Community: Additional Views of Senator Richard C. Shelby, Vice Chairman, Senate Select Committee on Intelligence.* United States Senate Select Committee on Intelligence. Washington, DC: US Government Printing Office. This report is also available on the Internet at http://intelligence.senate.gov/pubs107.htlm

Shenon, P. 2003. 'Domestic Security: The Line Starts Here.' *New York Times.* 6 Mar.

Skolnick, J. 1966. *Justice without Trial: Law Enforcement in a Democratic Society.* New York: John Wiley.

United States Senate, Appropriations Committee. 2003. *Full Text of Testimony before the Senate Appropriations Committee.* http://www.nytimes.com/2003/03/27/international/worldspecial/28FTEX-RUMS.htlm.

United States Congress. 2002. *Joint Inquiry Conducted by the Senate Select Committee on Intelligence and the House Permanent Select Committee on Intelligence. Findings and Recommendations.* Washington, DC: U.S. Government Printing Office. This report is also available on the Internet at http://intelligence.senate.gov/pubs107.htlm.

United States Congress House, Select Committee on Intelligence. 1977. *CIA: The Pike Report.* Nottingham: Spokesman Books for the Bertrand Russell Peace Foundation.

United States Congress, Senate, Select Committee to Study Governmental Operations with Respect to Intelligence Activities 94th Congress, 2nd Session. 1976. *Intelligence Activities and the Rights of Americans* (The Church Report). Washington, DC: US Government Printing Office.

Websites

EPIC (Electronic Privacy Information Centre): http://www.epic.org/privacy/profiling/tia

FERET and Face Recognition Vendor Test: http://www.frvt.org/

Government exec.com: http://www.govexec.com/dailyfed/1102/112002ti.html

IAO: http//www.darpa.mil/iao.index.html

New Tools for Domestic Spying and Qualms (Cryptome): http://cryptome.org/tia-balk.html

New York Times: http://www.nytimes.com/2002/11/09.politics.09COMP.html

Total Information Awareness Program (TIA) System Description Document (SDD) (Official Document, 150 p.): http://www.epic.org/privacy/profiling/tia/tiasystemdescription.pdf

Washington Post: http://www.washingtonpost.com/ac2/wp-dyn/A40942-2002Nov11

8 An Alternative Current in Surveillance and Control: Broadcasting Surveillance Footage of Crimes

AARON DOYLE

In February 1974, a crude but dramatic piece of surveillance camera footage captured the television news spotlight. The segment, from the camera of the Hibernia bank in San Francisco, featured kidnapped heiress Patty Hearst brandishing a gun as she accompanied her Symbionese Liberation Army captors during a hold-up. Surveillance cameras were rare then, but the televising of this halting, flickering piece of black-and-white footage marked the beginnings of a trend. As public and private surveillance cameras have multiplied, authorities have taken the initiative more and more in providing broadcast television with footage of many crimes, either for the news, or for reality programs like *World's Wildest Police Videos* and *Crimewatch*. As Norris and Armstrong (1999: 67) note, broadcast TV and video surveillance were made for each other, especially when crime is added to the mix. Other broadcast surveillance footage has since taken on iconic status, for example that showing two-year-old James Bulger being led from a shopping mall to be killed, that showing the rampage by two teenaged gunmen at Columbine High School, or recently the segments showing some of the final moments of murdered television presenter Jill Dando, who was, in a terrible irony, herself a *Crimewatch* host. Sometimes TV audiences are enlisted to identify unknown suspects appearing on the surveillance footage, a technique I will refer to as the 'video wanted poster.' Otherwise, the footage is handed over to the media simply for various promotional reasons, including promoting the surveillance technology itself. Alternatively, footage originally recorded in other ways, say by news cameras or home camcorders during a riot, is given to the media so TV audiences can identify the unknown suspects shown. The increasing role of webcams and Internet videoplayers as a vehicle for capture and broadcast of surveillance video will take the phenomenon to another level.

Broadcasting surveillance camera footage is anything but routine. In fact, it runs counter in various ways to most of the surveillance practices considered elsewhere in this book: it is visual surveillance rather than simply collecting data, it relies on human informants (the TV audience), and it is an extremely visible process rather than being surreptitious. Broadcast of the footage also occurs unsystematically, in isolated cases, in contrast to the routine monitoring of large numbers of people characteristic of most forms of surveillance.

This chapter will argue that, although unusual, the broadcast of the footage has important impacts. It is also a highly suggestive example of some broader processes, so I will use it to make wider points about how we think about media influence, and how we theorize trends in surveillance and control.

As a media product, surveillance camera footage has various qualities that make it resonate in the wider culture, for example, its often-horrifying 'realness.' The footage is important in reinforcing a particular, prominent system of meaning about crime, one in which, for example, crime is portrayed as random, inexplicable, violent acts by strangers. Yet this cultural impact is only part of the story. We tend to think of media influence rather narrowly in terms of this kind of effect on the understandings and views of audience members, and we miss other impacts. Surveillance camera footage has additional effects at the institutional level, impacts on various criminal justice players and processes, and a fuller account of media influence must include theorizing such impacts.

The images of criminal justice presented by these slices of footage have a lot in common with how crime is portrayed in the news and entertainment media more generally. Yet they are not simply more of the same. Much of what is unique about such footage stems from the particular properties of TV and video as media, especially when they directly record the events in question. 'Medium theory' (Meyrowitz 1994) can begin to help us understand how the properties of TV and video as media shape the various influences, both cultural and institutional, of broadcasting this footage.

Of course, the properties of TV and video are not the whole story. Medium theory helps us, but it needs to be reworked into something more critical, to show how these properties of TV and video work in interaction with the key institutional players, in this case the police and media organizations, and how the properties of TV and video also operate in interaction with the broader culture. Media impacts are fundamentally shaped by patterns of domination in both our institutions and our wider culture.

Finally, I will argue in the last section of the chapter that all these various effects mean that broadcasting surveillance camera footage is a key instance of something wider, of a significant alternative current in contemporary control. This alternative current is quite at odds with surveillance through routine data gathering, and at odds with the current tendency towards control through dispassionate, amoral, technical management of populations at a distance, that is, at odds with what I will call 'actuarialism' (Feeley and Simon 1994). Broadcasting surveillance footage is 'anti-actuarial.' It moves us towards a more emotionally charged, punitive, vengeance-oriented approach to crime and control, parallel to past spectacles of punishment, but also new in key ways.

I begin by presenting a little more empirical detail on the phenomenon and on the factors that led to it.

Explanations for the Trend

Although analysis of surveillance is usually couched in language and metaphors of vision and visibility, most contemporary surveillance does not involve literally watching the subject. Instead, when talking of surveillance, vision becomes a 'master metaphor' for the other senses (Marx 2002). Nevertheless, enabled by the miniaturization and proliferation of video cameras, more and more of social activity is visually recorded, and there has been an increase in direct visual surveillance as well as in other forms of monitoring.

Closed circuit television (CCTV) surveillance of public spaces has become pervasive in Britain with the installation of more than 1.5 million cameras. Publicly owned and operated surveillance cameras are also spreading in North America, although less rapidly. By the late 1990s, they had come to cities such as Tacoma, Newark, and Baltimore (Surette 1998: 175). Public cameras were given a further boost as part of the wave of new security measures after 9/11. Canadians have, in typical fashion, adopted a more cautious stance in relation to the cameras. In Vancouver, a proposal has been under debate for some time to mount twenty-three surveillance cameras to monitor segments of the downtown area. Cameras mounted in police cruisers have become another key source of footage for broadcast. Private surveillance cameras have also spread rapidly: the Security Industry Association recently estimated there were over two million surveillance cameras in operation in the United States (Murphy, *National Post*, 30 Sept. 2002, A13), the majority privately owned.

The spread of the cameras has converged with several other factors. Broadcasting surveillance footage so that the public can identify suspects is also part of a rising culture of informing. Ironically, contemporary forms of surveillance are in part a response to the rise of privacy (Nock 1993) and the breakdown of local trust and of informal social control that previously existed through local social ties. For example, insurance investigators must go to databases now for information about people they might once have obtained from neighbours, because these days neighbours are more and more often strangers (Ericson, Doyle, and Barry 2003). Gary Marx (1988: 207) has suggested that negative attitudes towards informing that were prevalent in the eighteenth and nineteenth centuries shifted, and informing came to be seen more as an act of good citizenship during the twentieth century. The culture of informing was further fuelled in the aftermath of 9/11. Within two months after 9/11, the FBI's Internet Fraud Complaint Center, which invited browsers to report terrorist activity, received approximately 150,000 tips (Gormley 2003). Informing is paid relatively little attention in surveillance studies, but what Marx (1988: 208) suggested in the 1980s appears to remain the case today: 'the scale of human informing is ... dwarfed by electronic informers.' However, the rise of human informing, if a small part of the whole surveillance picture, also tells us something about the shifting cultural place of surveillance more generally.

Trends in the television industry have also played a role in the rise of surveillance footage as broadcast content. Although many people think of television news as image-driven, most of the time, TV news has not literally shown us what it is telling us about; it has featured after-the-fact recounting of events by 'talking heads' rather than actual footage of the events in question (Ericson 1998). However, video records of daily events are increasingly available, and television news is offering more and more footage of such 'real' occurrences (Doyle 2003). Surveillance footage is picked up not only by news, but also by various other TV formats such as Crimestoppers ads, reality series like *Crimewatch*, *Crime Beat*, *Eye Spy*, and *Police! Camera! Action!* and videos like *Police Stop!*, and *Caught in the Act!*

The fragmentation of the mass audience with the proliferation of channels led producers to branch out, searching for cheap and innovative programming that would appeal to narrower audience segments. Starting in the late 1980s, Rupert Murdoch's Fox network began to focus on young, low-income earners who had not previously been targeted by the three major U.S. networks (Fishman 1998: 66). This segment has been a key audience for Fox reality programs that use surveillance footage, like *World's Wildest Police Videos* and *Real TV*.

Another contributing factor is the cooperation of numerous police forces in providing the media with surveillance footage. Police are the central source of crime news more generally, often dominating its production (Chibnall 1977; Ericson, Baranek, and Chan 1989; Doyle 2003). In recent decades, police have increasingly taken the initiative in seeking media attention for their own political ends. Other possible news sources, such as victims, witnesses, the accused, academics, or citizens' groups, may not be as accessible or cooperative. Police have a taken-for-granted authority, and conferring with them is built into the daily routines of news production.

Certainly, not all surveillance footage that is broadcast comes from police. Sometimes how the media obtain the footage remains secret. In some instances it is definitely obtained without police help, as with the Columbine footage that was surreptitiously rerecorded for local TV news during a training session for police and emergency workers (Associated Press, 13 Oct. 1999). Yet it is clear that in many cases surveillance footage is passed along directly by police to television outlets, for a variety of reasons. Firstly, police use 'video wanted posters' on TV news or programs like *Crimestoppers* to call on the TV audience to help with investigations. There have been a number of success stories for the video wanted poster. A man who robbed a Louisiana convenience store was caught on camera banging the clerk's head on the cash register. His father saw the footage on local news and turned him in (*The Rivera Show*, 30 May 1998). Advocates claim video wanted posters helped to capture a number of important fugitives: for example, the young killers in the Bulger case, a suspect in the bombing of a government building in Oklahoma (Graham 1998: 90) and a British subway bomber (Freeman 1999).

Enterprising police also use video wanted posters to identify suspects in the aftermath of riots, employing footage from surveillance cameras or confiscated from television news cameras. One example comes from the so-called Stanley Cup riot that occurred in Vancouver after the last game of the 1994 hockey final. Police executed warrants at three local TV stations, confiscating hundreds of hours of footage of the action to create video wanted posters. These were broadcast on two Vancouver television stations to solicit public help to identify selected rioters. Police also set up a computerized public video booth in local malls, car shows, and other locations which displayed selected video clips of the riot and asked the public to type in information identifying particular rioters. Charges were then laid against more than a hundred people (Doyle 2003).

Video wanted posters have been used in many other crowd situations, most notoriously after the 1989 Tiananmen Square uprising. Chinese authorities had massive surveillance footage of the student-led protest for democracy, footage from what was nominally a traffic control system. Images of the protesters were repeatedly broadcast over Chinese television with an offer of a reward for information. Many participants were identified, captured, and punished. Police used TV footage in a similar way after the Los Angeles riot of 1992, plus riots in Alberta, Ontario, Ohio, Colorado, and Illinois.

Police give footage to TV even when it does not portray the suspects, for example, footage of murder victims from shortly before they were killed, which may jog the memory of potential witnesses. Other surveillance video is broadcast even when there is no investigative reason: in the Latasha Harlins case, for example, the news repeatedly showed footage of a Korean shop-owner in Los Angeles shooting a young black woman in the back of the head and killing her (Fiske 1996). Other criminal justice footage that is televised comes from camcorders mounted in police cruisers. The reality program *I Witness Video* featured footage from a Texas police officer's car of suspects gunning down the officer, followed by footage from a second police car of one suspect being killed in a subsequent shoot-out with police (*LA Times*, 22 Feb. 1992).

Why would police give surveillance footage to the media? Firstly, the footage is used to publicize the technology itself. A British police chief thus made available to the media a 'greatest hits' tape to promote local CCTV (Groombridge 2002). Police also use the media to promote their own successes, for example, releasing footage of successful sting operations. Police may also turn over footage as a quid pro quo to maintain favourable relations with journalists. Finally, in some cases, police may be paid for footage. Labyrinth Video confirmed that it had made 'donations' to some police forces that had given footage to *Police Stop!* (*Daily Telegraph*, 2 July 1994).

Surveillance footage that is broadcast not only spotlights crimes that are already high profile, as in the Patty Hearst case, but often features more routine and prosaic incidents. These are crimes that make the news because they happen to be caught on camera and are dramatic visually, even in the absence of any other obvious news value. I will call these 'found television crimes.' A prominent example came in 2002: an Indiana woman received national media attention after a surveillance camera in a supermarket parking lot taped her striking her child. In an earlier case, a secret camera captured a nanny in Tennessee slapping the infant under her care. While one might argue that both individuals should have been charged, normally, these types of crime, while repugnant, would receive little media attention. They only became notorious because dramatic surveillance footage was broadcast.

Rethinking Media Influences

Now I will move on to discuss the various influences of broadcasting surveillance camera footage. One tool to help us understand these influences is 'medium theory' (Meyrowitz 1994). Medium theory analyses the social impact of the relatively fixed properties of communications technologies such as broadcast television, the video camera, and the Internet. It examines, for example, the speed, direction, and kind of encoding that characterize a particular medium, and which of our senses the medium engages.

Medium theory's most prominent advocate is Joshua Meyrowitz[1]. Meyrowitz's celebrated book *No Sense of Place* (1985) argued that because TV and other electronic media made different social situations more visible, they had a liberating effect. Television and other electronic media broke down barriers between various social groups by including many people in new 'information systems,' moving us out of a more segregated world based on print communication. Thus, TV and other electronic media had massive effects of social levelling, for example, between politicians and ordinary people.

In a broad way, Meyrowitz's argument parallels writing about sur-

---

1 Many of medium theory's key ideas were pioneered by Harold Innis (1950) and popularized (although in sometimes cryptic fashion) by Marshall McLuhan (1964). More recent social theorists, for example, Mark Poster (1990), John Thompson (1994), and Jean Baudrillard (1988), have made broad arguments about the theoretical importance of shifts in the dominant media of our era. Meyrowitz's work is most immediately applicable in this case.

veillance, such as *The Transparent Society* by David Brin (1998). Like Brin, Meyrowitz argues that new communications technology has the potential to enhance democracy through making everyone more visible and thus promoting equality. As Brin rightly points out, however, one key to any such transformation will be the power relations around the technology and whether or not there is full democratic participation in its use. In his medium theory analysis, Meyrowitz does not pay sufficient attention to the social contexts of media production and consumption and the power relations in those contexts. We need to resituate medium theory in a more critical vein. The properties of media reshape social situations, but the power relations in particular contexts largely dictate how this reshaping occurs. The surveillance produced by the interaction of the cameras, authorities, broadcast television, and the public will be selective. Broadcasting surveillance footage has broad ideological effects as well as more specific effects on criminal justice processes and institutions, but both kinds of effects will tend to work to the advantage of dominant institutions and groups, and against the less powerful. We can divide the reasons for this into four categories:

a) police institutional factors;
b) media institutional factors;
c) existing broader cultural understandings of crime; and
d) properties of visual media themselves.

The images of crime resulting from broadcasting the footage have much in common with how criminal justice is represented in news and entertainment more generally. These media images tend to fit with dominant cultural understandings of criminal justice, and to reinforce existing power relations, for a number of reasons. One is the role of police as the key source of crime news. For example, extensive content analysis shows that when police sources make attributions concerning the causes of crime, they seldom mention broader social causes and focus instead on individual evil and pathology (Ericson, Baranek, and Chan 1991). If police tend to offer a particular, ideological view of crime, surveillance cameras supplement the package with dramatic visuals. Police are not only key gatekeepers concerning which surveillance footage is broadcast, they also play a vital role in providing the official definitions of what is on the footage.

Secondly, the news also offers an ideological vision of criminal justice because of various properties of the news media as an institution. While media outlets and formats vary somewhat (Ericson, Baranek, and Chan 1991), in general the media focus heavily on individual crimes rather than on broader situations in criminal justice. Journalists normally report on day-to-day events that have occurred since the last deadline rather than examining underlying issues – a tendency known as 'event orientation' (Ericson, Baranek, and Chan 1987). The news media are also oriented towards violence, especially homicides and sexual assaults. The use of surveillance footage fits with these tendencies, focusing on dramatic, individual crimes, most of which are violent.

A key way in which inequality is reproduced in criminal justice is by making the crimes of the lower classes are more visible than those of elites. This is also true with broadcasting surveillance footage. Firstly, public surveillance cameras are more likely to be present in poorer areas (Davies 1998: 270). Secondly, camera operators cannot monitor every camera all the time. A further kind of bias occurs with respect to precisely who the camera operators choose to monitor. In a study of British surveillance camera operators in three areas covered by 148 cameras, Norris and Armstrong (1999) found that the young, the male, and the black were systematically and disproportionately targeted.

Police also dictate which footage from public surveillance cameras is available to the news. Not surprisingly, as with the reality-TV show *Cops*, for example (Doyle 2003), the broadcast of footage from police surveillance cameras showing police in a bad light is relatively rare. As Norris and Armstrong (1999: 70) point out, 'the more contentious, yet routine, aspects such as the surveillance of political protesters, the deployment of police officers to move on "troublesome" youth, or the use of the system by private security guards to exclude children from shopping malls are of course not shown.' Similarly, the footage broadcast for surveillance purposes after Vancouver's Stanley Cup riot predictably contained no examples of possible police misconduct, focusing only on the activities of rioters (Doyle 2003: chap. 5).

Of course, some surveillance video makes it onto the news despite the police rather than because of them. The news media are sometimes resistant to and critical of police surveillance efforts (Doyle 2003: chap. 5; Gilliom this volume). The use of home video on the news also opens up new avenues of resistance, epitomized by the Rodney King case. Yet, as I elaborate elsewhere (Doyle 2003: chap. 4), the use of home camcorder footage on the news will necessarily be limited compared to the use of surveillance footage. Indeed, the Rodney King case itself achieved such prominence not simply because of the dramatic visuals but because they arose at a particular moment in the political history of Los Angeles (Lawrence 2000).

In sum, surveillance footage that is broadcast features a structured bias towards reporting certain types of crimes. In particular, surveillance footage may tend towards street crimes committed in poorer urban areas, and by populations that are visibly different, such as visible minorities and certain youth subcultures. Surveillance footage that is broadcast will also tend to be of crimes committed by strangers, because they are the crimes for which police investigations will use video wanted posters. Such representations of crime have an ideological impact.

We certainly should not assume, however, that media images simply make people support a particular view of criminal justice, or support surveillance. Reception analysis shows us that audiences often actively resist dominant representations and create their own meanings from television texts (Fiske 1987). The public is more educated and more critical of major social institutions than ever before (see, e.g., Jones 2004). They also have a number of alternative sources of knowledge besides the media.

Surveillance studies tend to focus on forms and techniques of surveillance, just as media and cultural studies tends to focus on media products and texts; we have to be very careful about the inferences we make concerning how things are perceived and reacted to by various audiences or publics. As John Gilliom suggests in chapter 5 of this volume, we still know surprisingly little about what everyday people make of surveillance. If there is anything that decades of study of mass media tell us, it is that we must not overread and oversimplify their ideological force. Public understandings of crime and surveillance evolve in a complex interaction of peoples' direct experience, popular wisdom, and broader sets of beliefs about society, and peoples' interactions with their immediate social networks, the media, and other sources of information such as police and politicians (Sasson 1995). In short, media do not simply create public beliefs and attitudes.

Even so, general patterns of crime news and of broadcast surveillance footage in particular – for example, a focus on individual crimes without broader context, on violent crime, on crimes committed by strangers, or by those who are visibly different – reinforce one system of understandings about crime that is fairly pervasive in our culture. Because they interact with this broader set of meanings about crime, the various influences we have discussed here will tend to reproduce the status quo.

A surface adherence to crime control as a rational, utilitarian aim seems bound up with unconscious currents of fear and anger and identification with powerful authority. Some sources have suggested punitiveness involves displacement of anxieties and angers from other sources (Garland 1990; Sparks 1992; Scheingold 1995). If such displacement occurs, it may be acute among those who experience a broader sense of being under threat from various social and economic pressures.

This way of thinking about crime is far from universal, yet the images of crime discussed here fit with pre-existing understandings of certain target audiences. In the American context, for example, there is substantial research evidence indicating that several overlapping categories of audiences may be more inclined to hold a constellation of punitive views that fit with this way of understanding crime. These overlapping categories include heavy television viewers in general, the demographic who are viewers of Fox reality programs in particular, and the 'Reagan Democrats' with conservative views on crime who have come to exert a disproportionate influence on American crime policy because they are swing voters (Beckett 1997: 85, 88).

The status quo is reinforced as new kinds of media products like broadcast surveillance footage interact with older ways of thinking about crime and punishment that are deeply rooted in our culture. Television presents crime in new ways, using new technologies to deliver these stories to new audiences. However, these are often the 'same old stories,' following enduring cultural scripts that predate the advent of TV.

Televising of surveillance footage thus has much in common with conventional media visions of crime. Yet it is a mistake to see the televising of surveillance footage as simply 'more of the same.' The properties of television itself as a medium also play a key role in the shaping of impacts. I will argue that we must turn to a critically reformulated medium theory to understand both key particularities of this phenomenon and its cultural and institutional effects. I will now make some general comments about the properties of broadcast television, especially when it captures 'real' social activity.

First, like any technology, media simply do not have inherent formal properties outside of the social and cultural contexts of production and consumption. These formal properties are not inherent in the medium, as medium theory would have it, but are instead socially and culturally determined. It follows from this that the properties that media take on in particular contexts are shaped by the power relations in those contexts, a key dimension missing from Meyrowitz's (1985, 1994) account, which consequently veers towards technological determinism.

In broadcasting surveillance footage of crimes, TV has the following important properties which shape both its wider cultural effects and its effects on criminal justice processes and institutions:

1) television is emotive,
2) television is embodying,
3) television is epistemologically forceful,
4) television is both collectivizing and individualizing, and
5) television is voyeuristic.

Each of these properties is elaborated upon below.

Television is an emotive medium. As Meyrowitz (1994: 58) argues, 'while written and printed words emphasize ideas, most electronic media [he refers centrally to television] emphasize feeling, appearance, mood ... There is a retreat from distant analysis and a dive into emotional and sensory involvement.'

Surveillance footage in particular has certain emotive aesthetic properties. The immediacy of footage of actual events – seeing 'real' footage of a violent crime or of the last moments of a person about to be murdered – adds to this emotive effect. The crudeness, starkness, and graininess of surveillance video suggest in particular a grim, harsh, street-level 'reality,' evoking a 'gritty realism' which audiences may actually have learned from viewing fictional crime programming (Cavender and Bond-Maupin 1993). The grey and black palette of surveillance footage has a kind of 'film noir' quality. These visual properties fit with a 'common sense' view of crime – committed on dark, 'mean' [i.e., poor] streets at night by strangers – a vision of crime that is so naturalized it may take the critical observer a while to realize that this is a particular, ideological way of understanding crime. Footage from public surveillance cameras of crime on urban streetscapes combines several potent symbols that have come to stand in for broader anxieties about modernity (Sparks 1992): crime, the city, television itself. The footage is just a brief fragment, so the crime, often violent, comes out of nowhere, emphasizing the notion of 'random violence' (Best 1999). Other aesthetic properties contribute: the footage is murky and inchoate, and has a dreamlike or liminal property that makes it resonant, with suggestions of the unconscious. It is darkly disturbing and thus contributes to a vision of crime that is passionate and irrational.

Much of contemporary surveillance has a disembodying quality (see Lyon 2001: chap.1), focusing on personal data in the absence of the body itself. Although televisual surveillance does not occur 'in person,' tele-vision at least visualizes the body and thus cuts somewhat against this tendency. Because of this embodying quality, television and visual me-dia have a particular affinity for sex and violence, and for violent crime in particular. Television is visceral and passionate and thus also has an affinity for passionate punitive sentiments that feature an undercurrent of wishing violence against the body of the criminal. Television culture is saturated with visions of violence against the criminal meted out as informal punishment.

Television is also epistemologically forceful. Different forms of media have different epistemologies linked with them. Like surveillance cam-eras, television, particularly that which uses 'real' footage, often relies on the particular epistemology that 'seeing is believing.' Thus, for ex-ample, the anchor of television's news magazine *Inside Edition* (5 June 1999) described the video camera as 'the truth machine' and stated, 'It never lies.' The notion that 'seeing is believing' is also captured in the point that TV news has greater powers of validation than do radio or newspapers (Ericson, Baranek, and Chan 1991: 23). These increased powers of validation account for the consistent research finding that TV is the most trusted and relied-on source of news, in comparison with print media and radio (24). Ericson et al. suggest that TV news derives these powers of validation from its ability to present sources making their statements directly to the camera in appropriate social contexts that convey the sources' authority.

In part, the truth-telling power of TV resides with those, like police, who are the gatekeepers of what is broadcast. However, this power also falls to a great degree to those who are authorized to provide verbal scripting of TV's images (Doyle 2003). Because we think 'seeing is believing,' TV's actual effect is often to validate the words of those who hold the upper hand. We are told what we are seeing; the powerful players are the ones who tell us.

Surveillance is not simply the capacity to watch but rather the capac-ity to define what is under consideration. The knowledge produced by surveillance cameras always involves interpretation. Those who are authorized to give verbal interpretations of surveillance images – who produce what I will call the 'official definition of the situation' – are the ones who hold the upper hand.

The truth-telling power of the televisual in general is supplemented by the crude underproduced quality of surveillance footage in particu-lar, which gives it a strengthened claim to authenticity. This property works to deny the existence of artifice, suggesting the tapes have come, undoctored, from a 'real' source.

A fourth property of television that is important for my analysis is broadcast TV's property of collectivizing and individualizing. As David Lyon emphasizes in chapter 2 of this volume, broadcast TV shows the few to the many. On the one hand, it reaches a vast and diverse audience and involves them in a collective experience, seeking a common denominator. While this is the case with all mass media, television epitomizes this tendency. Even though the television audience is fragmenting somewhat, nevertheless, compared to print media, the medium of television tends to include large diverse groups of people in the same 'information systems' (Meyrowitz 1985).

This property of TV has implications for the kind of collective constituted through surveillance by broadcast television. Surveillance as conducted through broadcast TV is not surreptitious, a situation of someone watched in isolation by a sole observer, but rather places the individual in front of a huge public. If old-style spectacles of punishment featured an assembled crowd of onlookers, these new forms of surveillance through broadcast TV create a new type of watching collective or public. The social process of informing is now different: once secret, it now becomes a shared, collective, ritualized experience among audiences that reinforces collective sentiments about crime and punishment.

If it reaches a vast audience, on the other hand, broadcast TV singles out what it shows, zeroing in. Particular situations or events that are televised or otherwise mass-mediated thus become 'bigger than life' in various ways (Altheide and Snow 1979). This property is 'anti-actuarial' and has implications for how we theorize control as it occurs through television, as I discuss in the closing section.

A fifth property of television in these contexts is that it is voyeuristic. Voyeurism means getting pleasure from viewing the forbidden, or in viewing without being viewed. Viewing may thus be experienced as an act of domination. Being the watcher rather than the watched puts one in a position of power. Voyeurism may not involve literally watching the subject – one might experience the pleasures of voyeurism through reading a diary without permission, or looking at a secret file. Yet the experience of voyeurism may be most profound when one is literally viewing another person – seeing the 'real thing.' For some audiences the pleasures of one kind of power – voyeuristically intruding into the private – may become entwined with the seductions of another experience of domination – identifying with the authority of police.

Thus far I have focused on the potential impacts of broadcasting surveillance footage of crime on individual audience members. Yet a key point that we can draw from medium theory is that, as Meyrowitz (1985, 1994) and others have argued, the media do much more than simply influence the beliefs and attitudes of individual audience members. Media can affect justice institutions and the justice process more directly in a number of other ways. Possible media impacts can thus be divided for analytical purposes into, on the one hand, the cultural or ideological effects I have been discussing so far and, on the other hand, what we might call institutional effects. Here are six other possible, more particular, institutional effects of broadcasting surveillance footage of crime. A number of these institutional effects also fit with an emotional, vengeance-oriented approach to crime that David Garland has termed the 'criminology of the Other':

1) helping police investigations;
2) net-widening – criminalizing cases that otherwise would not be criminalized;
3) shaming of suspects;
4) trial by media – interfering with or pre-empting formal punishment in these cases;
5 increasing formal punishment in these cases; and
6) fuelling the spread of surveillance cameras themselves.

I have already suggested several examples of how the broadcast of surveillance footage has assisted police investigations, so I will not elaborate further here. Instead, I will point to one key possible problem. The massive publicity given to what I have called 'found television crimes' – otherwise undistinguished crimes that become notorious simply because they offer dramatic footage – presents the potential for an unfortunate kind of net-widening, in which acts are criminalized simply because of the publicity they recieve. Similarly, critics charged that in the case of the Stanley Cup riot, the post-riot surveillance program led to the unnecessary criminalization of numerous individuals long after the fact.

The media can also have a massive informal shaming effect that fits with the 'criminology of the Other.' After the Tennessee nanny mentioned above had already pleaded guilty in court, the footage of her slapping the baby was released to TV news and broadcast widely. That footage became so notorious it will probably haunt the woman for the rest of her life. Her lawyer said: 'It was like taking a sledgehammer to an ant' (*Newsweek*, 22 July 1991: 45).

While the media shaming applied in such cases may be quite punitive itself, television also may influence the formal justice process. Television broadcast of surveillance video may pre-empt the accused's right to a fair trial, resulting instead in 'trial by media.' In the Azscam episode in Arizona, a police sting operation videotaped state politicians taking bribes from an informant. Police released the incriminating surveillance camera footage well before any criminal proceedings could be initiated, prompting resignations and guilty pleas and effectively by-passing the criminal justice system (Altheide 1993).

Another consequence of such crimes becoming heavily publicized is a potential intensification of formal punishment, also fitting with the current trend towards a passionate, punitive, vengeance-oriented approach to crime. Some crimes that are captured directly on video and broadcast seem to draw down extra public and judicial wrath partly just because they are visually dramatic. For example, a group of American teenagers videotaped themselves driving around shooting at passersby with a paintball gun. When this video became available to the media, it captured the spotlight and aired nationally because the dramatic footage looked just like 'real' drive-by shootings. The paintball gunmen subsequently received prison sentences (Doyle 2003). In a somewhat similar case, a sixteen-year-old Nebraskan high school student videotaped a friend beating up one of his peers at school. After the footage was broadcast, the teenaged cameraman was himself arrested for assault for his role in wielding the camera. A reporter with the local NBC affiliate noted the impact the television images had on singling out this situation of schoolyard bullying and making it 'bigger than life' (Altheide and Snow 1979): 'The community in general seems to be pretty shocked about it ... the police don't want to minimize it ... because most people don't know what a fistfight looks like, and they have to see it firsthand ... that is shocking. It's as shocking as the Rodney King beating ...' (*Rivera Live*, 5 June 1996). The mother who beat her child in front of a surveillance camera in an Indiana supermarket parking lot had her child taken away by social services and faced a possible three-year sentence. In the sensational case of the killing of two-year-old James Bulger in Britain, which was high profile at least in part due to the frequent broadcast of surveillance footage of the two-year-old being led away to his death by older children, there was a massive public appeal for harsher sentences. The home secretary eventually intervened in the case, raising the sentences of the ten-year-old perpetrators to nearly double their original length (Young 1996: 126).

Finally, the broadcast of surveillance footage has helped fuel the further spread of surveillance cameras. This is one partial explanation for the rapid spread of CCTV after its initial introduction in Britain: once introduced, CCTV effectively provides its own dramatic visuals for media promotional material (Norris and Armstrong 1999: chap. 4). This is not simply another form of the media's ideological influence on audience members; often criminal justice policy makers react to media coverage rather than actual public input (Roberts 1992; Surette 1998). In other words, officials will simply infer from the fact that the media are spotlighting a criminal justice measure that it will also get public support.

In sum, then, medium theory helps us to understand the influence of television. Yet television as a technology does not, in isolation, cause cultural and institutional changes. The medium is not simply the message. Instead, powerful players often adapt the particular properties of broadcast TV to their own ends, playing off a broader culture that offers visions of crime that also tend to reproduce existing power relations.

Beyond all this, the resulting cultural and institutional impacts of broadcasting surveillance footage of crimes also show us something broader about social control, an alternative tendency that cuts against the main contemporary current, as I will now elaborate.

An Alternative Current in Surveillance and Control

The broadcast of surveillance footage marks an alternative tendency in surveillance and control. It runs counter to the predominant tendency towards 'actuarialism' (Feeley and Simon 1994). Actuarialism is characterized by low visibility, routinized and technicized procedures such as dataveillance, and control of categories of people at a distance. The various impacts here fit instead with an expressive, vengeance-oriented mode of control, which has affinities with Foucault's (1977) notion of 'spectacle' as well as with Mathiesen's (1997) 'synopticism' and Garland's (2001) 'criminology of the Other.'

If actuarialism is epitomized by the computer database, this alternative mode is epitomized by broadcast television. Actuarialism is dispassionate; this mode is emotive. Actuarialism is categorical; this mode involves singling out, is individualized, and has a random property. Actuarialism is disembodied; this mode is physical. Actuarialism is rational; this mode is irrational. Actuarialism is focused on the future and prevention; this mode is focused on the past and punishment. Actuarialism is secretive; this mode is public and collective. If, in actuarialism, 'the few watch the many,' here 'the many watch the many'; if actuarialism is amoral, this alternative mode is moralizing and blaming. If actuarialism is calculative, this mode is non-calculative and features a 'precautionary' (Ewald 2002; Haggerty 2003), less rational approach that reacts to any risk, no matter how remote.

This is not simply a recurrence of old-style spectacle. Of course, spectacle is now greatly expanded across time and space. Spectacle is now combined with surveillance. It is found earlier in the criminal process, not in the formal administration of punishment. Finally, there is another key shift. Public spectacles of punishment were abandoned partially due to negative public reactions: they were dangerously unsettling. There is much more consensus here, more approval around this new kind of spectacle.

Michel Foucault (1977) famously argued that the rise of the modern prison marked a shift in predominant modes of control beginning in the early nineteenth century, from spectacles of criminal justice in which the 'many' saw the 'few,' to surveillance in which the 'few' saw the 'many.' Somewhat in parallel, various authors in different formulations portray a more recent movement in twentieth-century criminal justice and control from a highly passionately charged mode of control to a more dispassionate, technical, and instrumental one, a shift from moral outrage to a more utilitarian morality (see, e.g., Shearing and Stenning 1984; Lyon 1993; Feeley and Simon 1994). In particular, the widespread introduction of surveillance cameras or CCTV is seen by some (Norris and Armstrong 1998, 1999; McCahill 1998) as part of this broad trend towards dispassionate, managerial, technicized approaches to crime at the level of categorically regulating populations (Feeley and Simon 1994; Ericson and Haggerty 1997).

Despite this, it has also become relatively commonplace to point out that the analyses by Foucault and others neglect the persistence of somewhat similar spectacles of punishment in contemporary society. For example, Thomas Mathiesen (1997) sets up an opposition between Foucauldian panopticism, in which 'the few see the many,' and what Mathiesen calls 'synopticism,' in which 'the many see the few,' and argues that Foucault neglects the persistence of synopticism. That is, Foucault ignores how spectacles persist in contemporary society as a complement to the kind of control in which 'the few see the many' (see also, e.g., Thompson 1994: 42–3; Garland 1990: 61, 163; Sparks 1992: 134; and David Lyon, chapter 2 in this volume).

Did Foucault 'really mean' that sovereign power as displayed in spectacles of punishment would be superseded, or simply complemented, by newer forms, disciplinary and bio-power? Certainly Foucault may be read instead as arguing that the three forms now co-exist. Contemporary Foucault scholars have generally focused on the more recent shift from disciplinary power to bio-power, but questions of the persistence of spectacular power also remain worthy of further consideration. Clearly a shift *has* occurred in the visibility of *formal* penal practices, which are now largely invisible to the general public (Garland 1990). However, rather than disappearing, it may be argued that spectacular power has simply shifted in form and location. The examples considered here show how a somewhat analogous kind of spectacular power continues – but it is now sometimes intertwined with surveillance in complex ways, so that these two forms of control may converge rather than exist in opposition (Donovan 1998).

Foucault's neglect of contemporary spectacles of control derives in part from the empirical focus of his analysis: the rise of the prison. However, another crucial shift would take place in Western criminal justice systems around the same time: the birth of the modern police institution. While Foucault's account focused on how punishment was increasingly made private and invisible in the prison, police have always been a much more public and visible institution. As Garland (1990) has observed, the locus of publicity in the criminal justice system has moved since the eighteenth century from the formal administration of punishment to earlier in the criminal justice process. Loader and Mulcahy (2003) argue that sociologists have neglected the broader cultural and communicative aspects of the police institution. As opposed to the dramatization of pre-modern sovereign power described by Foucault (1977), however, police came to symbolize a new kind of power and authority. If spectacle persists in criminal justice, I argue it is now much more the property of the police, as well as occurring in court; the connection between spectacle and the formal administration of punishment has been severed. The spectacle now lies in the crime, the investigation, the pursuit, the trial and sentence, sometimes even in the confession, as in the Court TV reality program 'Confessions' built around police interrogation videotape.

Of course, the panoptical metaphor did not necessarily mean that 'the few' were actually watching 'the many.' It conveyed instead simply that 'the many' did not know whether they were being watched or not, and understood they might be watched at any time. Foucault's use of the panopticon metaphor suggests surveillance was something that was built into a technological system, rather than necessarily being conducted by individuals. In fact, as empirical research on CCTV tends to show (Norris and Armstrong 1999), often there are no 'few' capable of monitoring the 'many,' limiting the surveillance potential of CCTV (at least until the advent of more effective facial recognition technology). It is impossible for a small number of watchers to be so omniscient and all knowing. Instead, surveillance is often embedded in technological systems themselves, as Foucault's reading of the panopticon suggests, rather than conducted by individuals. However, an alternative possible solution to this problem – how only a few may monitor a large population – may lie in broadcast television: the 'many' may be enlisted to watch each other in an informing culture.

The metaphor of the panopticon suggests that each individual is watched in isolation. In contrast, as surveillance has become greatly elaborated as one part of the increasing complexity of modern social relations across time and space (Giddens 1990), surveillance has sometimes become a more public and collective phenomenon. These new forms of surveillance through broadcast TV create a new type of watching public, in which the informing culture will become a shared experience and thus arouse collective sentiments. Surveillance becomes a media ritual (Couldry 2003) conjuring a shared illusion of the social. It moves us towards the electronic lynch mob.

While much of contemporary institutional practice is actuarial, focusing on categorical kinds of regulation at the level of the population, TV is profoundly anti-actuarial. Television exemplifies a countervailing tendency in which late-modern institutions, as they expand over time and space, have an unprecedented ability to zero in on and magnify the individual case: to isolate, single out, and (often) to visually dramatize and spectacularize frightening individual risky cases and situations drawn from huge pools. Because key institutions are now better at finding and communicating these individual cases, the resulting images of risks are more dramatic than ever before. Sometimes they represent the most extreme cases in pools of millions: winning the lottery or succumbing to the flesh-eating disease. At other times, they are cases of risks that are unexceptional but easy to dramatize: somewhat arbitrarily selected, relatively routine crimes for which dramatic visuals are available.

While surveillance tools like the database govern populations by slicing them into risk categories, the opposing tendency towards zeroing in on the individual case always lurks as a latent alternative potential in the database and other forms of contemporary surveillance. Public and private cameras and public informants join with police and media in a 'surveillant assemblage' (Haggerty and Ericson 2000) in which the tools are owned by unaligned public and private organizations with a variety of agendas, each targeting diverse populations. Nevertheless, the assemblage may be drawn together in a tight focus. The potency of the surveillant assemblage is thus in part its ability to single out and isolate. This parallels a tendency of the mass media, which also have a dynamic of singling out individual cases and are highly skilled at dramatizing them. Particular situations that are mass-mediated thus become 'bigger than life' in various ways (Altheide and Snow 1979). This 'zeroing in and singling out' tendency is epitomized by broadcast TV, especially when it uses footage of 'real' events, and particularly in the broadcast of surveillance footage.

Contrasting tendencies in the realm of criminal justice are captured well by David Garland's (2001) formulation, in which a kind of schism or bifurcation has developed in crime control. Garland argues that in the current justice systems of Western democracies there are two alternative ways of thinking about crime control, and correspondingly, two quite different types of strategies for dealing with crime: on the one hand, a collection of rather dispassionate, preventive, technical, and managerial strategies which Garland calls the 'criminology of the self,' and, on the other hand, a set of more expressive, emotionally charged, retributive, vengeance-oriented ways in which the justice system responds to crime, the 'criminology of the Other.' Routinized surveillance of categories of people fit with the former mode of justice; the singling out of particular cases when footage is broadcast fits with the latter.

Garland argues that the punitive public sentiments which fuel the retributive form of criminal justice are especially strong now in Britain and the United States, for various historically specific reasons (2001: 11). These reasons include the current media situation and particularly the rise of television:

Television viewing emerged as a mass phenomenon at much the same time that high crime rates began to become a normal social fact i.e. between about 1950 and 1970. TV's ... affinity for crime as a theme, its sympathetic portrayal of individual victims who have suffered at the

hands of criminals and been let down by an uncaring system, have transformed perceptions of crime and further reduced the sense of distance
from the problem that the middle classes once enjoyed ... This is not to say
that the media has *produced* our interest in crime, nor that it has *produced*
the popular punitiveness that appears as such a strong political current
today ... My point is rather that the mass media has tapped into, then
dramatized and reinforced a new public experience – an experience with
profound psychological resonance – and in doing so it has institutionalized that experience. (Garland 2001: 28–30)

Garland emphasizes how television may have contributed to the
schism in criminal justice by fostering more punitive public attitudes
towards crime, yet he focuses only on the broader cultural and ideological effects of television, and ignores institutional effects. Television has
contributed to an alternative, more expressive mode in contemporary
criminal justice not just by influencing individual audience members,
but also by directly influencing the justice system in a variety of other
ways that move it towards passionate, vengeance-oriented, punitive
practices: shaming, trial by media, intensification of punishment. The
television culture of crime does not simply affect audiences, but feeds
back into the actual day-to-day practices of criminal justice. For example, crimes that receive media attention may be less likely to be plea-
bargained and more likely to be pursued vigorously by media-conscious
prosecutors and judges and punished to the full extent of the law
(Pritchard 1986). Judges may devise spectacular sentences to achieve
media attention (Altheide 1995). Informal media rituals of shaming
develop: for example, the 'perp walk' in which authorities parade hand-
cuffed suspects in front of media so they can be photographed and
videotaped. The broadcast of surveillance footage is also part of this
tendency. As we have seen, particular individuals who are captured on
surveillance footage that is subsequently broadcast may not only be
shamed but then prosecuted and punished with extra vigour.

The concept of 'surveillance' uses a visual metaphor and visual language to encapsulate a variety of activities that often do not involve
literally watching the subject, but rather employ non-visual ways of
gathering knowledge to be used for control. However, medium theory
(Meyrowitz 1994) calls attention to the point that, with the advent of the
video camera, and even more so with the broadcast of surveillance
footage on television, surveillance has not only expanded its reach,
it has become literally visual again. Working through broadcast TV

surveillance takes on a cluster of related properties: for example, it is emotive, embodied, and voyeuristic in a way that may connect it with identification with powerful authority. Surveillance has also become intertwined with spectacle: it is more passionately charged and instrumental, and thereby fuels a retributive 'criminology of the Other' (Garland 2001).

The spread of surveillance is in part a quiet, almost invisible build-up in which practices and techniques of surveillance are unknowingly insinuated into our daily lives, resulting in a kind of 'quiet revolution' (Staples 1997: 128). But consider the alternative example of Fox's reality-TV program *America's Most Wanted*, which often broadcasts surveillance footage of crime. The program's host, John Walsh, enlists the audience's help in identifying the suspects and announces that criminals will be 'hunted down by millions of viewers.' In this alternative current, surveillance can also be a highly visible, shared public cultural phenomenon in which individual cases are singled out for vengeance.

REFERENCES

Altheide, D. 1993. 'Electronic Media and State Control: The Case of Azscam.' *Sociological Quarterly* 34:53–69.
– 1995. *An Ecology of Communication: Cultural Formats of Control.* New York: Aldine de Gruyter.
Altheide, D., and R. Snow. 1979. *Media Logic.* Beverly Hills, CA: Sage.
Baudrillard, J. 1988. *Selected Writings.* Cambridge: Polity Press.
Beckett, K. 1997. *Making Crime Pay: Law and Order in Contemporary American Politics.* Oxford: Oxford University Press.
Best, J. 1999. *Random Violence: How We Talk about New Crimes and New Victims.* Berkeley: University of California Press.
Brin, D. 1998. *The Transparent Society.* Reading, MA: Perseus Books.
Cavender, G., and L. Bond-Maupin. 1993. 'Fear and Loathing on Reality Television: An Analysis of "America's Most Wanted" and "Unsolved Mysteries."' *Sociological Inquiry* 63 (3): 305–17.
Chibnall, S. 1977. *Law and Order News: An Analysis of Crime Reporting in the British Press.* London: Tavistock.
Couldry, N. 2003. *Media Rituals: A Critical Approach.* London: Routledge.
Davies, S. 1998. 'CCTV: A New Battleground for Privacy.' In C. Norris, J. Moran, and G. Armstrong, eds., *Surveillance, Closed Circuit Television and Social Control.* Aldershot, England: Ashgate.

222   Aaron Doyle

Donovan, P. 1998. 'Armed with the Power of Television: Reality Crime Pro-
gramming and the Reconstruction of Law and Order in the United States.'
In M. Fishman and G. Cavender, eds., *Entertaining Crime: Television Reality
Programs*. New York: Aldine de Gruyter.

Doyle, A. 2003. *Arresting Images: Crime and Policing in Front of the Television
Camera*. Toronto: University of Toronto Press.

Ericson, R. 1998. 'How Journalists Visualize Fact.' *Annals*, AAPSS, November
560: 83–95.

Ericson, R., P. Baranek, and J. Chan. 1987. *Visualizing Deviance: A Study of News
Organization*. Toronto: University of Toronto Press.

– 1989. *Negotiating Control: A Study of News Sources*. Toronto: University of
Toronto Press.

– 1991. *Representing Order: Crime, Law and Justice in the News Media*. Toronto:
University of Toronto Press.

Ericson, R., A. Doyle, and D. Barry. 2003. *Insurance as Governance*. Toronto:
University of Toronto Press.

Ericson, R., and K. Haggerty. 1997. *Policing the Risk Society*. Toronto: University
of Toronto Press.

Ewald, F. 2002. 'The Return of Descartes's Malicious Demon: An Outline of a
Philosophy of Precaution.' In T. Baker and J. Simon, eds., *Embracing Risk:
The Changing Culture of Insurance and Responsibility*. Chicago: University of
Chicago Press.

Feeley, M., and J. Simon. 1994. 'Actuarial Justice: The Emerging New Criminal
Law.' In D. Nelken, ed., *The Futures of Criminology*. London: Sage.

Fishman, M. 1998. 'Ratings and Reality: The Persistence of the Reality Crime
Genre.' In M. Fishman and G. Cavender, eds., *Entertaining Crime: Television
Reality Programs*. New York: Aldine de Gruyter.

Fiske, J. 1987. *Television Culture*. London: Methuen.

– 1996. *Media Matters: Everyday Culture and Political Change*. Rev. ed. Minneapolis: University of Minnesota Press.

Foucault, M. 1977. *Discipline and Punish*. New York: Vintage.

Freeman, A. 1999. 'Closed Circuit TV Cameras Have Invaded Great Britain.'
Globe and Mail. 25 May.

Garland, D. 1990. *Punishment and Modern Society: A Study in Social Theory*.
Oxford and Chicago: University of Chicago Press.

– 2001. *The Culture of Control: Crime and Social Order in Contemporary Society*.
Chicago: University of Chicago Press.

Giddens, A. 1990. *The Consequences of Modernity*. Cambridge: Polity.

Gilliom, J. 2001. *Overseers of the Poor: Chicago*: University of Chicago Press.

Gormley, W.T. 2003. 'Reflections on Terrorism and Public Management.'

Governance and Public Affairs Symposium, Campbell Public Affairs Institute, Syracuse University.

Graham, S. 1998. 'Towards the Fifth Utility? On the Extension and Normalization of Public CCTV?' In C. Norris, Moran J. and G. Armstrong, eds., *Surveil-lance, Closed Circuit Television and Social Control*, Aldershot, England: Ashgate.

Groombridge, N. 2002. 'Crime Control or Crime Culture TV.' *Surveillance and Society* 1 (1): 30–46.

Haggerty, K. 2003. 'From Risk to Precaution.' In R. Ericson and A. Doyle, eds., *Risk and Morality*. Toronto: University of Toronto Press.

Haggerty, K., and R. Ericson. 2000. 'The Surveillant Assemblage.' *British Journal of Sociology* 51: 605–22.

Innis, H. 1950. *Empire and Communications*. London: Oxford University Press.

Jones, D. 2004. 'Why Americans Don't Trust the Media.' *Press/Politics* 9 (2): 60–5.

Lawrence, R. 2000. *The Politics of Force: Media and the Construction of Police Brutality*. Berkeley: University of California Press.

Loader, I., and A. Mulcahy. 2003. *Policing and the Condition of England: Memory, Politics and Culture*. Oxford: Oxford University Press.

Lyon, D. 1993. 'An Electronic Panopticon? A Sociological Critique of Surveillance Theory.' *Sociological Review* 41 (4): 653–78.

– 2001. *Surveillance Society: Monitoring Everyday Life*. Philiadelphia: Open University Press.

Marx, G.T. 1988. *Undercover: Police Surveillance in America*. Berkeley: University of California Press.

– 2002. 'What's New about the "New Surveillance"? Classifying for Change and Continuity.' *Surveillance and Society* 1 (1): 9–29.

Mathiesen, T. 1997. 'The Viewer Society: Michel Foucault's "Panopticon" Revisited.' *Theoretical Criminology* 1 (2): 215–33.

McCahill, M. 1998. 'Beyond Foucault: Towards a Contemporary Theory of Surveillance.' In C. Norris, J. Moran, and G. Armstrong, eds., *Surveillance, Closed Circuit Television and Social Control*. Aldershot, England: Ashgate.

McLuhan, M. 1964. *Understanding Media*. New York: McGraw-Hill.

Meyrowitz, J. 1985. *No Sense of Place: The Impact of Electronic Media on Social Behaviour*. New York: Oxford University Press.

– 1994. 'Medium Theory.' In D. Crowley and D. Mitchell, eds., *Communication Theory Today*. Stanford, CA: Stanford University Press

Nock, D. 1993. *The Costs of Privacy: Surveillance and Reputation in America*. New York: Aldine de Gruyter.

Norris, C., and G. Armstrong. 1998. 'Introduction: Power and Vision.' In

C. Norris, J. Moran, and G. Armstrong, eds., *Surveillance, Closed Circuit Television and Social Control*. Aldershot, England: Ashgate.

– 1999. *Maximum Surveillance Society: The Rise of CCTV*. New York: Berg.

Poster, M. 1990. *The Mode of Information: Post-Structuralism and Social Context*. Cambridge: Polity Press.

Pritchard, D. 1986. 'Homicide and Bargained Justice: The Agenda-Setting Effect of Crime News on Prosecutors.' *Public Opinion Quarterly* 50: 143–59.

Roberts, J. 1992. 'Public Opinion, Crime and Criminal Justice.' In *Crime and Justice: A Review of Research*, ed. M. Tonry. Chicago: University of Chicago Press.

Sasson, T. 1995. *Crime Talk: How Citizens Construct a Social Problem*. New York: Aldine de Gruyter.

Scheingold, S. 1995. 'Politics, Public Policy and Street Crime.' *Annals*, AAPSS, 539: 155–68.

Shearing, C., and P. Stenning. 1984. 'From the Panopticon to Disneyworld: The Development of Discipline.' In A. Doob and E. Greenspan, eds., *Perspectives in Criminal Law*. Toronto: Canada Law Book.

Sparks, R. 1992. *Television and the Drama of Crime: Moral Tales and the Place of Crime in Public Life*. Philadelphia: Open University Press.

Staples, W. 1997. *The Culture of Surveillance: Discipline and Social Control in the United States*. New York: St Martin's Press.

Surette, R. 1998. *Media, Crime and Criminal Justice: Images and Realities*. 2nd Pacific Grove: Brooks/Cole.

Thompson, J.B. 1994. 'Social Theory and the Media.' In D. Crowley and D. Mitchell, eds., *Communication Theory Today*. Stanford: Stanford University Press.

Young, A. 1996. *Imagining Crime: Textual Outlaws and Criminal Conversations*. London: Sage.

9 Surveillance and Military Transformation: Organizational Trends in Twenty-first-Century Armed Services

CHRISTOPHER DANDEKER

The concept of surveillance has served as a means of identifying the distinctive features of administrative power in modern society and, from that point of view, refers to a cluster of one or more of the following activities: (1) the collection and storage of information, presumed to be useful, about people or objects; (2) the supervision of the activities of people or objects through the issuing of instructions or the physical design of the natural and built environments; and (3) the application of information-gathering activities to the business of monitoring the behaviour of those under supervision and, in the case of subject populations, their compliance with instructions, or with non-subject populations, their compliance with agreements, or simply monitoring their behaviour from which, as in the control of disease, they may have expressed a wish to benefit. In reflecting on the social forms of surveillance, we should be aware of the variety that may be encountered: (1) the few monitoring the many (panoptical), (2) the many monitoring the few (synoptical), (3) the members of a group monitoring each other more or less simultaneously, and (4) processes of self-monitoring whether or not these are consciously following the agendas of other agents. Furthermore, we should bear in mind whether any of these processes of monitoring are designed to exercise power *over or on behalf* of human communities (see Dandeker 1990: 37–8).

Historically speaking, armed services have been in the vanguard of the development of surveillance as a means of administrative power in society. They have pioneered the means of hierarchical and panoptical forms of surveillance in which political and military elites, the few, have monitored the performance of their subordinates, the many, in order to achieve their objectives: the defeat of enemy forces. Military organizations, past and present, are also characterized by simultaneous and synoptical forms of surveillance, reflecting the fact that their hierarchies also include cooperative, horizontal networks of information, and that at least some of their operations are open to scrutiny by wider civilian society.

In eighteenth-century Europe, the military provided the most developed examples of surveillance with, for instance, the sailing line of battleship being the most advanced, complex machine of the pre-industrial age, requiring sophisticated forms of coordination, discipline, and the world's leading naval professional bureaucracy to manage it to repeated success in war (Dandeker 1990, Rodger 1986). During the nineteenth century, those engaged in new forms of capitalist industrial organization, especially those of a large scale such as the railway companies, looked to the military as a model of administrative control and indeed as a source of personnel to run their concerns: no other entity had the experience of organizing such large-scale, complex enterprises.

Looking back from the standpoint of the twentieth-century one could say that the military, with its system of collective discipline and subordination of the individual to a chain of command, served as a model for other areas of life in terms of administrative power. Of course, individual heroism and collective discipline stand in tension with each other. As Weber argued, while military discipline pioneered discipline *tout court*, the conflict between discipline and individual charisma 'has its classic seat in the development of the structure of warfare' (Weber 1977: 1150; Dandeker 1993).

However, by the end of the twentieth century, as business enterprises sought to make the most of the global economy and the revolution in information technology, an interesting question emerged: was the administrative learning relationship between military and society being reversed? Indeed, with the end of the Cold War, as the Western military sought to adjust to the new strategic environment, it realized that, in searching for new forms of flexibility, it had much to learn from the civilian business sector. This is the setting in which one can situate some of the debates about 'military transformation,' known more grandly as the 'revolution in military affairs' (RMA). Although focused on the United States as the world's leading military power, the RMA has relevance for any other power seeking to exercise military influence in international affairs. As we shall see, in the military environment, surveillance continues to be panoptical. At the same time it has become accompanied by other forms, for example, simultaneous kinds, linking military actors in networks that allow them access to a common picture of the battle space to achieve military objectives. There are also synoptical forms, whereby the media and public opinion can exercise a degree of scrutiny of some aspects of military operations. In the latter case, transparency of some operations is also combined with silence and secrecy concerning others. Thus tensions can arise between panoptical and synoptical forms of surveillance in the military context.

This paper explores the sociological implications of military transformation from the standpoint of surveillance, and analyses the tensions that this process is set to produce within the military organizations of the United States and other (mostly Western) powers that seek to be able to cooperate, or compete, with it. It also evaluates the extent to which military transformation – and not least the increase in surveillance capacity it brings in its train – is set to provide the step-change in military power that its enthusiastic promoters envisage.

The Promise of Military Transformation

Since the end of the Cold War, Western armed services have sought to develop more flexible structures to enable them to respond to the more uncertain strategic environment (Dandeker 2003). In that context, those using the term 'flexible forces' refer to armed services configured as follows: they are equipped with the appropriate hardware, force structures, and personnel policies to enable states to respond swiftly, in collaboration with allies and/or friends bonded in 'coalitions of the willing,' to a wide variety of crises, the precise nature of which is quite difficult to predict in advance. Consequently, military responses to such crises will increasingly have to be configured in 'force packets' drawing on a range of military elements to meet the particular needs of a specific crisis in ways that echo the long-standing distinction made in the world of business between mass and customized production (see Dandeker 1999, 2003; Boene et al. 2000; Segal 1993; Moskos, Williams, and Segal 2000; Kuhlmann and Callaghan 2000). Indeed, a key theme in organization theory since the influential study by Tom Burns and Graham Stalker (1994) has been the functional linkage between uncertainty in the business environment and the benefits attending the adoption of flexible organization structures capable of being agile and responsive to unforeseen shifts in market conditions.

When considering recent developments in military organization, from the point of view of flexibility, especially since the end of the Cold War, one is struck by parallels between the drivers of change facing armed forces and those encountered by private sector organizations. There are five key drivers, discussed below:

(1) The end of an immediate state-based and direct threat to national territorial sovereignty is paralleled by the lack of a stable market for business. We live in a more uncertain world, at least when compared with what may now be regarded as the relative certainties – indeed, even relative security – of the bipolar Cold War era. One must be cautious about drawing too firm conclusions from the terrorist outrages of 9/11 in the United States, which threaten states and values far beyond the boundaries of the state in which they were committed. Yet it does seem to be the case that these events have added further impetus to the perceived necessity of advancing the organizational trend of flexibility in the armed services.

(2) Military establishments such as those in the United Kingdom are at the lowest level since the Second World War, paralleling the general course of company downsizing since the 1980s. Over the past two years, while the United States has increased its defence budget significantly, Europe has yet to do so, although modest increases have occurred in the United Kingdom (Dandeker and Freedman 2002). While this may lead to increases in the numbers of uniformed personnel in the military establishment, the focus remains on exploiting the potential of technology: the balance of investment between capital and labour will remain in favour of the former, although, as witnessed in Iraq, the importance of 'boots on the ground' cannot be underestimated.

(3) The military has to address a range of missions involving operations other than (and in addition to) major war (OOTW), namely, interventions abroad in multinational contributions to international peace and stability. This focus on the projection of force to dispersed points on the globe from a home base parallels the ways in which companies have to respond to increasingly global markets.

(4) Armed services are having to think through the possibilities offered by the application of business models to the military, such as contracting out – or 'outsourcing' – of functions, restructuring of hierarchies, and so on, processes that echo civilian developments in the empowerment and restructuring of companies.

(5) Both sets of organizations are seeking to make best use of the new information technologies in enabling them to achieve flexibility and a competitive edge over their rivals. This is evident in all aspects of organizations, from the personnel functions (such as the administration

of pay, personnel records, and so on) to logistic, support, and operational areas – not least in the offensive and defensive aspects of 'information warfare,' which in turn connect with the broader phenomenon of military transformation and with the dynamics of surveillance in the contemporary military.

Discussions of military transformation in military and academic circles have been triggered by technological and strategic developments. First, there is the promise of information technology in terms of surveillance and coordination, together with the technical capacity to deliver precision-guided munitions (PGMs) to targets from long range. Second, there is a strategic imperative for the United States and other military powers to respond – alone or with allies or friends in coalitions of the willing – quickly to crises in an agile fashion, with forces equipped for war fighting, peace keeping, or missions that lie at some point between them (for good discussions see Freedman 1998: 11–32; Hirst 2001: 88–101).

As one commentator remarked recently, 'Advocates of military transformation have argued for years that the same information-technology revolution that so changed business could radically change warfare. With better surveillance and improved information technology, military commanders would have better information about the enemy's position, lifting what has been called the "fog of war."' (Jaffe 2002 R5; see also Freedman 1998, 11–13; Owens 1995). Another has observed, 'Since the gradual demise of the Soviet Union, certain scholars of combat had been arguing that the great lumbering military machine constructed for the cold war was stubbornly ill suited to the new threats of a disorderly world and slow to exploit the new technologies of the information age' (Keller 2002: 32).

A number of distinct, but interdependent themes are encapsulated within the idea of 'military transformation': the surveillance and information revolution should be regarded as one aspect of a broader cluster of developments. For convenience, these can be grouped in five broad, interdependent categories.

First, there are changes at the level of institutional organization, which include the modularization of the all-volunteer force in 'force packages' (on a mix-and-match basis depending upon the nature of the mission) together with an emphasis on 'jointery,' that is to say, an increased focus on interservice integration. In addition, interservice integration involves units drawn from the armed forces of different countries, posing further issues of 'cultural interoperability': how to ensure effective cooperation among different national traditions.

The task of building force modules has stimulated debates over whether such elements are likely to be more on the 'light' rather than 'heavy' side of military equipment: for example, light infantry and Special Forces working in close collaboration with airpower (as occurred, successfully, in Afghanistan), or whether such forces need the extra power provided by heavy armour (which in turn imposes some constraints in terms of rapid power projection). Such debates were very much to the fore in discussions about the evolution of U.S. war planning for the invasion of Iraq in 2003.

A second feature of military transformation is an emphasis on the use of panoptical and simultaneous surveillance to apply destructive power with fine calibration: information is gathered from networks of sensors and advanced communication systems to target munitions with great precision. For example, one of the differences between the first and second Gulf Wars was a decline in the proportion of dumb to smart munitions that were used, which is part of a longer phase of development since the Second World War (see United Kingdom 2003: 22–3).

Third, great effort is being expended on what has been called 'information superiority.' This provides the basis for '[d]ecision superiority and increased tempo through capabilities such as operational net assessment, common relevant picture and joint intelligence, surveillance and reconnaissance' (JDCC 2003: 8–14).

Advocates of transformation have been enthralled by the prospect of the combination of precision weapons and information superiority underpinning a 'new sort of fighting force able to zip intelligence from manned and unmanned sensors scattered around the battlefield to planes tanks and artillery pieces in a matter of seconds. "Our military must be able to identify targets by a variety of means – from a marine patrol to a satellite – then be able to destroy those targets almost instantly" Mr. Bush said at the Citadel' (Jaffe 2002: R5, referring to a speech by President George W. Bush – see below).

A key point here is that, in the targeting process, many vertical and lateral points in the command and control system have access to the *same* information picture. This complex linkage of panoptical and simultaneous forms of surveillance is an echo of the idea that surveillance should be construed not only as a means of the centralization of power but also as an assemblage of dispersed processes, which enable the coordination of a variety of actors operating in a network who are all able to access a common picture of the battle space (Haggerty and Ericson 2000).

A fourth element of military transformation has been termed 'full dimension protection,' which involves, 'avoiding close combat where possible and ... minimising operational risk by ensuring sufficient superiority of firepower and tempo to overwhelm the opponent comprehensively, so avoiding any possibility of matched combat' (JDCC 2003: 8–14).

This point raises a number of interesting issues. Emphasis is placed not just on people and platforms but on *networks of information* in military organization. Such networks are designed to produce not only accurate information but at great speed, which in turn allows for the compression of the *decision cycle*: this is the time it takes to collect intelligence about threats, identify potential targets, decide on courses of military action, implement them, assess their effectiveness, recalibrate action if required, and then repeat the cycle. The capacity to do this again and again at great speed is what is meant by tempo, and is especially effective if an enemy is attacked at a number of different points simultaneously, with all the effects of shock and surprise that such attacks engender. One important issue in all of this is whether an opponent will be able not only to avoid being placed in this position of vulnerability, but to inveigle an armed force designed on the basis of military transformation into forms of close combat in which the strengths of the players are more evenly matched, a point taken up in the next section.

Tempo can only be effective if the military system can deliver 'smart' munitions, which has implications for defence procurement and logistics, the fifth aspect of military transformation. The shift from dumb to smart munitions requires a restructuring of the defence industry. This is more of a challenge than would appear at first sight, because all aspects of military transformation need to be served by an industry that has the capacity to stay at the cutting edge of all military-related technologies and thus ahead of potential competitors. And the challenge is easier to set than to meet: technology changes significantly every eighteen months, yet on past experience the U.S. defence acquisition process, for example, takes ten years to change, posing a serious barrier to military transformation (Jones 2002).

Proponents of military transformation have envisaged a more 'focused' approach to logistics, designed to facilitate a more agile and effective projection of power (JDCC 2003: 8–14). In some ways this echoes systems of 'just in time' delivery practised in the civilian business sector, with the military advantage of lessening the dependence of the armed services on an unwieldy logistics chain. The more operations are organized jointly, the more possible it becomes to structure logistics on a tri- rather than a single-service basis. Every choice, however, has its costs. Should the light, agile approach to operations falter, supply chains can appear flimsy and unsustainable in their lightness, indicating that the logistics dimension connects with the earlier debate over how far armed services facing specific missions should be designed on a light or heavy basis. There are, therefore, both advantages and disadvantages to having an armed force that would be more 'agile, lethal and readily deployable' and able to fight wars on foreign territory with a 'minimum of logistical support' (Jaffe 2002). This writer was referring to President George W. Bush's now famous, but at the time rather ignored, speech at the Citadel, when he talked not only about military transformation but also about the foreign and security policy it was designed to serve: a more pre-emptive war fighting posture (see Bush 2001).

Four Limits on Military Transformation

Military transformation, in terms of the five overarching themes set out above, is set to define the evolution of the U.S. military – as well as that of the armed forces of less powerful states that aspire to be able to cooperate with it – for at least the next decade. However, military transformation is not something that just happens: it is a choice, and choices are made by states within the constraints set by domestic and international politics. These constraints are the first limit on the potential of military transformation.

Much depends on what states desire and consider being in their interests together with what capacities they already possess. Whatever doubts that other Western countries have or have had about current U.S. policy responses to 9/11, whether concerning Afghanistan or 'rogue states' such as Iraq, it is still in their interests for them to develop the capacity to be able to cooperate militarily with that country. Indeed, this is also in the interest of the United States because, notwithstanding its sole superpower status, it cannot do without the material, political, and symbolic support of its allies: it does not especially want to go to war alone for material as well as legitimacy reasons.

However, a major issue is what form that collaboration should take, and this has become much more complex as a result of the war in Iraq. Some states, like the United Kingdom, continue to see it as in their interest to be able to cooperate with the United States as a relatively small, junior military partner in combat operations. Cooperation brings some degree of influence over the United States, while enabling the United Kingdom's security to be protected from global threats.

Meanwhile, other states may decide that, for reasons of cost and political preference, they would prefer not to engage in forms of military transformation that would allow participation in military-led operations. Yet they might welcome being able to engage in activities that would enable them to assist the United States in ways other than lending political and diplomatic support, for example, providing bases, over flight access, medical services, and other specialist functions such as Nuclear, Chemical, and Biological (NBC) decontamination. Alternatively, they may wish to undertake missions such as stabilization and reconstruction after military operations have been completed, which some might view as a suitable broad division of labour between the United States and European powers, one that makes sense in terms of practical economics as well as political preferences (JDCC 2003: 8–11, 8–12). This process would allow national and historical differences as to where comparative advantage lies among national members of the NATO alliance, the European Union (EU) and 'coalitions of the willing' to come into play. For new as well as older members of NATO and the EU, there will be delicate issues of choice to be handled in regard to war fighting, peace support operations, gendarmerie functions, and other military roles (Dandeker and Freedman 2002: 469).

Other states, not least France, see this approach as a de facto role specialization *and* subordination to U.S. 'hyper-power.' They seem set on establishing the EU, led by the Franco-German axis, as not simply a counterweight to the United States in political and diplomatic terms, but also, in due course, in military terms as well. Yet whether the EU has the political cohesion and will to commit resources to defence to close the major technological gulf with the United States and thus to establish a Western but non-American regional power based on military transformation – a continuation of French strategic dreams from the past – remains debatable.

Three of the significant consequences of the war in Iraq are likely to be as follows: (1) the institutional expertise acquired by coalition forces will increase the gap in military capability between them and other states such as France and Germany; (2) this, in turn, will reinforce U.S. preferences for coalitions of the willing to provide political support and legitimacy for U.S.-led military operations but for the operations themselves to be conducted by a far smaller group comprising coalitions of the capable; and (3) in order to cooperate militarily with U.S.-led coalitions it is likely to be easy for very small and sympathetic powers (those in 'new Europe,' for instance) to identify and develop a niche capability that can dovetail neatly as a subordinate element in such a coalition. However, this approach will be much more difficult for larger powers such as France and Germany which – especially the former – baulk at being confined to such a role specialization within a U.S.-dominated framework. Yet the EU alternative looks quite flimsy both in economic terms and in respect of political cohesion, as pointed out above.[1]

A second limit on military transformation is that a focus on prevention by agile deployable forces abroad could lead to major problems if states take their eyes off defending the homeland. In adding momentum to the trend towards flexible, quick reaction forces with global and precise reach, 9/11 has also given states much to think about in terms of the balance between expeditionary operations and homeland defence and security. To take the case of the United Kingdom, in its Strategic Defence Review of 1998 the primary role of armed forces was conceived as 'going to the crisis before the crisis came to you,' which became a key theme in the current Labour government's foreign policy (see Centre for Defence Studies 1998, 34). Yet some modification of this kind of thinking is in order in light of the events of 9/11: the crisis can come to you before you go to the crisis. Indeed, even before these events, some commentators felt that if a state were to send its forces abroad to prevent a crisis, those hostile to the action might mount some kind of terrorist action within that state's homeland. September 11 has brought this scenario into stark relief. Consequently, defence planners now have to think even harder about 'securing the home base' while projecting power abroad or threatening to do so to deal with crises (Hoon 2002). This dual approach to security means that not only regular but also reserve forces must be organized and distributed appropriately to meet both power projection and homeland defence needs. In the case of reserve forces this may mean a reversal of past policy of force reductions. Against this background of the need to balance expeditionary efforts against the need for homeland defence, in the United States and other countries greater priority is being given to homeland defence. In the case of the United States, this means a twin effort; maintaining a technological edge in the military means of power projection abroad as

1 These points have emerged in conversation with Professor Peter D. Feaver from the Department of Political Science at Duke University, North Carolina.

well as in the means of detection and prevention of intrusions into the homeland.

Focusing on forces used abroad, there are other problems that need to be addressed. As mentioned earlier, those who have argued in favour of military transformation have been enthralled by the idea that 'with better surveillance and improved information technology, military commanders would have better information about the enemy's position, lifting what has been called the "fog of war"' (Jaffe 2002: R5).

However, a third set of limits on the potential of military transformation derives from the fact that the technology can lessen but not facilitate an escape from either the fog or the friction of war. This is not simply a technological issue but rather a result of the necessary part played by fallible human beings in the most 'high tech' of military operations. Mistakes are made by personnel even when aided by the best available equipment, as is illustrated by the occurrence of friendly fire or 'blue on blue' incidents, and by civilian casualties when smart munitions go astray (leaving aside the question of an opponent seeking to ensure such casualties occur by placing military targets close to or among civilians). As observed earlier, given the focus on increasing the tempo of operations, the strains on human fallibility must increase as personnel seek to cope with the sometimes bewildering number of hostile and friendly elements in and around the modern battlefield.

The fourth limit on military transformation concerns what might be called 'operational complexity.' In the first place, the idea might encourage overly optimistic judgments about the ease with which 'high tech' armed forces will be able to prevail over an opponent, together with an underestimation of the latter's capacity to use asymmetrical responses. The concept of military transformation can reinforce ideas based on a risky over-specialization in light agile forces assisted by airpower. Thus what appeared to work well in Afghanistan – and operations there are still ongoing – may work less well in other scenarios, such as Iraq, or only work in a campaign when the opponent's will to resist is weak or based on a preference to resist in the post-conflict phase through irregular operations. In that case, as in Iraq, there is no substitute for a large number of 'boots on the ground' to serve as a buttress for a stable, post-conflict, and reconstruction phase.

To be fair, for political reasons, it may well have been impractical to inform the public about the true scale of effort required to prosecute a successful war against Iraq. This point is rather different from the one made in the media that no plans had been made for this scenario: the media made too much of the idea that the military and political elite in the United States believed that light forces would 'do the job easily.' Furthermore, many of the thoughtful proponents and practitioners of military transformation have recognized this challenge from the outset. Indeed, they seek to focus on how military transformation can itself provide an asymmetrical counter-threat to the more publicized asymmetries deployed by states or non-state actors that are, in conventional military terms, outclassed by the United States. For example, these would include terrorist acts like suicide bombing and/or guerrilla tactics – including using the media to play on the sensibilities of public opinion – to disrupt and divert a conventional military from its prime objectives.

Asymmetrical strategy is thus *not* the sole property of the 'underdog,' and military transformation can serve as a counter to the goals of those who seek to subvert the strong by avoiding their favoured landscape: this is what Iraqi forces sought to do by learning from their errors in 1990–1. However, Iraqi tactics could be countered by intelligence, Special Forces, PGMs, and airpower, and they delayed rather than avoided the final outcome of defeat, even though much media reportage tended to convey a more pessimistic message, if one takes the standpoint of coalition forces after the first two weeks of the war. The key problem lies not in the defeat of Iraqi forces but in replacing the regime by a different one that functions effectively, and ensuring that the tactics of armed resistance focused on inflicting a steady stream of U.S. casualties do not sap the coalition's will to prevail in its mission of reconstruction.

One lesson from all of this is that there is a serious risk in viewing light and agile versus heavy as an either-or choice: it is possible to integrate both within a military based on transformation, especially from the point of view of information superiority, precision firepower, and command and control. Then, based on a flexible posture, it would be possible for a state to use a modular approach to produce a mix of light and heavy forces depending upon the needs of the operation: sometimes this will involve relatively small numbers of Special Forces – some on horseback(!) – with the support of airpower and high technology surveillance and reconnaissance.[2] On other occasions, a broader

2 Note that the combination of Special Forces and airpower in Afghanistan only worked to the extent it did because substantial ground forces were provided by the Northern Alliance. Airpower with or without Special Forces can achieve only modest results unless backed up by 'boots on the ground,' *or the threat of them*, as was proved in Kosovo.

panoply of firepower, including both light and heavy armoured forces, would be appropriate, as in the campaign in Iraq.

There is a second aspect of operational complexity, which has two different strands. First, prevailing militarily over an opponent does not exhaust the missions that the armed services need to undertake: force with other levers can be used for a wide range of purposes, including stabilization and reconstruction, as in Afghanistan and Iraq. Second, while the first Gulf war was the last war of the twentieth century, the war in Iraq was the first of the twenty-first. Here was a mission intended to combine destruction and a degree of 'shock and awe' achieved through a sophisticated blend of force and information operations with reconstruction, reassurance, and regime change. It was and remains a quintessentially complex operation, some might say postmodern in its fragmented narrative and in the way in which the protagonists seek to convey their image of what the war is and means through the world-wide media. It is worth dwelling on the political implications of this strand of operational complexity.

While the largely U.S. literature on military transformation has focused on new ways of prevailing over an opponent, relatively little attention has been paid to these political issues. The idea of political complexity highlights continuities and discontinuities in the use of armed forces. While the purpose of the military lies outside of itself – to achieve the political goals set for it by its client, the state – the way in which it does so depends upon the historical context. The division of labour between political ends and military means and the separation, through political filters, of the spaces in which each of these actors operates has become more and more blurred in modern warfare (Dandeker and Gow 2000: 70–3; Prins 2003: 221–3). As will become apparent below, the factors that have caused this process of blurring include: the creation of more complex and ambiguous end states at the political level; the political imperative to monitor lower-level military decision making and tactics in light of the changing logic of the political as well as military situation, including a desire to limit both enemy and home side casualties; the ways in which public opinion and the media are able to scrutinize some aspects of military operations, which in turn leads to politicians being reluctant to consider the space of military affairs as outside their purview; technological developments, which not only facilitate such scrutiny but also increase the pace of political and military events; and finally, the ways in which technologies of military power and communications facilitate the compression of the levels of war – that is to say, allowing small-scale tactical events to have major strategic consequences, as when individual military personnel become entangled in highly publicized events such as the shooting of civilians and other aspects of 'collateral damage.'

The United Kingdom's Chief of Defence Staff gave a speech in which he reflected on the military and political dynamics of what has been called 'achieving effect,' and the links with contemporary military surveillance capacities. He argued that a key priority for the armed services is their

> need to move towards effects-based campaigning by undergoing transformation and developing a Networked Enabled Capability. (Boyce 2003: 30)

> Crucial to effects-based campaigning, the Network Enabled Capability intends to link sensors, decision makers and weapons systems so that information can be translated into synchronised military effect. (Ibid., 36)

Emphasis is placed not only on the main features of military transformation, discussed earlier, including the need for a joined-up or service-integrated military, but also on the use of *asymmetrical responses to asymmetrical strategies*: for example, the use of propaganda through information operations and disruption of the opponent's communications with precision weaponry, including their own propaganda efforts.

Interestingly, when referring to quite modest-sized U.K. operations such as in Sierra Leone or ISAF in Afghanistan, relative to the size of force deployed, as in ISAF, for example, the size and significance of the HQ deployed is much more substantial than would have been expected in the past. This is due to the political sensitivity of the operation and the need for command and control of the political effect to be achieved. All concerned with making a success of operations today must be able to work in complex, fast-moving environments. Yet this raises the question of how far such precisely focused force requires the discretion of the soldier on the ground, and how far that individual is the problem or the solution to the ways in which military transformation might overcome the limits of operational complexity. Here, as discussed below, different forms of surveillance are at work in military systems and can give rise to organizational and wider tensions.

Surveillance and the Dialectic of Control in the Military

Military transformation needs to be placed in its historical context. It might be viewed as significant a turning point as previous waves of

change in the field of warfare: the use of fighting columns in Napole-
onic warfare; the mass industrial armies reliant on artillery that fought
in the First World War; and the war of manoeuvre pioneered by the
Germans in the build-up to the Second World War and taken further –
indeed to its apex – by the United States with its military success in the
last war of the twentieth century as it defeated the Warsaw Pact –
equipped and trained Iraq forces in the open desert.

From the standpoint of the individual soldier, through these successive waves of change there has been a long-term process of dispersion
of military authority to lower levels of the command chain.[3] This dispersion has been driven by developments in war and military technology, together with the development of the citizenship state. Thus, as a
result of the importance of the consent of the governed to the legitimacy
of political elites, persuasive forms of authority and 'group consensus' –
to use Morris Janowitz's formulation – have become significant features
of the military command system (Janowitz 1960). Yet this system re-
tains coercive and hierarchical elements that mark it out from civilian
systems. These retained elements are due to the functional imperatives
of a war-fighting organization, but also reflect the residue of tradition.
How much further the military can be made even flatter and more akin
to some civilian business organizations and lose its coercive elements
remains a moot point.

Military transformation is leading to an increase in the speed of
information flow, and thus to a major expansion in the quantity of
information that military and political elites have to consider when
formulating decisions. It has also made more acute the problem of
determining the quality of this information: what is true, who can be
believed, how does one respond to propagandist uses of information on
TV or the Internet? This adds to the pressures on political elites, their
advisers, and military commanders. In the fast-moving events of modern war, leaders have less and less time to digest the increased information at their disposal, to assess its quality, and to make decisions. These
decisions and their outcomes will be scrutinized just as quickly, thus
adding to the 'telescoping' pressures. The performance of NATO in
Kosovo, and even more of the coalition forces in Iraq, are testament to
this process.

Dispersion is, however, connected with a counter-development: a
tendency towards the centralization of control and what has been termed

---

3 I would like to acknowledge Bernard Boene, with whom I have developed some of
the arguments in this section.

the micro-management of military operations. These two conflicting trends constitute in the military context what might, following a formulation of Anthony Giddens, be called a 'dialectic of control' (Giddens 1977). The drivers of this counter-trend are first, the new technologies of communication which provide the means of installing systems of micro-management. Second, political leaders and their advisers are increasingly aware that quite small scale events at the tactical or sub-tactical level can have a major impact at the strategic level. Such events are likely to have both military and non-military dimensions, for example, involving implications for refugees, human rights, and relations between military and other organizations. In addition, the media that report the events magnify their impact by emphasizing their likely consequences through often graphic images. Third, as mentioned above, the very tempo of events, and the consequent need to adapt strategy and tactics in the light of fast-moving situations, or at least to evaluate the need for such adaptation, leads to a tendency to increase knowledge and control at the centre.

The dialectic of control – that is, between dispersion and micro-management – leads to a compression of the three main levels of war: these are the strategic (where the political and military objectives of the overall campaign are established), the operational (where a specific campaign is conducted in a 'theatre of operations' and within the framework of strategic plans), and the tactical (where units and higher formations or combinations thereof of the armed forces carry out plans by engaging the enemy). This is despite the fact that, objectively, the complexity and pace of events indicate a need for dispersion. Thus, the fact that the lowest level events can have major consequences encourages the political centre to become 'control freaks.' This is especially so when they feel under pressure and become aware of the need to, for example, keep public opinion, as well as the leadership and publics of potentially wavering allies, committed to the operation. The 'control-freak' tendency can also lead to tensions within the top echelons of the military command system at the planning, deployment, and operational phases of military activities. That is to say, the political level might draw on a narrower rather than a broader range of military expertise in formulating its decisions (e.g., in the case of the United Kingdom, turning to the chief of defence staff and not the individual service chiefs). It might narrow its circle of advisers and thus lessen the chance to hear critical but constructive criticism of plans. Finally, it might seek to manage operations in such a way that the military chain of command extending from the operational level to tactical matters is subjected to political monitoring. One possible type of monitoring is target sets, for example, in terms of sensitivities about civilian casualties, especially in complex operations that combine destruction and reconstruction as in Iraq. Another example is the minimum altitude of military aircraft.

The inherent tensions within the dialectic of control can only be mitigated satisfactorily by trust- and confidence-building measures being installed at the political-military interface. This points to the need for appropriate education and training for personnel on both sides of what has been and will continue to be a blurred dividing line (on fusionism see Huntington 1957, 1961; Janowitz 1960; Perlmutter 1977; Feaver 1996; Luttwak 1999; Mackenzie (Maj. Gen.) 1993; Sarkesian 1998; see also *Orbis* 1999; Davis 1996; Avant 1996; Snider and Carlton-Carew 1995; Barnes 1996; Betts 1977; Danopoulos and Watson 1996).

At the lower (even lowest) levels of chain of command, soldiers need to be politically aware of the broader framework in which their actions occur. In particular, they must be able to place the objectives of an operation in the context of the contingencies of the situations they confront. As an ex-NATO commander put it, 'the ordinary soldier has to be educated to understand that his actions can have as large an impact on events as Madeleine Albright.'[4] Observation of the conduct of young soldiers at checkpoints in Iraq under media scrutiny and the global repercussions of events at Abu Graib prison leads one to concur with this statement. This contextualization can only occur through trust and a doctrine of mission tactics. Successful application of these tactics requires all levels to appreciate the doctrinal basis on which they depend. In addition, the highest political and military levels need to be made aware that, understandable though it may be for them to press for the centralization of control, the logic of the situation points to the need for dispersion. By the same token, higher levels must recognize the damage that can be caused by second-guessing those situations from locations far removed from the action on the ground. It is here that Moskos's formulation of the 'soldier-statesman' and 'soldier-scholar' becomes pertinent (Moskos, Williams, and Segal 2000; Boene and Dandeker 1998).

The soldier-scholar is required to think through the conditions for applying force in the new security context: for example, in those opera-

---

4 In an off-the-record briefing to the author.

tions which lie midway between classical peacekeeping operations and all-out war, where the defeat of an enemy is sought (Dandeker 2000). The most likely military operations probably lie at this midpoint; experience and doctrine were relatively undeveloped here, although much has been learned during the last few years (Gow and Dandeker 1995; Dandeker and Gow 1997; Williams 1998). The complex mission of regime change and reconstruction in Iraq provides further related experience.

The more military operations become politically complex and militarily unconventional, the more compelling this argument becomes. In set-piece battles in high-intensity war fighting, the pressures on commanders and personnel at unit and higher level formations, such as brigades and divisions, can be intense, as they seek to manoeuvre their forces in order to fight and defeat the enemy. For the lower sub-units involved – battalion and below – so long as they do their jobs, their work is relatively straightforward. They follow set procedures and routines in which they have been thoroughly trained. However, in more complex, 'broken backed' operations – let us say around Basra, compared with U.S. encounters with the Republican Guard near Baghdad – the lower formations face a rapidly changing operational and political environment in which success depends upon creative responses to novel problems thrown up by the situation. This is not to say that one kind of operation cannot shift into another, or that political effect is not the desired outcome in the latter, as we witnessed with the U.S. attempt to take Baghdad airport. It has strategic value, but its capture conveyed a political message. In doing so, at least three considerations were at work: forcing military defenders to give up, winning the hearts and minds of Iraqis, and keeping the support of both world opinion and the public in coalition states. In all of this, a number of interesting questions arise: What will be the balance between these two modes of warfare of the future? Which national military cultures are most attuned to their demands? Is it possible and desirable to train the same formations for both modes?

The soldier-scholar's role is promoted not only by new strategic circumstances, but also by political and technological conditions. With the decline in the military experience of the political elite, both inside and outside government, politicians are less well versed than they used to be in the conditions under which force can usefully be applied in pursuit of security policy. Yet the situation is complicated not only by the need to deal with new types of mission, but also by the effects, alluded to above, of the revolution in communications. Thus it is increasingly risky to give the armed forces missions without the appropriate means or to use technology to micro-manage operations: the consequences can undermine the operation as well as civil-military relations.

The new security climate has promoted another professional military role: that of the soldier-statesman who is adept at handling the media and international diplomacy. Political skills are becoming increasingly important. In connection with the 1990–1 Gulf conflict, for example, General Sir Peter De la Billiere remarked on: 'one of the basic principles of high command, which I was learning as I went along: that a senior commander must bring together everyone concerned, not only in theatre, but outside as well, and that often he must act almost more as a diplomat than as a soldier.' (De La Billiere 1992: 104)

The soldier-statesman's role is becoming more significant because of the political complexity of coalition warfare. This is especially the case in missions where threats to national interests fall well short of the threat to national survival that characterized war planning in the Cold War. Also, as mentioned above, the pace of events and their reporting made possible by the modern electronic media telescopes the decision time available to political and military decision makers. Much closer cooperation between them is required, and the result is a blurring of the divide between political and military skills and a challenge to traditional ideas of the military professional as an apolitical technician. Finally, because of the delicate nature of a mission, mandates may well change during an operation. Again in such a case, military commanders must be politically sensitive to the rapidly evolving diplomatic context. For example, although it is reasonable for the military to request clear objectives and rules of engagement, it is now increasingly unrealistic to ask that these change as little as possible while an operation is in progress.

The soldier-statesman occupies a key role in the processes whereby contemporary military operations are observed by political elites, the mass media, and the public. With the integration of the media into the very texture of the battle space, the objective of the armed forces is not only to use force to achieve a precise effect but to use information to convey a particular message: to enter into the opponent's decision-making loops and to undermine the enemy's will to resist; to convey a sense of success and momentum of the mission to domestic and international audiences (see Prins 2003 for an analysis of these processes in U.K. operations in Sierra Leone). In all of this, the soldier-statesman must be able to move with subtlety and agility up and down the levels of war.

These processes involve a complex dynamic between panoptical and synoptical processes of surveillance (Lyon, chapter 2 this volume; Hier 2003; Mathiesen 1997). The media and public opinion are able to access in real time some aspects of military operations in ways that are dramatically different from earlier periods (for example in the Crimean, Boer, or even Vietnam Wars). Here, the many survey the few; and as we have seen in respect of the compression of the levels of war, this is why the few at quite low levels can have such an effect at the higher levels. At the same time, political and military elites seek to ensure that the public and the media are 'on message': that the image and narrative of the war conforms to their logic, and that the many follow the view of the few. We may refer to two examples of this phenomenon from the recent Iraq war. First was the way in which coverage of the events around Baghdad and Basra was matched by a complete absence of news about highly significant but secret operations in the western desert. Second, the action of U.S. soldiers in symbolically placing an American flag over the head of the statue of Saddam Hussein that was about to be pulled down, only for them to be ordered, from what point in the chain of command is unclear, to take it down. Embedded journalists, as they experienced in Iraq, occupy an often difficult position in terms of the conflicting forces of panoptical and synoptical surveillance.

Conclusion

Military power has been a major focus of surveillance during the development of the modern state. It is a key component of the nation-state as a power container, and the armed services have constituted a site from which surveillance techniques and practices have spread to other social contexts. Many of those who have discussed and practised military transformation are using models from business enterprises to describe the recent past and to project a course of development for the next fifteen years or so. Military reliance on lessons from the civilian business world leads one to ask whether the future will see a process of two-way learning or whether, perhaps, the balance will shift again, with business seeking to learn new ways and means of conducting its affairs from the military context.

The process of military transformation raises the familiar question of the role of the individual in the bureaucratic machine. Transformation produces a 'high tech' surveillance machine. For the soldier on the ground and his or her superiors engaged in the dialectic of control, there is a need for great skill in managing the complex environments of contemporary military operations as well as for discretion and trust, despite the technical means that are available to micro-manage activities. And no matter how well that machine works, there will also be fallible human beings making errors that lead to loss of life and military and political embarrassment.

Success in military operations depends upon the military achieving the right political effect. Increasingly, this involves ensuring that the intent behind that political effect is communicated through the media, which have increasingly become integrated into the battle space. The war in Iraq has led to more intense media coverage than before. This coverage was not simply a technical achievement; it was just as much a political result of states seeking to ensure that the media were present to achieve one aspect of a campaign that included a dimension of 'shock and awe' through the real-time reporting of embedded journalists. That things turned out rather differently does not detract from the importance of this point. More significantly, we have to ask, as governments and commentators have done, whether the intense media coverage of operations means that the public *see* more about war than before but are in fact *learning* much less about it because the stream of tactical and sub-tactical information is not located in a broader strategic context – something that is difficult for the logic of twenty-four-hour news coverage and the fast-forward culture of wider society to deal with. We also have to ask whether the public's role in monitoring what the armed services do on their behalf is limited by the complex tensions between panoptical and synoptic forms of surveillance that characterize contemporary military operations (Hoon 2003).

Finally, while military transformation focuses, in the main, on the use of the military as a surveillance machine to deal with external threats and, as we have seen, has become harnessed to a pre-emptive foreign and defence policy, it occurs at the same time as homeland security becomes tightened. In so far as military operations involve the more secretive side of the military – Special Forces and the like – and the security services abroad and at home, then for the twenty-first century we have to be concerned about the tension between national security and democracy. How do we oversee the surveillance machine at home and abroad without compromising the secrecy it needs to protect the citizens it serves?

REFERENCES

Avant, D. 1996. 'Military Reluctance to Intervene in Low-Level Conflicts: A Crisis.' In V. Davis, ed., *Civil-Military Relations and the Not-Quite Wars of the Present and Future*, 25–32. Lexington: U.S. Army War College, Strategic Studies Institute.

Barnes, R.C. 1996. *Military Legitimacy: Might and Right in the New Millennium.* London: Cass.

Betts, R.K. 1977. *Soldiers, Statesmen and Cold War Crises.* Cambridge: Harvard University Press.

Boene, B., and C. Dandeker. 1998. *Les armées en Europe.* Paris: Éditions La Decouverte.

Boene, B., C. Dandeker, J. Kuhlmann, and J. Van Der Meulen. 2000. *Facing Uncertainty, Report No. 2, Flexible Forces for the Twenty-First Century.* Sweden: National Defence College, Department of Leadership, Klaria Tryckeri, Karlstad.

Boyce, Admiral Sir M. 2003. 'Achieving Effect.' The Annual Chief of Defence Staff Lecture. *RUSI Journal* (Feb.): 30–7.

Burns, T., and G.M. Stalker. 1994. *The Management of Innovation.* New ed. Oxford: Oxford University Press.

Bush, G.W. 2001. Speech of the President Delivered to Military Cadets at the Citadel December 11. http://www.whitehouse.gov/news/releases/2001/12/20011211-6.html

Centre for Defence Studies. 1998. 'The Strategic Defence Review: How Strategic? How Much of a Review?' *London Defence Studies* No. 46. (July).

Cohen, E.A. 2002. *Supreme Command: Soldiers, Statesmen and Leadership in Wartime.* New York: Free Press.

Dandeker, C. 1990. *Surveillance Power and Modernity: Bureaucracy and Discipline from 1700 to the Present Day.* Polity: Blackwell.

– 1993. 'Surveillance, Liberte et Modernite.' In M. Audet and H. Bouchikhi, eds., *Structuration de social modernité avancée: Autour des travaux d'Anthony Giddens*, 233–66. Laval: Les Presses de l'Université Laval.

– 1999. *Facing Uncertainty, Report No. 1 Flexible Forces for the Twenty-First Century.* Sweden: National Defence College, Department of Leadership, Klaria Tryckeri, Karlstad.

– 2000. 'The United Kingdom: The Overstretched Military.' In Moskos, Williams, and Segal, eds., *The Postmodern Military*, 32–50.

– 2003. 'Flexible Forces for the 21st Century.' In G. Caforio, ed., *Handbook of Military Sociology*, 407–21. Dordrecht: Kluwer Press.

Dandeker, C., and L. Freedman. 2002. 'The British Armed Services.' *Political Quarterly* 73 (4): 465–75.

Dandeker, C., and J. Gow. 1997. 'The Future of Peace Support Operations: Strategic Peacekeeping and Success.' *Armed Forces and Society* 23 (3): 327–48.

– 2000. 'Military Culture and Strategic Peacekeeping.' In E. Schmidt, ed., *Peace Operations between War and Peace*, 58–79. London: Cass.

Danopoulos, C.P., and C. Watson, eds. 1996. *The Political Role of the Military: An International Handbook*. Westport, CT: Greenwood.

Davis, V. 1996. *Civil-Military Relations and the Not-Quite Wars of the Present and Future*. Lexington: US Army War College, Strategic Studies Institute University of Kentucky.

De La Billiere, General Sir P. 1992. *Storm Command: A Personal Account of the Gulf War*. New York: Harper Collins.

Feaver, P.D. 1996. 'The Civil-Military Problematique: Huntington, Janowitz and the Question of Civilian Control.' *Armed Forces and Society* 23 (Winter): 149–78.

Freedman, L. 1998. 'The Revolution in Strategic Affairs.' *Adelphi Paper* 318, International Institute for Strategic Studies. Oxford: Oxford University Press.

– 2002. 'Calling the Shots: Should Politicians or Generals Run Our Wars?' *Foreign Affairs* (Sept./Oct.)

Giddens, A. 1977. *Central Problems in Social Theory*. London: Hutchinson.

Gow, J. and C. Dandeker. 1995. 'Peace Support Operations: The Problem of Legitimation.' *The World Today* 51:171–4.

Haggerty, K., and R. Ericson. 2000. 'The Surveillant Assemblage.' *British Journal of Sociology* 51 (4): 605–22.

Hier, S.P. 2003. 'Probing the Surveillant Assemblage: On the Dialectics of Surveillance Practices as Processes of Social Control.' *Surveillance and Society* 1 (3): 399–411. http://www.surveillance-and-society.org/articles1(3)/probing.pdf

Hirst, P. 2001. *War and Power in the 21st Century*. Polity: Blackwell.

Hoon, Rt. Hon. G. 2002. UK Secretary of State for Defence. *11 September – A New Chapter for the Strategic Defence Review* (text can be downloaded from the Ministry of Defence): http://moddev.dera.gov.uk/news/press/news_press_notice.asp?newsItem_id=1247

– 2003. UK Secretary of State for Defence. 'No Lens Is Wide Enough to Show the Big Picture.' *The Times*. Mar. 28. http://www.timesonline.co.uk/printFriendly/0,,1-152-626097,00.html

Huntington, S.P. 1957. *The Soldier and the State: The Theory and Politics of Civil-Military Relations*. Cambridge, MA: Harvard University Press.
– 1961. *The Common Defence: Strategic Programs in National Politics*. New York: Columbia University Press.
Jaffe, G. 2002. 'New and Improved? Wall Street Journal Report, Spending for Defense.' 28 March.
Janowitz, M. 1960. *The Professional Soldier*. Glencoe, IL: Free Press.
JDDC (Joint Doctrine and Concepts Centre UK). 2003. *Strategic Trends: The Military Dimension*. Ministry of Defence, UK. www.jdcc.mod.uk/trends.htm
Jones, General J. 2002. Commandant of USMC. Televised Speech to US Naval Institute Ap. 4.
Keller, B. 2002. 'The Fighting Next Time.' *New York Times Magazine*. 10 Mar.
Kuhlmann, J., and J. Callaghan. 2000. *Military and Society in 21st Century Europe*. Verlag: Munster.
Luttwak, E. 1999. 'From Vietnam to Desert Fox: Civil-Military Relations in Modern Democracies.' *Survival* 41 (1): 99–112.
Mackenzie, Maj. Gen. L. 1993. 'Military Realities of UN Peacekeeping Operations.' *RUSI Journal* (Feb.): 21–4.
Mathiesen, T. 1997. 'The Viewer Society: Michel Foucault's Panopticon Revisited.' *Theoretical Criminology* 1 (2): 215–34.
Moskos, C., J. Williams, and D.R. Segal. 2000. *The Postmodern Military*. New York and Oxford: Oxford University Press.
Orbis. 1999. 'The Future of Military Culture.' Special issue, Jan.
Owens, Admiral W. 1995. 'The Emerging System of Systems.' *US Naval Institute Proceedings* 121 (5): 35–9.
Perlmutter, A. 1977. *The Military and Politics in Modern Times*. New Haven: Yale University Press.
Prins, G. 2003. *The Heart of War: On Power, Conflict and Obligation in the Twenty-First Century*. London: Routledge.
Rodger, N.A.M. 1986. *The Wooden World: An Anatomy of the Georgian Navy*. New York: Collins.
Sarkesian, S.C. 1998. 'The US Military Must Find Its Voice.' *Orbis* 42 (3): 423–37.
Segal, D.R. 1993. *Organizational Designs for the Future Army*. Alexandria: VA. US Army Research Institute for the Behavioral and Social Sciences, special Report 20.
Snider, D., and M.A. Carlton-Carew, eds., 1995. *U.S. Civil-Military Relations: In Crisis or Transition?* Washington, DC: Center for Strategic and International Studies.

Strategic Defence Review. 1998. 'How Strategic? How Much of a Review.' *London Defence Studies*, no. 46 (July).

United Kingdom. 2003. Ministry of Defence, *Operations in Iraq: First Reflections Report*. London.

Weber, M. 1977. 'The Meaning of Discipline.' In G. Roth and C. Wittich, eds., *Economy and Society* 2:1148–57. Berkeley: University of California Press.

Williams, M.C. 1998. 'Civil-Military Relations and Peacekeeping.' Adelphi Paper 321, International Institute for Strategic Studies. Oxford: Oxford University Press.

10 Visible War: Surveillance, Speed, and Information War

KEVIN D. HAGGERTY

It was not until some months after the fighting had ceased that I saw what became for me the most lasting image of the 1991 Gulf War. A televised documentary aired a brief video, filmed in the ghostly green glow of night-vision equipment. In it, an Iraqi personnel carrier rests on the ground just a few yards below, as we hover above in an American Apache attack helicopter. Soldiers stumble out of their vehicle in a futile attempt to escape their unseen adversary by fleeing into the desert. None advance more than a few yards before they and their vehicle are shredded by the helicopter-mounted machine guns. This video is one small but powerful demonstration of how new visualization tools have been incorporated into the U.S. military, and the combat advantages that such devices can afford.

This paper explores the place of surveillance and visibility in the contemporary U.S. military. Military surveillance involves efforts to monitor different populations, places, and practices for strategic, tactical, and operational purposes. While human agents still conduct military surveillance, it is increasingly accomplished with the aid of a plethora of technological devices. I focus on developments in the United States, as it now possesses an unparalleled martial superiority that has been secured to no small extent by its embrace of new visualization devices.

The paper briefly outlines the current revolution in military affairs that is transforming the U.S. armed forces. It then concentrates on two issues related to the expansion and intensification of military optics. First, I consider how new informational and visualization technologies alter the speed of war. In some contexts new technologies accelerate the speed of combat, but in others they can have a deceleration effect.

Second, I address how the fixation on informational war can detract from other military objectives, and how other nations are apt to respond to this new mode of combat.

Information War

It has become routine to point out that the U.S. military is undergoing a revolution in military affairs. This revolution is expected to fundamentally alter almost all aspects of the armed forces, including tactics, doctrine, training, warrior skills, and execution. Although this new form of combat has been called many things, I prefer the expression 'information war,' as it best encompasses the type of conflict that the United States military is fostering.

Information war is distinguished by the prominent place of 'information' as a tactical and strategic resource. Information war is based on myriad technological enhancements of the military's computational, communications, and visualization abilities. The clearest articulation of this model can be found in the Department of Defense's *Joint Vision 2010*, the core premise of which is that emerging technology will grant U.S. forces 'information superiority,' where information is conceived of as a 'force multiplier.' Although the 1991 Gulf War is regularly seen as the moment when this form of warfare first coalesced, several dimensions of information war have predecessors in other conflicts, most prominently the war in Vietnam.

Below, I set out some of the main dimensions of information war and the larger context that has helped make such warfare possible (see generally Adams 1998; Arquilla and Ronfeldt 1993; Denning 1998; Ek 2000; Friedman and Friedman 1998; Gray 1997; Robins and Webster 1999; Van Creveld 1989; 1991). Several of these developments are interrelated and mutually reinforcing, and have been intensified by recent efforts to respond militarily to global terrorism.

Permanent War

Over the past half-century war has become a permanent condition in the United States. Virilo refers to this situation as 'pure war,' which involves coordinating political and scientific institutions with military agendas (Virilio and Lotringer 1997). It originated in the years following the Second World War, when politicians and military strategists recognized that any future wars would be fought with whatever weapons and troops were available at the outbreak of hostilities. As it would simply take too long to transform a peace economy onto a war footing, American leaders chose to position the nation in a situation of constant heightened military preparedness. This involves incessant military gaming and mock-ups of potential future wars through increasingly sophisticated simulation technologies (Bogard 1996; Der Derian 1990), and the development of an extensive military industrial complex that has profoundly shaped the evolution and practice of science (Cohen 1988; Leslie 1993; Mendelsohn 1997).

The blurring of the line between periods of peace and war has only been exacerbated in recent years with the rise of 'cyberwar' (Adams 1998; Arquilla and Ronfeldt 1993). Subsumed under the broader category of information war, cyberwar involves computerized attacks on, and defence of, the vast electronic infrastructure of the state, commerce, and military. Adversaries try to disrupt electronic systems by prodding computer networks for weakness, hacking into them, and occasionally infecting them with viruses. All of this usually occurs in the absence of officially declared hostilities. Rather than a fanciful future scenario, cyberwar is understood to be already ongoing. Many of these 'attacks' are little more than uncoordinated pranks, such as when hackers shut down the web page of Qatar's Al-Jazeera satellite TV station during the Iraq war after it broadcast images of American prisoners of war. Patrons were instead directed to a web page containing patriotic American images and slogans. That said, some cyberwar attacks involve more serious and coordinated efforts to disrupt a nation's political and military capabilities, as was done by the United States in Afghanistan and Kosovo.

Notwithstanding the ongoing military gamesmanship of cyberspace, military officials in the United States approach war as a permanent condition, where combat involves more temporally limited engagements. It is tempting to say that combat can be reduced to efforts to kill an adversary, but the broadened scope of information war places ever greater emphasis on destroying an enemy's command and control system, which can sometimes be accomplished without the loss of life.

While war has become a permanent condition in the United States, combat is now expected to involve comparatively brief engagements. Manual Castells (1996) refers to these as 'instant wars,' involving a combination of special forces, advanced technology, and overwhelmingly superior airpower. Conscious of the legacy of Vietnam, senior military personnel are also demanding explicit 'exit plans' to establish the conditions for the removal of U.S. troops. The hope is that such plans will avoid miring American troops in drawn-out civil wars, or serving as de facto police during lengthy and uncertain processes of 'nation building,' while at the same time pre-empting accusations that the United States is involved in colonial occupations of foreign lands.

The collapse of the Soviet Union produced a series of attendant uncertainties about the form and location of future wars. Military planners were forced to move away from their previous fixation on anticipating a clash of superpowers. They have responded with a series of more flexible strategies designed to allow the United States to rapidly exert military force in just about any region on earth. The perceived adversaries of these conflicts are no longer exclusively nation states, but now include assorted clans, warlords, drug cartels, and terrorist groups that threaten U.S. interests.

Advanced Technology and Intensified Visibilities

'We have an unblinking eye over the enemy formation.'

(General John Keane, U.S. Army, 2002)

These diverse developments point to a growing commitment to use advanced technologies to secure military supremacy (Gray 1997). While new technologies have historically produced military advantages (McNeill 1982; Van Creveld 1989), they have tended to emerge slowly, and in a haphazard fashion. Today, however, the United States' military establishment has embraced the idea that new developments in technoscience constitute the principle factor in securing military superiority. Almost all new scientific discoveries are scrutinized by Pentagon officials to determine whether they have potential military applications. Indeed, following the Second World War the United States embarked on a campaign to direct science to securing military advantage, a process that accelerated in the ensuing years. The number of contemporary technologies and sciences whose origins and development have been shaped by military concerns is too lengthy to list, but it includes lasers, pharmacology, computers, surgery, chemistry, electronics, satellites, sensors, atomic energy, space exploration, and the Internet.

New surveillance devices have been particularly important in establishing the current American global military dominance, and, in the process, have helped to transform the dynamics of combat. Here 'surveillance' is understood broadly to include all of the ways that populations, processes, and places are rendered knowable. Information war incorporates photographic images, geographic positioning systems, communication networks, and bureaucratic reporting systems, all of which help to make phenomena amenable to military monitoring, reflection, calculation, and action.

A major chapter in the history of warfare could be devoted to successive innovations in visualization tactics and technologies (Dandeker 1990). In antiquity, advantages in visibility were most commonly secured by positioning troops and fortifications on high ground with clear lines of sight. Industrialization rapidly enhanced military optics. During the American Civil War, for example, generals scrutinized enemy positions from hot air balloons. The subsequent trench warfare of the First World War ushered in numerous changes to military visualization. One of the most famous of these was the ill-fated communication cables laid between the trenches and into no-man's land in an effort to allow the general staff to maintain contact with their troops from behind the lines. Although buried six feet underground, these cables were regularly destroyed during the massive artillery barrages accompanying major campaigns, leaving the command structure completely in the dark about how battles were progressing, often with tragic consequences (Keegan 1976).

More recently, the American military used the war in Vietnam as an opportunity to experiment with cutting-edge surveillance devices. New reconnaissance aircraft, communication devices, and an extensive bureaucratic and statistical reporting structure were all introduced in Vietnam. More fanciful efforts were employed to detect North Vietnamese troops, including Agent Orange to defoliate the impenetrable Vietnamese jungles and helicopter-mounted 'people sniffers' to try and identify the ammonia traces left by human urine. Sensors shaped like twigs, plants, and animal droppings were scattered along the Ho Chi Minh trail in an effort to detect vehicle noises or body heat (Edwards 1995).

Today, the operation of surveillance in the U.S. armed forces involves an assemblage of visualization technologies. The sheer range, scope, and intensity of combat visualization devices means that the battlefield is being turned into a surveillant assemblage that aims to seamlessly connect sensors, fighters, and decision makers (Murray 2002), and allow for 'total transparency across the battlespace of air, land, sea and space' (Harknett 2000: 129).

We can gain an appreciation of these visualization practices by looking below the battlefield where caves, tunnels, bunkers, cables, and pipelines have been mapped for military purposes. Moving above ground one encounters a military cartographic frenzy, consisting of a multitude of minutely calibrated and informationally embedded inter- active maps. Geographic positioning systems allow for precise situ- ational awareness and contribute to the increasing digitization of the battlefield where soldiers and equipment are monitored through as- sorted sensors. Combat vehicles are equipped with night-vision equip- ment and thermal sensors that can see behind walls. Cameras are mounted on weapons systems, allowing combat to be monitored in real time and munitions to be guided remotely. Field commanders augment all of this with a steady stream of reports to central command, where statistics are produced and analysed. Overhead, airplanes, helicopters, and unmanned drones monitor ground and air space with the assis- tance of radar, night vision equipment, video, and reconnaissance cameras. Spy planes capture telephone, satellite, and microwave com- munications. Moving higher still, satellites encircle the globe, conduct- ing twenty-four-hour-a-day surveillance through darkness, cloud, and smoke. In the near future, military satellites are expected to be able to direct artillery rounds and pinpoint every friendly soldier and machine on the battlefield (Anderson 2001).

One of the more interesting visibility-related developments on the near horizon that will undoubtedly shape the execution and politics of combat will be the introduction of a host of mass-marketed visualiza- tion devices into warfare. Combatants and citizens of technologically advanced nations will increasingly have their own digital cameras, satellite camera phones, camcorders, and web-connected micro-com- puters. As images of combat disseminate more widely and quickly combat promises to become yet more transparent. Even this partial list indicates that surveillance is now a multi-layered component of the U.S. military and has become vital to strategy and tactics. Indeed, the primacy of visibility in combat situations is revealed in Murray's assessment of contemporary theories of battle command, which, he suggests, all begin 'with one's ability to see, visualize, observe or find' (Murray 2002: 49).

Visibility and the Speed of War

French cultural critic Paul Virilio has done much to advance theoretical understanding of these developments in military technology and visualization.[1] He has risen in prominence as recent technological advancements have caught up with long-standing themes in his writings. Inspired by Virilio's work other cultural scholars repeatedly emphasize how information technology has accelerated the speed of warfare.

Changes to the speed of combat is an essential theme in Virilio's 'dromology,' or the historical study of speed. Dromology involves an analysis of how speed functions in different social spheres, with the suggestion that speed operates on par with wealth in terms of its ability to create and reinforce social hierarchies. Following Virilio, 'speed is power' (Virilio and Armitage 1999: 35). His work is replete with historical examples of how the greatest speeds have been reserved for the political, economic, and military elite. Industrial societies have produced new vectors of speed which derive from new technologies of transportation, communication, and visualization, the last of which he refers to as 'vision machines.'

A second related theme in Virilio's work is the suggestion that new technologies inevitably bring with them new forms of accident. Hence, the automobile produces the car accident, and airplanes generate plane crashes. Most recently, he has been concerned with the prospect of an 'integral accident' resulting from the development of new information and genetic technologies (Virilio 2000).

Surveillance studies has demonstrated a long-standing emphasis on optics: how new technologies and practices can alter what is seen and the way it is understood. More recently, there have been efforts to connect practices of visualization with the creation and operation of space, where visualizing devices are recognized as a means to transcend distance. Virilio's work encourages us to contemplate yet another dimension: the relationship between visibility and speed.

Virilio's analysis of speed is germane to our examination of military surveillance, as one of the effects of new military visualization technologies has been to accelerate the speed of combat. Following Virilio, the history of warfare involves a series of transformations to the means of perceiving battle, each of which has tended to accelerate the speed of war. To see an enemy while remaining undetected is to gain a temporal

---

1 There is now a voluminous English-language Virilio literature, including an edited collection (Der Derian 1990), a volume by Redhead (2004), and a 1999 double issue of the prominent journal *Theory, Culture and Society* dedicated to his work. For our purposes, some of his most important texts include *Speed and Politics*, *War and Cinema*, and *Pure War*, although the themes of war, speed, and technology run throughout Virilio's writings.

advantage in the ability to reflect on, and react to, tactical and strategic developments. Increasingly, these advantages are measured in fractions of seconds. A host of almost instantaneous communication and visualization technologies have accelerated the speed of combat to the point that the ambit of human decision making has been markedly reduced or eliminated entirely. In the process, modes of human perception are surpassed as combat decisions become automated, a trend that brings the speed of war close to the speed of light. Virilio is particularly alarmed by how the accelerated speed of new technologies now surpasses and confounds human perception and sense-making abilities. On these points Virlio has been particularly influenced by the Cold War nuclear stand-off between the United States and the Soviet Union, which saw the increasing automation of nuclear launch decisions. Comparable dynamics are at work in other forms of combat. As Rochlin (1997: 204) observes: 'In an era of supersonic aircraft armed with high-speed missiles, quick-reacting radar-directed gun and missile batteries, and tank battles that may be won or lost on the first shot, there is simply not the time for centralized command systems to exercise real-time control over battlefield events.'

Virilio's insights are important for analysts of surveillance because he encourages them to go beyond their usual fixation on optics to contemplate the relationship between visibility and speed. That said, his own fixation on the technological acceleration of the speed of combat is limited by virtue of his tendency to ignore how the exact same technologies can also decelerate the speed of war. As such, his analysis of the speed of war represents only one side of the temporal coin. Looking to the other side, one finds a series of attributes of military technologies that can decelerate the speed of combat by producing information overload, centralization, and elaborate, prolonged war preparations.

Information Overload

New military technologies risk producing a form of informational overload that, notwithstanding the enhancements to the military's technological ability to see, can produce opaque and indecipherable command and control systems. This, in turn, can foster a corresponding tentativeness in decision making. The war in Vietnam provides perhaps the most telling examples of such tendencies. There, despite the United States' decided advantage in informational and surveillance capabilities, the ostensible benefits of such tools rarely materialized. In fact, several attributes of the U.S. military's information systems often helped to subvert their stated goals (Van Creveld 1985; 1989).

In Vietnam, the U.S. military approached information as a vital strategic and tactical resource, establishing remarkably detailed reporting structures to try and capture the minutiae of combat and administrative activity. It quickly became apparent, however, that rather than help to ensure decisive action, the sheer volume of information produced by these systems could breed a form of command paralysis. In the ensuing years the problem of information overload has only intensified. In the Gulf War, for example, satellites generated an unmanageable stream of information that delayed communication with commanders, meaning that information was often irrelevant by the time it arrived (Friedman and Friedman 1998: 324). The totality of information produced by military surveillance tools in the Gulf 'simply swamped the tactical intelligence system, and came near to paralyzing other systems for electronic integration and command-and-control' (Rochlin 1997: 180).

In Vietnam the military embraced new computing and communications systems to help alleviate their information management problems. An unforeseen consequence of this development was that such tools encouraged a tendency towards greater specialization and organizational complexity, factors which themselves tend to produce yet more information. In an escalating spiral, administrators then sought to rectify these new information management problems by introducing yet more information technologies. Rather than rendering decisions more rapid and efficient, these tools helped to produce 'an inordinate increase in the amount of information needed to make any given kind of decision at any given level' (Van Creveld 1985: 237).

Faced with unwieldy information systems containing data that were of increasingly questionable quality, commanders sought out alternative, more timely and less opaque, informational sources (van Creveld 1985: 258). Washington officials supplemented imprecise official reports with journalists' accounts of the conflict. Commanders abandoned the progressively useless official reports to personally oversee battles in helicopters (see below), a technological update of the Civil War strategy of monitoring combat via hot air balloon.

Vietnam presents a picture of a military information structure that produced a series of paradoxical feedback loops. As information technology was introduced to help resolve the problems associated with system complexity, specialization, and information overload, these selfsame systems contributed to the problems they sought to solve, in the process often further slowing and complicating decision making.

Command and Control

Contrary to Virilio's image of warfare, not all military decisions have been embedded in computerized forms of artificial intelligence. Soldiers and politicians continue to evaluate tactics, strategy, and logistics, and this often involves a host of deferrals and delays. Van Creveld (1985, 1989), again concentrating on the war in Vietnam, provides several examples of how an informationally intensive command and control structure can impede decision making and produce extraordinarily long periods of planning and preparation.

The most infamous example of this tendency was the U.S. raid on the Son Tay prisoner of war (POW) camp (Van Creveld 1985: 249). In May 1970 a U.S. intelligence unit initiated a remarkably extensive and complicated procession of operational planning when they discovered the Son Tay camp. Despite the primacy that had been placed on rescuing POWs, the ensuing drama in U.S. command amounted to a drawn-out and ultimately tragic series of briefings, planning sessions, feasibility groups, formal recommendations, more planning sessions, Presidential briefings, and operation training. All of this occurred over a six-month period, culminating in the 21 November raid that liberated an empty camp, the prisoners having been moved on 14 July. Van Creveld presents this as one example of the general tendency for complex military organizations, permeated with modern communication systems, to produce decision cycles that are longer than those of an enemy, who is far less sophisticated on such matters.

There are more recent examples of such delays. Although early reports are inevitably suspect, it has been suggested that the first 'decapitation' strikes on Saddam Hussein's compound in the Iraq War were delayed four hours while the plan was discussed at the White House. The attack was then delayed another hour while the target coordinates were programmed into the cruise missiles. Had the order come more quickly, Hussein might not have escaped (Cheney 2003).

That the decision to strike Hussein's compound appears to have originated in Washington accentuates another attribute of information war that can decelerate the speed of combat. Military visualization devices often operate as a form of distanciation. Rather than intensifying a commander's view of something that is already in his field of vision, distanciating technologies transmit representations to centralized locations of command and control over great distances. Remote commanders can now, as never before, watch combat unfold before their eyes. In the U.S.- led war in Afghanistan, officials in Washington were purportedly awed by their ability to monitor attacks on the Taliban in real time. Not long afterwards, CNN viewers shared a comparable experience, as they watched unprecedented videophone images of the 7th Calvary's assault across Iraq's desert.

As impressive as such distanciation can be, it introduces new concerns about the site of effective tactical decision making. To date, military doctrine has held that combat decisions are best made by field commanders who have the greatest appreciation for situational nuances. Senior officers have historically set broad parameters for how engagements should proceed, recognizing that as the fog of war rolls in, their subordinates should have maximum latitude to decide on different courses of action. New real-time visualization and communication abilities raise the prospect of combat micromanagement by senior officials who are considerably removed from the fighting (Harknett 2000: 135). Early evidence of this was seen in Vietnam, where the United State's command and communications technologies fostered 'the constant intervention of remote commanders into even the smallest details of battles on the ground' (Rochlin 1997: 139).

The tendency towards a centralized micromanagement of the battlefield is not an inevitable outcome of new information technologies. These self-same technologies also allow for still greater decentralization of decision making, as is evidenced from the prominent discourses about new military networks, non-hierarchical models, and a 'system of systems.' So, while there is nothing technologically deterministic about a movement towards greater centralization, such a propensity has been in evidence during the twentieth century as command and its associated decision-making authority have migrated upward with the expansion of communications capacities (Murray 2002).

This propensity for distant commanders to intervene in combat decision making reinforces another tendency fostered by new technologies. Field commanders become attuned to the fact that their actions are intensely monitored by different levels of the command structure. This multi-faceted monitoring reached absurd extremes in Vietnam, where officers took to the skies in helicopters to watch battles: 'A hapless company commander engaged in a firefight on the ground was subjected to direct observation by the battalion commander circling above, who was in turn supervised by the brigade commander circling a thousand or so feet higher up, who in his turn was monitored by the division commander in the next highest chopper, who might even be so unlucky as to have his own performance watched by the Field Force (corps) commander' (Van Creveld 1985: 255).

This preposterous layering of helicopters brings to mind Foucault's (1977) point about how disciplinary forms of surveillance, exemplified by Bentham's panopticon prison, also simultaneously allow for the scrutiny of the watchers. In combat situations, such monitoring is reinforced and intensified by new technologies. In the process, new delays can be introduced into combat decision making, as subordinate commanders defer to more senior commanders, and distant officers are more tempted to second guess a subordinate's actions (Murray 2002: 47). Such tendencies are mutually reinforcing, working together to occasionally reduce the speed of effective combat decision making.

The primacy accorded to information in contemporary warfare can slow decision making in still other ways. For example, the need for additional information can occasionally become an excuse for delay and inaction. Greater information is routinely equated with increased understating, but the exact opposite is often true. As military intelligence is always deficient in some important ways, there is the risk that, rather than acting boldly and decisively, commanders will be tempted to await yet more and more intelligence. Technologies do not produce this tendency, but new information technologies, combined with a military doctrine that valorizes information, can exacerbate the propensity to defer decisions until more timely intelligence or detailed images arrive.

Simply seeing an enemy does not suffice. Environments of enhanced military visibility produce heightened efforts to deceive an adversary (Latimer 2001). Hence, commanders do not simply see an enemy, but are always trying to discern whether the images and information they have is evidence of a feint, bluff, display, diversion or is, in fact, an important development (Keegan 2003). Such interpretive work cannot be quickly or easily 'read off' an image or report, but often requires careful, time-consuming debatable analysis, a situation that encourages commanders to seek out yet more information.

Marshalling

The theoretical emphasis on the speed of combat partially derives from a singular concentration on the flash point of conflict. While this is undeniably the most dramatic moment in warfare, focusing exclusively on the clash of arms ignores the remarkably complicated and measured process of military marshalling which is necessary before technologically sophisticated weapons can demonstrate their impressive speeds.

Troop ships and airplanes now allow armies to deploy at any place on earth in record time. When they arrive, however, the infrastructure necessary to fight an information war is not necessarily in place. Marshalling troops for combat still involves moving mountains of materiel, but now also includes establishing, calibrating, and aligning diverse visualization and informational technologies. The sheer scope and complexity of current marshalling practices is itself related to the use of advanced technologies which have helped produce a military with an unprecedented 'tooth to tail' ratio. Here the 'teeth' are the fighting soldiers and the 'tail' refers to the ever-expanding support structure of technicians and administrators. As the number of fighting soldiers has decreased, there has been a corresponding growth in 'tail' structure, which can require an inordinate amount of time to organize. Indeed, Rochlin (1997: 179) suggests that the 'single greatest lesson' of the Gulf War was the sheer scope of support necessary for an advanced military organization equipped with complex, and often fragile, machines.

The remarkable speed and success of the Gulf War relied on a meticulous organization of information systems and visualization devices during the six-month build-up to the war. The timing of the assault was itself partially dictated by the need to coordinate visualization systems, as the initial campaign was delayed to ensure that satellites would be overhead when the war commenced. The months preceding hostilities involved frantic efforts to collect data, coordinate intelligence, and establish systems: 'That time was needed to test, adjust, maintain, and make fully operational a large number of the high-tech platforms and systems. It was also needed to collect data and develop intelligence information. At the beginning of the buildup there were almost no up-to-date photos of the combat area. In fact, it took almost all of the six months to acquire, analyze, digitize, and program the key terrain and target information needed for programming the Tomahawk cruise missile's guidance computers' (Rochlin 1997: 178).

Chinese philosopher Sun Tzu may have been correct to claim that speed is the essence of war, but the current drive to maximize speed through new informational and visualization technologies produces its own unique delays and hesitations. This is evidenced by the tendency towards an increasingly unwieldy support 'tail' of technicians required to establish and maintain complicated technological systems. Before soldiers can wage informational war these systems must be installed, calibrated, and coordinated, all of which can be very time consuming. The volume of information these systems produce can overwhelm commanders, introducing delays and a form of paralysis in the execution of command and control. This is reinforced by new distanciating technologies that encourage remote commanders to micromanage battles, while prompting field officers, attuned to the various ways they are being monitored, to await direction from afar.

It is curious that Virilio has failed to engage with such limitations and contradictions of new military technologies. At times, he appears to take the proclamations of military officials about the capabilities and unqualified successes of such systems at face value, leading to his characteristically hyperbolic statements about the new speed of war. This uncritical embrace of official accounts is all the more peculiar given that Virilio is critical of most new technologies (Kellner 1999). One explanation for this oversight is that while Virilio is intimately concerned with the unintended consequences of military (and other) technologies, his focus is on the cataclysmic accidents that such technologies produce. Hence, his concentration on how the invention of airplanes simultaneously invented the plane crash, or how our newly networked society promises to produce a form of systemic global information failure (Virilio 2000). Consequently, he ignores the more mundane paradoxes of technological systems that routinely shape and limit the operation of military systems, including the simultaneous deceleration as well as acceleration of different components of warfare.

Conclusion

The fact that analysts, myself included, tend to concentrate on developments in the United States risks overlooking more general truths about the execution of combat. High technology developments in the United States are not necessarily indicative of how combat is conducted globally. Most martial technologies currently in use throughout the world are not markedly different from those employed during the Second World War. Casualties are still disproportionately produced by small arms and landmines, weapons that have not significantly changed in close to a century. The emphasis on the 'revolutionary' character of the current military transformation therefore downplays the many continuities connecting contemporary warfare with its predecessors. In previous epochs military analysts have accentuated the importance of individual initiative, superior strategy, logistics, brilliant commanders, or superior weaponry in bringing military success. Today, all of these factors are being re-coded as part of 'information war.' Therefore, what is particularly new is the perception within the American military establishment that information, acquired through greater technological ability, is *the* fundamental variable in warfare, and how this emphasis refocuses a host of traditional military concerns.

Information warfare is not without its own unique risks. America's informational infrastructure is vulnerable. Indeed, even simple component failure brought on by the sheer complexity of military systems can pose dangers. There is also the possibility that the remarkable successes demonstrated in Iraq might not have been the result of the inherent superiority of informational war, but were due instead to the fortuitous combination of informationally intensive and highly trained American troops pitted against an unmotivated and poorly equipped enemy. Not all engagements are fought in such optimal conditions. It remains to be seen how information war might operate in engagements with adversaries who are also embedded in an informational war paradigm. Or, alternatively, what advantages will information war bring to engagements against highly trained and ideologically committed adversaries such as the Japanese soldiers of the Second World War who dug into the caves of Peleliu and Okinawa and fought tenaciously against American troops (Sledge 1981). Were a comparable situation to arise today, it is likely that the level of American casualties would quickly surpass the comparatively low levels that the American public now seems willing to tolerate.

The commitment to high-technology also risks transforming military command into a version of the man with the hammer who approaches every problem as a form of nail. A fixation on high technology can restrict a broader appreciation of the requirements of a contemporary military, downplaying other important roles. Evidence of such tunnel vision can be seen in the reduced place for peacekeeping and post-conflict operations in the U.S. Army. Despite the fact that the history of colonial conflicts over the past century suggests the importance of being prepared to fight unconventional battles during transitions from war to peace, there is a general disdain for such operations in the American military. This attitude appears to be related to their 'low technology' status. General Wesley Clark, former Supreme Allied Commander, Europe, has suggested that part of the reason for the neglect of military post-conflict abilities can be attributed to the cultural and political value of informational technology in the U.S. military. As Clark notes: 'the Army, like the other services, had made its existence dependent on high-tech innovation and the creation of impressive, far-sighted procurement programs designed for high-intensity combat in the Middle East or in Korea. In view of overall US defense priorities, these programs were seen as more likely to compete successfully for funding. And, once funded, they would get important backing from contractors and subcontractors in many congressional districts' (Clark 2003).

These cultural and political priorities combined in such a way as to undermine any support for low technology post-conflict operations. The embrace of new informational warfare thus appears to be one example of what Steven Weinberg (2003) suggests is a tendency for officials to embrace military systems because of their ability to provide organizations with opportunities for glory, rather than a sober calculus of military requirements. In the process, less glamorous but nonetheless urgently needed options – such as post-conflict operations – are ignored.

Notwithstanding such reservations, information war is poised to shape the armed forces and geopolitics for years to come. Foreign nations have taken important lessons and formulated new courses of action in light of military capacities demonstrated in Afghanistan and Iraq. For the small handful of countries with the financial means to fundamentally re-tool their military to an informational footing, recent conflicts served as extended advertisements for U.S.-style informational warfare. Even these nations, however, must recognize that they are apt to remain behind the U.S. capacities in this regard for the foreseeable future. Other countries, like Iran and perhaps North Korea, appear to have concluded that they must accelerate their nuclear weapons program to maintain some semblance of independence from U.S. interests.

Still other nations will undoubtedly conclude that the days of massed, uniformed armies is drawing to a close. Fighting a war within this convention is not to fight fairly, but to commit suicide. Faced with few other palatable options, some nations will undoubtedly be driven to embrace more unconventional forms of warfare involving guerilla or terrorist tactics. This would be a profoundly paradoxical development, given that the United States' current martial efforts are justified as anti-terrorism initiatives.

If, as seems plausible, the new international imbalance in military capabilities encourages some nations or sub-national groups to embrace or intensify terrorist efforts, domestic repercussions in the United States will undoubtedly follow. In an environment of heightened concerns about terrorism increasing efforts are being made to integrate military strategies and technologies into domestic policing. As Richard Ericson and I have noted (Haggerty and Ericson 2001), there were already strong political and market pressures fostering the movement of military informational and visualization technologies into domestic policing prior to 9/11 (see also Nunn 2001). Such efforts have been redoubled following 9/11. The recent use of military spy planes over Washington in search of the Beltway snipers, and Coast Guard helicopters circling the Statue of Liberty, points to a future where it will not only be foreign soldiers who scramble madly for cover from American military surveillance devices.

REFERENCES

Adams, J. 1998. *The Next World War: Computers Are the Weapons and the Front Line Is Everywhere*. London: Hutchinson.

Anderson, Lieutenant General E.G. 2001. 'U.S. Space Command: Warfighters Supporting Warfighters in the 21st Century.' *Military Review* (Nov./Dec.): 11–17.

Arquilla, J., and D. Ronfeldt. 1993. 'Cyberwar Is Coming!' *Comparative Strategy* 12:141–65.

Bogard, W. 1996. *The Simulation of Surveillance: Hypercontrol in Telematic Societies*. Cambridge: Cambridge University Press.

Castells, M. 1996. *The Rise of the Network Society*. Oxford: Blackwell.

Cheney, P. 2003. 'Signs Point to Iraqi Role in Strike on Hussein.' *Globe and Mail*. 21 Mar.

Clark, W.K. 2003. 'Iraq: What Went Wrong.' *New York Review of Books*. 23 Oct.

Cohen, B. 1988. 'The Computer: A Case Study of Support by Government, Especially the Military, of a New Science and Technology.' In E. Mendelsohn, M. Smith, and P. Weingart, eds., *Science, Technology and the Military*, 119–54. Dordrecht: Kluwer.

Dandeker, C. 1990. *Surveillance, Power and Modernity: Bureaucracy and Discipline from 1700 to the Present Day*. New York: St Martin's.

Denning, D. 1998. *Information Warfare and Security*. New York: Addison-Wesley.

Der Derian, J. 1990. 'The (S)pace of International Relations: Simulation, Surveillance and Speed.' *International Studies Quarterly* 34:295–310.

Edwards, P.N. 1995. 'Cyberpunks in Cyberspace.' In S.L. Star, ed., *Cultures of Computing*, 69–84. Cambridge, MA: Harvard University Press.

Ek, R. 2000. 'A Revolution in Military Geopolitics?' *Political Geography* 19: 841–74.

Foucault, M. 1977. *Discipline and Punish: The Birth of the Prison*. New York: Vintage.

Friedman, G., and M. Friedman. 1998. *The Future of War: Power, Technology and American World Dominance in the Twenty-First Century*. New York: St Martin's Griffin.

Gray, C.H. 1997. *Postmodern War*. New York: Guilford.

Haggerty, K.D., and R.V. Ericson. 2001. 'The Military Technostructures of Policing.' In P. Kraska, ed., *Militarizing the American Criminal Justice System: The Changing Roles of the Armed Forces and the Police*, 43–64. Boston: North-eastern University Press.

Harknett, R.J. 2000. 'The Risks of a Networked Military.' *Orbis* 44 (1): 127–43.

Keegan, J. 1976. *The Face of Battle*. London: Pimlico.

– 2003. *Intelligence in War: Knowledge of the Enemy from Napoleon to Al-Qaeda*. Toronto: Key Porter.

Kellner, D. 1999. 'Virilio, War and Technology: Some Critical Reflections.' *Theory, Culture and Society* 16: 103–24.

Latimer, J. 2001. *Deception in War*. New York: Overlook Press.

Leslie, S.W. 1993. *The Cold War and American Science: The Military-Industrial Academic Complex at MIT and Stanford*. New York: Columbia University Press.

McNeill, W.H. 1982. *The Pursuit of Power: Technology, Armed Force, and Society since A.D. 1000*. Chicago: University of Chicago Press.

Mendelsohn, E. 1997. 'Science, Scientists, and the Military.' In J. Krige and D. Pestre, eds., *Science in the Twentieth Century*, 175–202. Amsterdam: Harwood Academic Publishers.

Murray, Major S.F. 2002. 'Battle Command: Decisionmaking, and the Battle-field Panopticon.' *Military Review* (July/Aug.): 46–51.

Nunn, S. 2001. 'Police Technology in Cities: Changes and Challenges.' *Technology in Society* 23:11–27.

Redhead, S. 2004. *Paul Virilio: Theorist for an Accelerated Culture*. Toronto: University of Toronto Press.

Robins, K., and F. Webster. 1999. 'Cyberwars: The Military Information Revolution.' In *Times of the Technoculture*, 149–67. London: Routledge.

Rochlin, G.I. 1997. *Trapped in the Net: The Unanticipated Consequences of Computerization*. Princeton: Princeton University Press.

Sledge, E.B. 1981. *With the Old Breed at Peleliu and Okinawa*. Novato: Presidio.

Van Creveld, M. 1985. *Command in War*. Cambridge, MA: Harvard University Press.

– 1989. *Technology and War: From 2000 B.C. to the Present*. New York: Free Press.
– 1991. *The Transformation of War*. New York: Free Press.
Virilio, P. 2000. *The Information Bomb*. London: Verso.
Virilio, P., and J. Armitage. 1999. 'From Modernism to Hypermodernism.' *Theory, Culture and Society* 16:25–55.
Virilio, P., and S. Lotringer. 1997. *Pure War*. New York: Semiotext(e).
Weinberg, S. 2003. 'What Price Glory?' *New York Review of Books*. 6 Nov.: 55.

PART THREE

Surveillance, Electronic Media, and Consumer Culture

In Part III, Joseph Turow, Serra Tinic, Emily Martin, David Wall, and Oscar Gandy Jr examine how surveillance technologies are transforming commerce, conceptions of health, and identity.

Surveillance is embedded in business enterprise. Joseph Turow considers how advertisers now conduct surveillance of potential customers. He describes a growth in such surveillance as part of a response to anxieties about how new information technologies are transforming long-established relationships between marketers and consumers. For example, alarmed by the capacity of digital interactive technologies to rupture their contact with potential consumers, advertisers have reconfigured traditional strategies such as product placement and direct marketing. Part of this reconfiguration involves the use of new and enhanced surveillance technologies.

Product placement integrates products into the content of the show itself. The demonstrated successes of such efforts in movies such as *ET* and *Crocodile Dundee*, combined with a relaxed regulatory environment, have encouraged advertisers to embrace this practice more thoroughly. Direct marketing, in contrast, is characterized by the ability to acquire clear feedback on advertising initiatives though such things as catalogues and coupons. While direct marketing was criticized by elite marketers during the 1970s, it regained respect through the use of computers to fine-tune the targeting of niche markets. Direct marketing is evident in television infomercials and shopping channels, although most targeted approaches during the 1980s and 1990s were conducted via the mail and telephone. Such initiatives combine precise audience identification, individualized communication, and speedy order fulfilment.

To connect marketers directly with desired targets requires a good list of prospects and it is here that direct advertising intersects with new forms of surveillance. The demand for such lists has been serviced by an expanding database business that culls information from a broad scope of universal and transactional databases.

Corporations, recognizing the importance of the repeat business that their existing customers provide, have tried to foster ongoing relationships by identifying loyal customers through such things as 800 telephone numbers, rebates, product registration, and sweepstakes. Such contact is designed to reinforce repeat buying with signs that the company is responsive to the consumer's needs. One of the most important ways in which this is being accomplished is through loyalty programs that provide consumers with benefits for increased business in exchange for allowing companies to collect data about their purchasing and lifestyle habits. This data is used to discern various market segments. Individuals then receive promotions that are directed exclusively at their particular market niche. Ultimately, the aim of this data collection exercise is to discriminate among customers, rewarding a company's best customers but also discouraging or even repelling customers who add no value.

In this context, Turow describes the emergence of 'customer relationship media' (CRM). The logic of CRM is that companies must reach desired consumers in ways that makes them feel connected to the company and its products. Customers must then be encouraged to surrender information that can help marketing efforts. Finally, the selling environment must be structured to ensure that consumers attend to advertisements. Two prominent examples of customer relationship media are 1) customized media and 2) interactive television.

Customized media apply the concepts of loyalty programs to gain the fidelity of desirable audience members. Evidence of this approach can be seen in the magazine industry where certain magazines customize their content to meet different subscription profiles. Interactive TV provides customers with the opportunity to interact with televised programming through personalized scheduling or, in some scenarios, allowing viewers to purchase products shown in televised programs. Each of these techniques involves the minute tracking of the viewing habits of customers. The data is then sold to potential advertisers.

This intense use of market surveillance intersects with two dimensions of trust. On the one hand, it is a response to the advertiser's declining trust in the ability of marketing corporations to place their messages in front of desired consumers using traditional advertising structures. At the same time, corporations undertake efforts to enhance the public's trust in the soundness of corporate activities, even as corporations engage in market-driven practices that risk undermining that trust.

Ironically, customer relationship media can induce anxiety and distrust in advertisers. There are growing public concerns about information profiles falling into the wrong hands. There are also complaints about the distastefulness of being treated as a member of a category. Customers are also apt to become more anxious about their ability to secure the right or best deal, as they become attuned to the fact that other market segments are offered better deals, and that choices they may have made long ago can structure their future opportunities in wildly unpredictable ways. Moreover, as customers become more aware that they are receiving solicitations based on their specified or assumed needs and preferences, their relationship to advertisements can profoundly change. Receiving a solicitation from a cancer clinic or divorce lawyer in this new marketing environment can foster a sense of paranoia, raising questions about what the marketers might *know* about the individual.

The ability of new interactive television (ITV) formats to allow individuals to personalize, individualize, and customize electronic communications technology has prompted a series of heady proclamations about how audiences will now be empowered against the controls of the broadcast industry. Serra Tinic takes exception to such utopian accounts because they overlook the social, political, and economic context in which information technologies are converging.

ITV involves the integration of broadcasting and computer Internet capabilities through two-way communications. It allows the audience to personalize their TV watching practices and schedule customized viewing. Viewers can use set-top boxes to record a set number of hours of programming, and additional features allow the system to anticipate viewing preferences and send audience-members programs that meet their viewing profile. A key selling feature of ITV is the ability to skip advertisements, making traditional advertising an uncertain venture and fuelling claims about a new era of audience empowerment.

The true revolutionary attribute of ITV, however, derives from the degree of domestic surveillance it allows for advertising interests. As such, ITV fulfils the industry logic of television as an advertising platform as it seamlessly integrates synoptic and panoptic dimensions of surveillance. These efforts are simply the most recent development in a long history of audience measurement initiatives that aim to gather information on the link between the type of programs watched and consumption patterns. Audiences are thereby transformed into a commodity to be sold to advertisers. ITV, however, takes such long-standing measurement initiatives a quantum leap forward. As individuals start to log on to their television, marketers will receive immediate detailed knowledge of relationships between programming, advertisements, and composite profiles of TV surfing patterns. This knowledge facilitates one-to-one targeted marketing and raises the prospect of customized advertising and program content designed to match audience profiles.

One unique attribute of ITV is how it brings an extensive data collection venture into the literal and symbolic heart of the home. Some consumer groups see this as a form of privacy violation and have lobbied for procedures requiring that customers explicitly 'opt in' to such monitoring. The advertising industry has consequently turned its attention to strategies to encourage individuals to opt in. This includes placing some of the most appealing programming outside of the reach of those who do not opt in, or charging a higher premium for those who do not opt in. They have also targeted those consumer demographics with few compunctions about surrendering information, such as children who are being socialized into a world of television and interactive technologies.

Tinic draws from Turow to distinguish between society-making media and segment-making media. Where the first can contribute to the formation of a broad public sphere, the latter is primarily concerned with creating a market. The rise of ITV diminishes the prospects for a broad television space, as its programming speaks only to those viewers who share its interests and market positioning. The corporate interests behind ITV contribute to a greater fragmentation along class, gender and ethnic lines, as they ensure that only select few – those with money – can fully participate in this new medium.

Ultimately, ITV fulfils the logic of television as a medium of consumption. Notions of empowerment are naively positioned within the liberal market realm of individual choice, ignoring how the power to shape the parameters of decision making remains in the hand of the industry.

New electronic surveillance devices not only make us knowable to others, but also allow for new ways to know and govern ourselves. Emily Martin shows how interior psychic states are being visualized and used for culturally specific purposes. She focuses on the mundane technology of the 'mood chart,' used by individuals diagnosed as manic depressive. This technology elicits information regarding a subjective interior domain and publicizes it for the use of assorted experts and institutions. It monitors and optimizes interior mood states, encouraging the rational management of a subjective domain.

To accentuate what is new and interesting about such charts, and the types of governance they encourage, Martin compares them with Emil Kraepelin's famous charts of moods originally published in 1921. For Kraepelin, manic depression was a natural disease that had an accepted and inevitable progression. Although he believed that individuals who had been institutionalized as manic depressive could not be cured, he held out hope that prevention was possible. Kraepelin developed a special form to categorize the information on each patient to be completed by physicians. The charts represented a linked number of mood states existing within the category of manic depressive, each of which was colour coded. Individuals were understood as being capable of cycling through mood states in an immense variety of ways over varying periods of time, from a period of years, to months, or even portions of months. The charts simply documented how one subtype of mood succeeds another, often demonstrating that patients could enjoy frequent and prolonged symptom-free periods.

Contemporary mood charts are radically different. Maintained by individual patients for the benefit of government, health organizations, patients, and families, these charts are now ubiquitous in books, magazines, and on the Internet. Individuals are required to list simple everyday feelings and behaviours. These are assumed to be self-evident, hence there is no need for expert mediation or diagnosis of these emotional states. The charts themselves have been intensified, now displaying only one day per page, which is divided into hours and minutes. The forms also contain a place to record efforts to ameliorate mood disorders, typically by encouraging individuals to chart their medication in relation to their mood, a process that transforms patients into a form of biofeedback mechanism in relation to their medications. The charts themselves provide comparatively little space for having 'normal' moods and individuals seldom self-identify as being normal, factors which tend to expand the experience of being 'abnormal.'

This type of personal accounting regime raises intriguing questions about resistance. The literature on resisting accounting systems tends to be rather despairing, suggesting that accounting regimes are only ever overturned in favour of new systems designed to address the gaps and inefficiencies of the previous regime. As such, 'resistance' to such regimes typically does not entail questioning the appropriateness or the meaning of the 'thing' being measured.

Martin analyses the broader social effects of such measurement systems. She demonstrates how the rise of a new accounting regime can stimulate activities that go beyond efforts to remedy the apparent shortcomings of the existing system. For example, new mood charts can change the definition of moods or alter the divisions between what information is recorded. As personal information becomes easier to compare, new axes of comparison and intervention become apparent, and individuals can be prompted to question the official criteria for mood disorders. By connecting an internal experience with numerical abstractions, mood charts transform moods into a social entity. At the same time they socialize moods, making moods putatively applicable to everyone, meaning that all moods can be compared. As Martin notes, for individuals who are categorized as mentally ill, having your experiences count as part of the social can itself be very significant.

Where engaging in acts of personal accounting is often seen as a quintessential act of conformity, the type of hyper-rationality performed by patients completing mood charts can be viewed as a form of resistance. Mentally ill people are so categorized because their ability to act rationally has been called into question. Completing the charts is a way to demonstrate rationality.

The cumulative effect of this record keeping is the emergence of new populations and new cultural relationships to mental diseases. A bifurcation between the manic and depressive extremes of bipolar disorder is articulated in public culture. Mania is now presented as something to be maximized in order to enhance productively. In contrast, depression is seen exclusively as a liability. In this imagining it would be optimal for depression to simply disappear so that a form of manic energy could be cultivated in the interests of commerce and industry.

Like Emily Martin, David Wall is concerned with how apparently mundane technologies can be connected with more monumental social processes. In particular, Wall examines the everyday phenomenon of unsolicited bulk e-mail, or spam, to demonstrate how a familiar nuisance of electronic communication raises issues that strike at the heart of informational capitalism.

Surveillance practices have helped expand and intensify informational capitalism. These practices are particularly apparent on the Internet, a transparent media that allows for remarkably detailed forms of monitoring. The information flows on the Internet are multi-directional, involving both panoptic and synoptic dimensions. Hence, like the panopticon, there is the ability for the powerful to monitor others, but not necessarily in a unidirectional manner.

Computerized personal surveillance involves one-to-one monitoring that can be accomplished through new forms of spyware. Spyware provides remote computers privileged access to a distant computer, and programs that record keystrokes or take a snapshot-like recording of a computer. New 'spam spider bots' crawl through 'www' sites collecting information about people, while small programs stored on a user's computer, called 'cookies,' allow owners of 'www' sites to surveil Internet users. Decentralized peer-to-peer technologies also allow others to monitory a person's 'swap' directories as part of file-sharing practices.

Mass surveillance, in contrast, typically involves scrutiny of a complete population. This can involve, for example, third-party cookies which compile information about users, matching them with e-mail addresses as part of mass profiling efforts. Another example is the electronic interception system known as ECHELON which extracts messages of interest from bulk or unwanted messages. The Total Information Awareness system exemplifies the rise of data-matching strategies which integrate a range of databases to monitor and produce data (see chapter 6 in this volume).

The spam industry approaches information as a form of capital. Spamming is primarily concerned with the compilation of bulk e-mail lists. Online users' private information is acquired through inducement, deception, or coercion. Individually, each discrete bit of information is not very useful, but when combined with other data it becomes valuable, particularly when sold to companies as lists of potential prospects. Most of these lists are not confirmed, and hence do not provide the types of intended responses and audiences that advertisers might desire. Confirmed lists are therefore much more valuable and expensive, and as result electronic solicitations will go to great lengths to try and entice a response from an e-mail recipient as a way of confirming their list.

Wall points to a general public misperception about spam. While the actual text messages of most spam involves solicitations for pornographic web sites, gambling opportunities, new surveillance devices, and pyramid schemes, the true aim of most spam is simply to elicit a response as a way to confirm a list. As few as 10 per cent of spams are legitimate attempts to sell products or services, and as many as one third of spam messages are a form of 'spoof spam' designed to elicit a response. This elicitation can be accomplished by programs that generate an automatic response when the message is opened, or text messages that are so inflammatory that they prompt a passionate response. Invitations to deregister from future unsolicited mailings can also confirm that a particular e-mail address is valid. When a person responds to any of these messages they have effectively confirmed their e-mail address, thereby enhancing the financial value of their address to marketers, and paradoxically increasing the volume of bulk mail that they will likely receive in the future.

The impact of spam can be approached at the level of the individual computer user or at a broader societal level. Individually, spam poses several different types of dangers, the most troubling being the risk that computer users will become embroiled in a direct relationship with a fraudster. Spam can also involve entrapment schemes that operate through embarrassment or blackmail. That said, unsolicited e-mail seems to pose less harm to individuals than might be assumed. While those who are not Internet savvy certainly face risks, increasingly sophisticated spam filters, combined with greater user awareness, seem to be reducing spam to the status of unwanted detritus of the information highway.

At the same time, however, spam risks undermine the investment of basic trust in these new informational technologies. Trust is crucial for a thriving informational economy. Genuine efforts to engage in e-commerce face the prospect of becoming lost amongst all of the noise, spams, and fraud.

Such developments in surveillance underscore the point that surveillance is not necessarily concerned with optics. For example, one of the most important surveillance techniques involves the little understood practice of 'data mining.' In his contribution, Oscar Gandy Jr explores the operation of data mining and related forms of knowledge production. He argues that such techniques both transform social relations and reinforce social cleavages.

Data mining is a specialized area in the field of applied mathematics that attempts to derive meaningful patterns from large sets of data. It differs from ordinary information retrieval by virtue of the fact that the information to be acquired by data mining must first be discovered within the data itself. Data mining seeks to transform raw data into strategic intelligence, and to generate hypotheses that can be explored in greater detail through other means. While data mining is a fairly new development, it relies on long-standing practices of assigning individuals to classes, categories, or groups on the basis of seemingly independent activities.

Data mining is augmented by techniques that allow analysts to visualize patterns of relationships within data. Some of the more sophisticated techniques include the use of 'neural networks' that can develop fine-tuned predictive algorithms even when working with a larger number of variables. Other techniques can help identify the variables or attributes of objects that are the most influential in predicting outcomes.

Data mining is now done by more people at more organizational levels and using more data than ever before. This expansion in dataveillance has been enabled by the diffusion of relatively inexpensive computers and the greater availability of data. Both private and public sector databases have increased along with the growing capacity of information technology to capture, store, and retrieve information. Databases have become more efficient, less expensive, and less demanding of specialist training. Moreover, there is now a greater ability to integrate previously independent data systems.

One of the most prominent commercial uses of data mining is target marketing. Target marketing evolved out of the opportunities for informational profiling that came with the advent of postal zip codes in the 1960s. Some database vendors now provide clients with lists from government databases that can specify, for example, the types of voters/consumers an advertiser or politician might want to reach. These types can be broken down according to remarkably fine-grained discriminations among market and population segments.

Data mining has also been recognized as a useful tool for security applications. For example, it has been used to try and identify attacks on computer systems based on deviations from the normal flow of server traffic. The terrorist attacks of 9/11 have provided greater impetus and official support for using new data-mining strategies for security efforts.

With all of the apparently beneficial uses of data mining come a series of deficits or troubling tendencies. Data mining in both commercial applications and security provision can reinforce disparities among populations. It also raises questions about privacy rights. In particular, the individualized focus of existing privacy rights makes them poorly suited to deal with the population-level group identities formed through data-mining exercises. Indeed, questions of equality, rights, and accountability are central to the new politics of surveillance and visibility.

11 Cracking the Consumer Code: Advertisers, Anxiety, and Surveillance in the Digital Age

JOSEPH TUROW

Critics have noted that a central strategy of the modern advertising industry centres on working for consumer trust while evoking consumer anxiety. They point out that ads aim to stir their targets' deepest angst about love, social acceptance, and success, and to then tie solutions to certain products and services. They also realize that in addition to playing on consumer anxiety, they must achieve successful sales by gaining consumers' confidence that the products will do what the ads say they will.

While certainly important, the long-standing critical focus on ad content's relation to anxiety and trust takes attention from two other strategic aspects of consumer-related trust. One relates to the media that carry the ads. Advertisers and their agencies must have confidence in the ability of those companies to place commercial messages in front of the right consumers at the right times. The second key area of trust involves audiences that the media firms try to reach. Ad executives must have confidence that substantial proportions of the target public will actually pay attention and look at the commercial messages as a prelude to being persuaded.

Critical observers of the ad industry take for granted advertisers' ability to ensure both these activities. They consider it obvious that over the past century, a phalanx of research companies has evolved to give the ad industry confidence that media firms generally do reach their audiences and that large numbers of people do attend to commercial messages. During the past few years, however, these two precepts have come under fundamental challenge, particularly when it comes to advertising aimed at the home. Advertisers look with consternation at the escalating number of technological and legal roadblocks that anti-ad enthusiasts have been erecting to mass marketing in the digital world. Media planners face spam filters, pop-up filters, new anti-ad laws for cell phones, and ad-skipping buttons in personal video recorders. Marketers fear that these instruments encourage increasing proportions of the public to abrogate an implicit contract with media firms that promises attention to ads in return for free or discounted media material. By extension, some fear that the new circumstances will make electronic media firms unable to ensure that ad messages actually reach their audiences.

Advertisers are engaged in a welter of rhetoric and actions that reflect a hope that they can 'crack the code,' in the words of one executive – that is, figure out ways to use media and audiences to achieve the product attention the advertisers need. This essay melds 'contextualist' and 'resource-dependence' perspectives into a framework for identifying activities that will bring the most changes to marketing in the digital world, and for understanding their social implications. The activities are customized media and interactive television. They trace back to a couple of longtime initiatives to grab consumer attention, product placement and direct marketing.

At first glance these businesses may seem unrelated to one another and peripheral to mainstream advertising. A closer look, however, reveals a converging strategic logic towards encouraging what might be called *customer relationship media*. The aim is to fundamentally change the ways advertisers reach out to consumers through media at home and even out of home. At heart, they involve surveillance of consumers in the name of trust. Moreover, the activities actively encourage new forms of consumer anxiety under the rubric of customer satisfaction.

Analysing Strategic Institutional Change

Organizations engage in strategic changes when they depart from customary activities to reshape basic relationships with their environments. As sociologist Andrew Pettigrew (1985) points out, attempts to track the logic, trajectories, and social implications of these activities are vulnerable to two traps. One is the dilemma of trying to see everything about the organizations that are being studied and getting muddled in irrelevant areas. The other, opposite, predicament involves trying to explain complex changes by focusing on only one or two influences and so appearing mechanistic.

To avoid these difficulties, Pettigrew makes a persuasive case for what he calls contextualism. This view insists on keeping the focus on the organizations' strategic activities in relation to their environment. It posits, however, that the exploration be carried out by examining the multiple organizational, social, economic, and political influences that ignited and structured the changes.

Pettigrew usefully points to the importance of multiple dimensions of context on organizations. Especially important is his suggestion that prominent strategies to face complex new environmental challenges often evolve from organizational activities that have not seemed at all central. A corollary to this idea is the possibility that seemingly unrelated responses to the challenges by different organizations might actually share a strategic logic – that is, assumptions by their creators about the best ways to move forward. If leaders across the industry begin to recognize these commonalities, and if technologies allow the interpenetration of the activities, the level of change within the industry may well be profound. So might be the social implications of that change.

This perspective has proven central to an understanding of the way media and advertising executives have been responding to the increasing ability of audiences to eliminate ads to the home. What Pettigrew's approach does not provide, however, is a framework for understanding specifically what about organizations and their environments to study. Here a resource dependence perspective is useful.

According to the resource dependence perspective, major goals of organizational leaders are avoiding dependence on other organizations and making others dependent on them. The general picture, one which meshes with the contextualist framework, is of decision makers attempting to manage their environments as well as their organizations. Power, in this view, is equivalent to the possession of resources. Emerson (1962: 32) adds that 'influence [perhaps a better word is leverage] is the use of resources in attempts to gain compliance of others.' It is through this struggle over resources on several organizational and interorganizational levels – and across time – that industry strategies develop.

Media organizations must depend upon entities in their environments for resources that help production, distribution, and exhibition. Those entities might demand substantial recompense for their help, especially if other organizations are competing for the same resources. Firms must cope with such demands by exchanging their own resources. They also try to manage their dependence on organizations and publics in their environment so as to be flexible in getting resources.

Advertisers and Resource Dependence

Advertisers occupy a key role in this institutional struggle. The reason is that in many media industries advertising money is most directly responsible for the solvency of producers, distributors, and even exhibitors. The interconnection of advertisers, ad agencies, media-buying companies, and market research firms in supporting production, distribution, and exhibition in media industries underscores that advertising should be seen as an industry in itself, rather than a single type of organization. The captains of the ad industry are huge marketing communication conglomerates (MCCs) such as Omnicom, WPP, and Publicis, which hold under their umbrellas several ad-agency networks, media-buying firms, public relations companies, and various kinds of research outfits. While advertisers put up the money to proclaim their products and services, ad agencies, buying companies, research firms, and increasingly, MCCs, often set the agenda for using the funds.

Advertising involves payment for calling attention to a product, service, or need. Whether the subject is Coca Cola, a cleaning store, or the United Way, the goal of advertising is straightforward: to persuade people to purchase or otherwise support the product, service, or need.

It is in the need to get audiences to believe in these stories and the sales messages that underlie them that advertising storytelling and audience trust intersect. As Anthony Giddens (1990) points out, a trusting public is a critical resource for sustaining the organizations of an institution. To keep their authority within the institution, these organizations need to continually convince the public of the soundness of the basic values and activities. At the same time, the resource dependence perspective points out that the limits of pursuing public trust can go only so far. Organizational leaders come to realize that while public allegiance is important, they must meet other demands on them that often conflict with the public's perceptions of its interests. Organizations may sometimes need to surreptitiously betray public trust to meet other resource goals. Advertisers, for example, will put the best public face on a product because revealing its flaws may lead to economic loss.

Despite its immense power over the media system, advertisers are in most cases dependent on media producers to deliver their ads to the right number and kind of people. Advertisers also depend on media firms to lead audience members to believe that the presence of commercial messages is legitimate and ought to be a taken-for-granted fact of life. In turn, advertisers depend on their audiences to trust the basic importance of advertising and the need to attend to it.

But advertisers' need to cultivate public attention, trust, and legitimacy by relying on other institutional actors raises the following question: What strategic actions toward media firms and its audiences would ad industry firms take if they perceived that a fundamental breakdown in the organizational actors' ability to satisfy advertisers' needs is leading to a major failure in the advertising industry's reason for being? This paper sketches early conclusions in an ongoing project that aims to address this and related issues. Following one of Pettigrew's exhortations, the emphasis here is on the ongoing (historical and contemporary) context of change. Specifically, the paper focuses on the ways that the forces influencing current strategic changes by the ad industry towards media and audiences developed through the decades. The findings suggest a sobering trajectory regarding advertisers' exploitation of trust, surveillance, and anxiety to ensure attention to their messages in the evolving digital environment.

A Breakdown in Trust

No one can fully assure attention to commercial messages. From the beginning of the first sponsored domestic electronic medium, radio, stories in the trade and out were of people talking, visiting the kitchen, or using the bathroom during commercial breaks. An extreme example of audience inattention in *television's* early days relates to the wild popularity of Milton Berle's program. It seems that so many people flushed the toilet simultaneously during the sponsors' segment of that show that the water pressure in some cities actually dipped noticeably (see Barnouw 1970).

Advertisers tended to view this sort of distraction as nothing new. They knew that for centuries newspapers and magazines could not guarantee that their readers would actually look at, let alone read, every print ad when they turned pages. Similarly, advertisers understood that radio and television stations and networks had no way to force people to stay in their seats when the commercials came on. Advertising agencies, in fact, saw garnering people's attention through compelling ads as one of their primary challenges.

Reactions by advertising practitioners to the remote control during the 1980s and 1990s reflected a mixed sense of blaming themselves and the television industry. Ad executives who looked to their own industry argued that audiences simply did not find commercials interesting enough. They exhorted creative types that making commercials that people would want to watch would go a long way towards meeting this technological challenge. Others blamed the television industry for creating so cluttered a TV environment that viewers would try to escape it now that they had the technology to do so.

The rise of digital interactive media in U.S. homes during the 1990s added a new spin to the problem of ad skipping, which some in the ad industry saw as potentially far more dangerous than the earlier analogue technology. Many in the ad industry began to worry that the digital world was generating a different level of consumer anger towards commercial messages in domestic media. More important, they worried that it creates a different level of ability to act on that anger and so sever the ad industry's basic bonds of trust in mass media organizations and audiences.

Two angles of attack were especially disconcerting. One involved attempts to get rid of unwanted Internet advertising, especial pop-ups and spam. The other involved the presence of personal video recorders that allowed viewers to ignore commercials. Both seemed to foreshadow a new era that would encourage consumers to eliminate, not just ignore, ads.

Pop-ups are magazine-like ads that thrust themselves onto a person's computer screen before or after a person views a web page. Nielsen/ NetRatings estimated that from the first to the second quarter of 2002, the number of pop-up ads grew from about 3.9 billion to nearly 5 billion impressions (Hansen and Stefanie Olsen 2002). Spam, a direct descendent of commercial junk mail, imposes itself into an Internet user's e-mail in-box without prior permission. Various web-auditing firms estimated that in 2003 over 40 per cent of all email fit this characterization.

Mainstream marketers distanced themselves from spammers, insisting that there is a big difference between the e-mail they send to known customers (even if the customer hasn't asked for it) and the barrages of messages that spammers supposedly sent to anyone. At the same time, mainstream marketers justified their active use of pop-ups as a vehicle that helped to sponsor the websites that their target audiences visited. The broad marketing industry watched in consternation, however, as angry technologists and nervous Internet service providers (ISPs) developed technologies that promised eventually to disable all pop-ups and eliminate all unwanted e-mail, not just spam.

Free distribution of 'pop-up killers' was already common on the web by 2003. In that year, AOL announced that it would no longer accept such ads for its 34 million customers and would give its users software to block pop-up ads on other websites. A prior champion of this ad form, AOL changed its tune in a bid to hold onto subscribers who were in danger of leaving for Internet service providers such as Earthlink that had less commercial clutter. (Earthlink had earlier begun promoting pop-up blocking as a feature of its service; see Hansen and Olsen 2002 and Hansell 2003.)

Spam was a harder creature to kill, but it was also a more expensive one for ISPs to carry. AOL also trumpeted its presence at the forefront of trying to eliminate spam, which it defined as 'unsolicited bulk e-mail' (see http://www.aol.com/info/bulkemail.adp). In early 2003, the company won a major $6.9-million judgment against a firm accused of sending more than a billion junk e-mail messages promoting sexually explicit websites to AOL customers. A website announcement of its victory promised more court activity and added that it would improve its automatic spam mail filters (Weiss 2003). In the popular press and among Internet specialists, better filters meant a better Internet life. Direct marketers saw the situation quite differently. They understood that preventing unsolicited contact of people via the Internet posed a danger to them and to legitimate advertising. Traditionally, after all, much salesmanship depended on reaching out to new, unsuspecting contacts. Increasingly, technology specialists such as Paul Graham (2002) were promoting techniques such as Bayesian filtering which, they hoped, would expunge from their e-mail boxes not just unsolicited bulk e-mail but all commercial messages.

Efforts to technologically disable web ads and commercial e-mail were not the only activities that distressed marketers. They also became nervous that digital technology could fundamentally disrupt media attempts to deliver messages to consumers at the centre of the marketing world: network television. At issue was what the digital video recorder, also called the personal video recorder (PVR), would do to television advertising. Essentially a Linux-driven computer with a large hard drive, the PVR is like a VCR in enabling its owners to record programs and view them at other times. Unlike a VCR, the technology marketed to the public by TiVo and Replay is connected to an updatable guide that made finding programs across more than one hundred channels easy. Also unlike a VCR, in some versions made by Replay (and in hacked versions of TiVo) it allowed viewers to skip ahead thirty seconds at a time without at all viewing what was skipped. That, advertisers knew, would be commercials. In fact, Replay used its PVR's facility for skipping over commercials as a selling point in its ads.

Both TiVo and Replay tried to assure advertisers and media firms that they were not fundamentally out to destroy commercials. Others disagreed strongly. They pointed out that the sales rate of branded PVRs was increasing and that home satellite firms and cable systems were beginning to integrate unbranded versions into set-top boxes. They noted Tivo's admission that 60 per cent to 70 per cent of people watching via its technology were skipping commercials. And they admonished that whatever accommodation advertisers would make with PVR firms, it would undercut the by-then traditional approach of mounting fifteen or thirty second commercials within shows.

Jamie Kellner, chair-CEO of Turner Broadcasting, warned that PVRs were destructive to the TV business, contributed to lower ratings, lower ad revenue, and fewer quality programs for TV distributors. 'What drives our business,' noted Kellner, 'is people selling bulbs and vacuum cleaners in Salt Lake City. If you take even a small percentage away, you are going to push this business under profitability' (Friedman 2003: 13). Agreeing, a Hollywood talent agent urged marketers to understand that they could no longer present ads through home-based electronic media in traditional ways. 'The genie is out of the bottle,' he asserted (Friedman 2003: 13).

A New Strategic Logic

Clearly, traditional commercial forms in traditional media still represent by far the most prevalent approaches in the first decade of the twenty-first century. Nevertheless, advertising practitioners firmly believe that the genie *is* out of the bottle. They insist that new digital technologies that allow for the elimination of commercials mean they must be prepared to use new ways to ensure that consumers attend to their electronic solicitations in the home. Increasingly, they are turning to alternatives to standard advertising as instruments to force consumer attention. As these separate sets of activities develop, they are coming together in an industry strategy for reaching the public that holds new implications for information privacy and ad-induced anxieties. To understand the trajectory of this strategic change it is useful to sketch the background of two approaches from which these activities evolved: product placement and direct marketing.

The Rise of Product Placement

As one firm involved in the business defines it, 'product placement is the process which integrates an advertiser's product into movies and

TV shows for clear, on-screen visibility' (MMI website, March 2003). The description might have added that radio had audio versions of this 'clear' placement. During the first four decades of that medium, from the 1920s through much of the 1950s, many programs integrated the sponsors' products into skits, jokes, songs, and other material that was part of the entertainment. Television entertainment programs of the 1950s continued this activity. The reason for it was straightforward. In those days, advertisers owned the programs, their ad agencies produced the shows, and they were quite happy to allow their stars to show the audience that they cared about the product.

As commercial television moved away from radio's mode of full sponsorship towards the selling of scattered sponsorship spots beginning in the late 1950s, the direct connections between the program content and advertisers was loosened considerably. Some avenues for product placement remained. Game shows paraded prizes that their producers had traded for a mention and cash. Also notable was the Ford Motor Company's provision of vehicles to prime time dramatic programs in exchange for the publicity. Through much of the 1980s, though, the three major television broadcast networks did not allow producers of prime time programs to present products in exchange for cash. One reason was network executives' fear that when added to the time for traditional commercials, product insertions would exceed the maximum time for commercials that the broadcasters had announced in response to pressure from the Federal Communications Commission.

It was the success of product placement in movie theatres during the 1980s that got television producers thinking of re-introducing the form to the home tube. In 1982, *E.T.: The Extra Terrestrial* jump-started the lucrative modern placement business when the film increased the sales of Hershey's Reeses Pieces after the candies made up a key plot point. (Mars had refused to pay the producers' asking price to get M&Ms into the movie, so the firm had then approached Hershey). Research companies began to audit audience recall of product placements much as they assessed commercials. The 1988 movie *Crocodile Dundee II*, for example, included the prominent handling of a Nikon camera. An exit survey of audience members included questions about the presence of Nikon and Cannon, even though only the first was placed in the movie. Nikon, which was held by a cast member in the film, had an overall recall level of 48 per cent. The recall on Cannon, which was not in the film, was 10 per cent (Sharkey 1988).

Such findings excited marketers, who were looking for ways to get their products to stand out in a cluttered ad environment. With an increasing number of film companies integrating products into scenes to help defray costs, television producers (whose firms were often owned by the same conglomerates) moved to do the same. They were pushed into doing it by a confluence of factors: substantially higher production costs; program licence payments from the TV networks that no longer covered production costs (due to lower real ad revenues); and increasing uncertainty that they could find profits in the local and global syndication markets. They could get away with it by the late 1980s and early 1990s. One reason was that the National Association of Broadcasters abandoned its code that included limits on prime time commercial length. Another was that federal administrators of the period did not care to regulate commercialism and sympathized with the television industry's difficulties in making money in a multi-channel era.

By the early twenty-first century, product placement had become a business that ranged widely across all forms of television. The Big 4 broadcast networks (CBS, NBC, ABC, and Fox) were deeply involved, especially in 'reality' programs such as *Survivor*. According to *Advertising Age*, the activity was most fine-tuned on cable TV, especially cable networks heavy on sports programming. A growing number of cable networks were 'aggressively exploring ways of integrating sponsors' messages directly into programming as part of more complex media packages that also include ... traditional 30-second commercials' (Fitzgerald 2002: 518).

The Fox SportsNet's *Best Damn Sports Show, Period*, for example, became what *Advertising Age* called a 'laboratory for new sponsorship efforts.' That included converting the talk show's studio into a lemonade stand to promote Mike's Hard Lemonade. In another product promotion within the program, hosts Tom Arnold and Chris Rose described Dockers pants as 'stain proof' and used this as comedy material in the talk show. The Dockers jokes aimed to appear to be spontaneous, but they actually were part of Docker's media buy (Fitzgerald 2002).

It all seemed very much like the early years of television, except that the strategic purpose had changed drastically. Although some cable programmers intent on selling traditional time slots scoffed, others acknowledged that the surge of advertisers' interest in product placement was part of their attempts to learn how to place their products in front of consumers when cable and broadcast TV networks could not reliably present commercials to them. *Advertising Age* concluded that 'advertisers and networks are scrambling to integrate commercial messages into programming to blunt the effect of sophisticated replay devices, like TiVo, that allow consumers to skip over commercials more easily' (Fitzgerald 2002: 518).

Direct Marketers

Paralleling the growth of product placement on television in the 1980s and 1990s was a very different advertising business from image-oriented product placement: direct marketing. Split into two rather distinct trades around the 1920s, 'image' advertising and direct advertising had developed different cultures. As late as the mid-1970s, the elite of the ad world looked down on 'direct' work. They associated it with the selling of shoddy or quirky mail-order products – breast enlargers and Ginsu knives – by preying bluntly on the audience's anxieties about being left out socially. Image-ad practitioners mined the same psychological tactics, yet they saw themselves as doing it much more subtly than direct marketers, via campaigns that enhanced the long-term reputations of brands.

A common example of direct marketing was an ad in a magazine that encouraged a reader to write for a free catalogue. Another was a set of coupons mailed to the home. In both cases, sales could be tracked by the company giving the discounts. Direct marketers saw their ability to get clear feedback from their work as the heart of what they did. Mainstream agency people typically revered ads that encouraged their competitors and their audiences to wonder at the imagination that made them. Direct marketers snickered at that meaning of creativity and scorned prizes for it. They argued that kudos should be awarded only for campaigns that demonstrably increased sales. That, they contended, was something mainstream advertising had a hard time doing because it was often impossible to draw a causal link between a commercial message and a consumer's purchase response.

Despite these marked philosophical differences, by 1982 fifteen of the top twenty advertising agencies had bought or started a direct-marketing capability (*Advertising Age* 1982). It was primarily their clients' growing interest in computer-guided targeting of niche markets that brought them to look at direct practitioners with grudging respect. A Dun and Bradstreet executive saw the changes as paralleling the multiplication of U.S. media channels and the distribution of audiences across more outlets than ever. The traditional hallmarks of direct marketing – precise audience identification, individualized media communication and speedy full-satisfaction order fulfilment – meshed constructively, he said, with the new technologies (MacDonald 1983: M9). Futurist John Naisbitt (in Pagnetti 1983: M9) was even more blunt. 'Direct marketers are at the forefront of where everybody is going to be,' he predicted in 1983. 'We can all learn from them.'

The range of direct work grew wider in the 1980s and 1990s than ever before. The number of TV commercials that invited viewers to use 800 numbers and credit cards to buy products by mail increased dramatically, especially on cable. Cable and independent outlets were also vehicles for longform ads called infomercials. Taking the infomercial one step further were shopping channels that invited immediate purchases by phone twenty-four hours a day. Still, while TV-based platforms for consumer responses were especially visible in the 1980s and 1990s, it was via the mail and the telephone that most targeted advertising took place (see Turow 1997: 129).

Direct marketers' penchant for dividing consumers fit with the movement towards increased specificity about audiences in mainstream advertising. Nielsen TV audience data; MRI and Simmons syndicated data on demographics, psychographics, and purchasing patterns; geodemographic extrapolations from database firms Prizm and Conquest – these and other storehouses of information were scavenged by ad, promotion, and publicity people as well as by direct tacticians. In direct response work, as in standard advertising, public relations, and event marketing, the focus was on reaching out – signalling – to a particular population with certain categories.

Where the direct-response business diverged from the mainstream, and where marketers saw its greatest possibilities, was in the connections that its practitioners could make with their targets. The 800 number and the personal computer made it much easier than in the past to link up with potential customers quickly. The technologies also made it easier than before to track the 'pull' of an ad (Delay 1980). On a cost-per-thousand basis, this approach was clearly more expensive than using magazines or network television. Direct practitioners insisted that what they lost in efficiency of reach would be more than made up in the careful selection of people likely to act on the sales pitch. The trick was to get good lists of likely prospects.

That in itself was not a new challenge. Companies were purchasing names for marketing purposes in the nineteenth century (see Rapp and Collins 1990: 110). The computer's ability to store and sort the names of millions of people and their characteristics, however, moved the list business – now called the database business – into overdrive. By the

1990s the buying and selling of names and information about them had become a major industry. Broadly speaking, direct marketers used two resources for name gathering: universal databases and transactional databases. Universal databases are compilations of information on every individual and household in an area, even an entire nation. Inferring buying interest from a range of demographic, psychographic, and broad lifestyle information was often helpful as a starting point. Yet many direct practitioners of the 1980s and 1990s followed the long-held dictum of their business that led in a different direction: the best predictor of future behaviour is actual past behaviour. This proposition underscored the importance of transactional databases. A transactional database is a list of people who explicitly responded to a particular marketing or fund-raising appeal. Marketers would often purchase names from other marketers with products that reflected compatible lifestyles. So, for example, names of men who paid for season tickets to sporting events might attract a company trying to sell sports memorabilia. Similarly, a frequent-flyer list would likely draw a hotel chain with an eye on the business traveller.

Increasingly into the 1990s, however, advertisers began to believe that as important as prospecting for new customers was, marketers should pay more attention to the customers they already had. The reason was the finding that a high percentage of a company's profit comes from repeat purchasers and that it costs several times more to get a new customer as it does to retail a loyal one. The key to believing that it was worth the effort lay in recognizing a repeat customer's lifetime value. Such value was clearly evident to airlines, whose frequent flyer programs were emblematic of the new approach. Similarly, upscale store and hotel chains saw the utility of keeping updates about their customers and contacting them on regular bases.

As the 1990s progressed, manufacturers of inexpensive 'package goods' – diapers, cereals, soups, inexpensive cosmetics, over-the-counter pharmaceuticals – also moved towards tracking and wooing individuals one on one. Enormous competition for shelf space in supermarkets and department stores forced executives from even the largest manufacturers to find ways to explore the hypothesis that repeat customers, properly handled, could help keep brand prices up and niche brands on the shelves. The idea was that if a marketer could identify loyal consumers of corn flakes as people whose habits made them likely to buy a range of the firm's products over a large number of years, using direct channels to establish relationships with them might well be worth the cost.

The proposition presupposed getting information about those heavy users, and marketers were increasingly adopting a variety of methods to gather data about those they called their best customers. They were asking for information from consumers on sweepstakes forms, mail-in rebates, product-registration cards, and in-box coupons, as well through 800 numbers and at events aimed at their target audiences. According to a study commissioned by the Direct Marketing Association, by 1994 two-thirds of the leading retailers in the United States were using databases to build customer relations and store traffic by collecting and linking customer data with transactions (*Advanced Promotion Technology Vision News 1995*). Among package-goods firms, the ten most highly committed consumer database builders in 1995 were marketing giants: Ralston Purina, R.J. Reynolds, Quaker Oats, Gerber, Philip Morris, Dowbrands, American Tobacco, Kraft Foods, Sara Lee, and Kimberly Clark (*Progressive Grocer* 1995).

Keeping customers, direct marketers said, requires establishing 'dialogues' with the firm's consumers (Peppers and Rogers 1993: v–vi). The aim was to reinforce repeat purchasing with signs that the company was tailoring its activities to their needs and those of the people like them. 'What we're seeing are major attempts by package-goods companies to find a way to make this kind of relationship marketing work,' admitted a direct practitioner in 1993. 'Efficiency is the challenge – can we make it work so that it's profitable?' He also said: 'Package-goods marketers ... are enthused about the possibilities of database marketing, but no one has yet cracked the code (quoted in Freeman 1993: 52).

'Cracking the code' became the Holy Grail for advertisers and their agencies. A major attempt became known as loyalty programs. Loyalty programs involve firms giving customers increasing rewards or discounts based upon their increasing business with the firms. A subset of customer relationship marketing, the basic assumption of loyalty programs is, as one *Ad Age* writer put it, 'that companies must discriminate among customers.' The writer quoted an expert who put the issue bluntly: 'We argue strenuously, strenuously against naïve sentimentalism on the part of companies who insist 'We love our customers and we love all our customers the same."

This emphasis on individual differences leads to two corollary assumptions. One is that marketers should reward their very best customers. The second is that they should push away, even alienate, those who are less valuable.

Airline frequent flyer programs pointed the way, with American Airlines being the first, in 1981. The customers that an airline wants most to cultivate are naturally the ones that had the most business to offer. Airline computers single them out for special treatment by all employees who deal with them. They are given low-cost upgrades to First or Business Class, special reservations phone lines, special check-in counters, priority baggage handling, and private airport lounge privileges.

It is a lesson not lost on hotel, telephone, car-rental, banking, supermarket, electronics retailer, and even packaged goods companies with loyalty programs. 'Points' alone, loyalty experts agree, will not be enough to bind customers to a company. As one marketing trade writer put it, 'they are a means of assembling customer purchasing data, which can then be used to build real CRM.' That raises another assumption of loyalty programs – that customer loyalty cannot be secured without directly or indirectly eliciting personal information. 'Once these consumers have offered their contact information and some basic purchasing data,' the writer continues, 'the trick is to build a true loyalty program that incorporates customer-based segmentation with different strategies for each segment. A firm must also devise a progression of strategies that will change as the consumer matures' (Levey 2003b: 5).

Despite evidence that consumers hate some loyalty programs and critiques that some are simply not carried out well enough to instil devotion, the numbers and types have increased (Cyr 2003). A model that U.S. consultants have heralded is the cross-company Nectar loyalty marketing firm in the United Kingdom. With affiliates such as Ford and the Sainsbury supermarket chain, Nectar has managed to encourage millions of Brits to give up information about themselves for various purchase-driven rewards. Sainsbury creates purchasing segments from the customer demographic and shopping cart data that it feeds into a company-wide database. One, for example, is the 'Foodie' – an affluent consumer who enjoys serving and eating unusual premium foods. Another is the 'Starter Family,' less well off with one or two working adults and at least one small child.

Members of the groups receive incentives in the store and by mail that tie the rewards they get for increased shopping to the store's categorization of them. Foodies receive serving suggestions and mailings with images of gourmet-style products. Company executives believe that targeting this audience with cents-off coupons would be a waste. Instead, the company redesigned some of its private-label products to appeal to Foodies by creating elegant designs and covers on some of its lines. Starter families, by contrast, receive messages touting healthy meals that are easy to prepare. As consumers assumed from the collected data to be more cost-sensitive, they also receive ads that contain more price-based incentives (Levey 2003a).

An aspect of loyalty programs, and of relationship marketing generally, is trying to communicate with customers across a variety of venues and media. As the web grew in popularity, it became a place to connect interactively with loyal customers. Although building websites was de rigueur for marketers at the start of the new century, general agreement emerged that e-mail was the best way on the Internet to keep in touch with consumers, reward them with coupons, and send them to appropriate parts of company websites.

For large marketers, however, this approach raises problems about the digital age. One is the clutter caused by spam; customers might ignore or filter all commercial e-mail because of it. Another problem relates to the nature of the medium. Loyalty marketing as practised through postal mail, e-mail, product websites, infomercials, and telemarketing calls still stands at the psychological and physical periphery of media to the home. The context in which they appear is not associated with audiovisual news and entertainment that compels both attention and interest.

Reading through marketers' trade discourse and listening to them talk about the subject at industry meetings revealed a strong air of futility among some that any new technique or combination of techniques can make CRM and loyalty marketing a persuasive aspect of Americans' media lives. Those observers wonder why consumers would embrace alternatives to the thirty-second television commercial. Marketing consultant Erwin Ephron (2003: 30) has suggested acidly that many consumers would recognize the contradictions inherent in CRM when it came to consumer trust: '[H]ave you noticed how torn advertisers are between their warm and fuzzy strategies to build "relationships" and "trust" with consumers, and their hard-edge database and pricing programs to squeeze profits out of them? When "trust" is mouthed as a marketing strategy, God save us. It's as naive as it is cynical.' U.K. marketing expert Alan Mitchell (2003: 40) has similarly asked, 'how many CRM, product placement, point of sale, or viral messaging programs really offer big wins for consumers?' Clearly implying that few programs do, he contends that 'in the absence of alternative win-wins, the old model, creaking as it is, will probably remain superior.'

In March 2003, Rance Crain (2003), the editor-in-chief of *Advertising Age*, captured the nervousness swirling around advertisers' approaches to consumers and media in an editorial titled 'urgency of change is touted but marketers favor safe bets.' He noted that network and cable TV were raking in lots of money for traditional commercials even as marketing executives were saying that 'if the new model [based on nontraditional ways to reach consumers] isn't developed, the old one will simply collapse.' He ended his piece, however, with the observation that 'Some of our forward thinkers preach radical change. The reality is that forces embracing the status quo are fighting a very effective rear-guard action.'

It seems clear that many in the ad world see traditional media approaches as rear-guard actions, activities that will necessarily fade as they are proven less and less successful. Crain, Mitchel, Ephron, and others acknowledge that many in the marketing establishment are groping for perspectives that will allow them to trust that the emerging mainstream digital media environment is a vehicle for their commercial messages. These sceptics point out that no common roadmap to those perspectives has evolved.

Converging towards Customer Relationship Media

A roadmap is, however, emerging. It is developing out of the seemingly disparate activities of product placement (with its emphasis on mainstream audiovisual media) and direct marketing (with its emphasis on databases and interactivity). Advertisers are experimenting with activities that aim to insert themselves unfiltered into their desired customers' domestic lives in ways that encourage consumers to accept surveillance and relationships tailored to their personal characteristics. Two labels are increasingly central to marketers' discussions of the evolving ad world: customized media and interactive TV. Though the activities to which the labels refer increasingly overlap, ad-industry practitioners associate them with different ad vehicles and aims. Moreover, technological drawbacks now make it impossible for all the activities associated with each label to take place across all ad platforms.

Nevertheless, discussions of the labels share a strategic logic about the best ways to address marketers' declining trust in traditional media's ability to stop audiences from eliminating or abandoning ad messages. The logic can be parsed into three propositions that bring together the traditional goals of product placement and direct marketing. First, a firm must reach wanted consumers in the main stream of their media activities in ways that make them feel connected to the company and its products. Second, the connectivity must encourage consumers to give up information that can help the company fine-tune discriminatory sales pitches. Third, the selling environment must be structured to make sure that they pay attention – or at least do not eliminate – their commercial messages.

Taken together, the labels and the programs to which they refer point strongly to marketers' interest in the evolution of all home-based information, news, and entertainment vehicles into what might be called customer relationship media. A look at major assumptions underlying discussions of the programs underscores this interest. It suggests, too, that in the emerging digital world, the social message linking product placement to surveillance-driven CRM will be an exhortation to 'feel special, and anxious.'

Customized Media

Customized media are those that apply the concepts of loyalty programs to help themselves and their advertisers gain the fidelity of desirable audience members. Such media vehicles already exist, though with highly varied degrees of sophistication. Perhaps surprisingly, of the traditional print and electronic industries the magazine business was the quickest to adapt to growing advertiser interest in customization during the 1980s. Until recently, customization seemed a procedure that television and radio could not achieve.

In the vanguard of print media was a venerable agricultural business periodical, *Farm Journal*. In 1982, the invention of a new computer-driven binding system enabled the company to put together customized contents to match the special farming needs and interests of each of its 825,000 readers as noted in a database gathered in periodic telephone surveys. Advertisers had the option of matching their ads to the individual subscriber profile. Three consumer magazines, *Games*, *American Baby*, and *Modern Maturity*, soon followed *Farm Journal*'s lead, though not in quite so ambitious a manner. In *American Baby*, for example, the Gerber baby food company ran different personalized ad messages in the magazines for mothers of two- to three-month old infants and mothers whose babies had reached three to four months.

Signs that the industry was really moving in a new direction came when Time Magazines, Inc., the nation's largest publisher, made a decision to enter the business of selective binding and imaging. At a premium price, an advertiser could specify that its ads should go to certain individual subscribers and not to others, depending on the advertiser's needs and the information about the readers in a Time Magazines database. Moreover, using ink-jet printing technology, the advertiser could alter on-page copy to reflect the information in the database.

The operative phrase is 'at a premium price.' Even by 2003, producing truly separate versions for different audience segments, let alone individuals, increased costs dramatically. The costs were so high that advertisers were typically loath to support editions that were tailored to individuals or niche segments. Nevertheless, in an era where they had to offer advertisers something special to compete with television and radio, magazine firms – and newspapers – increasingly share the assumptions of loyalty programs. They realize that they need to entice desired customers and create databases that can aid in understanding readers, and in selling them to advertisers. Increasingly, they are turning to their own branded websites for customizing relationships with their readers – and readers' relationships with their advertisers. There they are finding television networks that are doing the same thing.

Discovery Networks, a cable network conglomerate, created a hybrid between a traditional loyalty program and one that points towards future possibilities. The company aimed to sell its branded merchandise, find out a lot about potentially high-spending viewers it can promote to advertisers, and reinforce loyalty to its channels and merchandise via a tailored e-newsletter. Discovery used its brick-and-mortar store loyalty program to get e-mail addresses; it adds to those people by asking customers at point of sale if they want to receive a free Discovery e-newsletter. During Christmas 2002, the newsletter program went on hyper-drive, increasing distribution from once or twice a month to once a week, and refocusing the creative effort from product offers to attention-getting special promotions. In line with the discrimination logic of loyalty programs, Discovery differentiated between products and promotions that customers saw depending on the recency of their store purchases, their frequency of purchases, and models of the amount of money they would be likely to spend. For instance, someone who had never ordered would see a cheaper item than the person who had ordered multiple times. 'If you show that person a product that's selling well with others [like them] and it's $9.99, it might do better than a $200 telescope,' an executive explained (Oser 2003).

A more common tack of media-based loyalty programs has been to focus on seducing web users to trade their data for useful personalization features. CNN.com, for example, exhorts users to 'personalize your CNN.com page experience today and receive breaking news in your e-mail inbox and on your cell phone, get your hometown weather on the home page and set your news edition to your world region' (CNN.com 2003). Such customization allows the site to cultivate a relationship with its audience and to develop data about audience members' interests and movements that it can use for targeting ads.

Using a somewhat different approach towards the same ends, the *New York Times* tries to cultivate loyalty by encouraging users to sign up for a 'news tracker' service, which sends them e-mail when articles are published on particular topics. Armed with registration, user-interest data, and information about readers that the *Times* purchases from database firms, the company touts a technology to force ads in front of desired consumers through 'surround sessions': the ability to 'own a consumer visit to nytimes.com' by having targeted visitors receive ads almost exclusively from one advertiser during individual sessions on the site (Taylor 2003).

As the decade moved on, attempts to increase loyalty to marketers on the internet via personalization grew markedly. Increasingly accepted was Amazon.com's 'collaborative filtering' approach for customizing the home page of site visitors. It matched the purchases and searches of visitors who had registered to the purchases of other people who exhibited similar interests. What those interests were, how they should be categorized, and when particular suggestions should be made were determined by the firm's data-miners and statisticians. Though rather basic as a result of computing demands, Amazon's success with collaborative filtering confirmed for marketers the logic of customizing the main page based on previously learned customer characteristics. It pointed to increasingly sophisticated future website personalizations.

Personalization was also developing quickly in the web advertising arena by mid-decade. The most widespread activity was contextual advertising via search engines. Led by Google, Yahoo, and a number of other firms, the basic activity triggers specific ads to appear based on the search word a person typed into a search box. The more elaborate version involves an advertising network of ad sites that launched ads on a website that matched content of the web page the person was reading.

A variation on these approaches, practised through technologies of companies such as Revenue Science, Tacoda, and 24/7 Real Media, is *behavioural targeting*. It uses cookies (small text files placed into individual computers upon their arrival to websites) to track the movements of users of those computers within and across a network of websites that recognize the cookies and to serve up ads to them on the sites based on their history of web activities. Such behavioural targeting typically took place without the firms knowing the identities of the people being followed, or even demographic characteristics about them. 'Adware' is a controversial variation on these activities. Practised most notably by Claria Corporation through its Gator program, the idea is to place software into a computer when the user is downloading attractive free software, such as the music file sharing program KaZaa. This not only allows for tracking an individual's web journeys, it makes possible the insertion of pop-up ads that reflect Claria's inferences about that person's interests and habits. The ads appear above the ads of the websites that the targeted individual is visiting.

Not surprisingly, adware activities have earned the wrath of website operators, who argue that companies should not be allowed to hijack their sites by placing ads atop the ones that website advertisers bought. The adware firms counter that the people who own the computers have given them permission to do just that in order to personalize their marketing information and offers. To privacy advocates, this claim of permission is what makes adware unacceptable. Claria and its counterparts claim that computer users know that when they download free software they are giving them licence to download adware, as well. Their critics retort that most people really have no idea that they have placed software into their computers that track them and report back to firms that will send them ads.

Several of these arguments were winding their way through the courts in 2005. Mainstream marketers' opinions of their validity varied greatly; Gator counted among its clients a number of major advertisers, while others refused to go near it. The logic of behavioural targeting remains unquestioned among marketers, however. Moreover, companies such as Microsoft and Yahoo are beginning to select ads for people based on combining the tracking of individuals' search or web activities with huge amounts of demographic and psychographic data they are collecting about them. Privacy advocates worry that people know little about how data are collected online, or about the factors that lead such firms to reach out to them with certain materials and not others (see Turow 2005). The companies respond that the collection of such data are noted in their privacy policies and that, in any event, people get great benefit from receiving targeted, even personalized, materials that match their particular needs.

The behavioural targeting and personalization strategies – and the debate about them – are slowly spilling out from the web across the growing range of digital media. Marketers are actively working on ways to send customized product placements, ads, and programs into PDAs, cell phones, Mp3 players, and other devices. The proposition is that surveillance of valued customers combined with getting useful materials to their consumers wherever they go is the key to maintaining loyalty in the twenty-first century.

Interactive Television

The emerging logic behind these activities has carried over to discussions of the least advanced area of electronic media, interactive television. In industry discussions, the term is quite elastic, ranging from already available video-on-demand cable, DVR, video via the internet (IPTV), transmissions to cell phones and streaming to video iPods to over-the-horizon technologies that encourage even more radical changes in the way people use TV. When it comes to the latter, advertiser and media executives have expressed consensus about the direction of major changes, if not the time it will take to see them implemented. They are pretty well unanimous that the future of television in the twenty-first century lies in presenting viewers with menus of programs that they can watch at times of their choosing – often for a small fee. Not only will the twentieth-century notion of linear network programming be challenged, but viewers will be guided into their own program choices and flows by 'intelligent navigators' that learn what they like and search for programs that fit those characteristics. Moreover, say the futurists, viewers who would like to interact with programs could do that, to change the plot, perhaps, or to find out about – and even purchase – the clothes or objects that the characters use on the programs.

Some of these capabilities are already here. Satellite firms use devices that allow viewers to interact with, and order from, programming. Television production firms hire a company that digitally places products into programs such as *Seinfeld* that have been placed into TV syndication. Comcast and other cable system owners use technology that can digitally reconfigure commercials based on information about the neighbourhood to which they will be sent and the time they are delivered. The custom-placement of placement of products into commercials and programs of specially targeted viewers is technically possible but still expensive and unlikely now because of technical complexities as well as because so few households subscribe to digital TV. That kind of activity – and especially the personalization of commercials – may be more easily carried out on the internet, via IPTV. Moreover, the basic infrastructure for garnering the required information for this sort of product placement is developing rapidly. TiVo already minutely tracks and categorizes the viewing habits of its users as part of its service contract. The company sells the data in aggregate to potential advertisers. For its own purposes, it analyses the habits of individual subscribing households. It then recommends, and even records, programs for the households based on preferences inferred from the computer data as well as deals with advertisers who want TiVo viewers to watch those shows and commercials (see chapter 12 of this volume).

Discourse about the future of interactive television suggests a movement from tracking and customizing for the household to doing that for the individual. The goal of media firms in this interactive environment will be not only to instil loyalty to their own brands, it will be to cultivate relationships with their advertisers who, despite the subscription nature of much TV programming, will pay a substantial part of the programming freight. Product placement through sponsorship and embedding will be one step in the process. Another step, being tried by content sites such as Salon.com on the web, is to give desirable viewers discounts to content if they view commercials. A third step – emerging as the most common – is to force the viewer to watch a commercial (of around fifteen seconds) if the viewer wants to view the video. Still another is simply to give them advertiser-sponsored discounts to premium content so as to emphasize goodwill and encourage the transfer of consumer data.

Converging Social Anxiety

It is in this convergence of strategic logic that the social rub lies. The combined language of direct marketing and product placement suffuses discussions of trusted media and interactive television. The digital world marketers and media firms are building consequently has as its core the belief that success will come from seducing customers to release their personal data in the interest of rewarding relationships with media and marketers. Marketers will claim to reward consumers' distinctiveness. Purposefully or not, they may well also encourage feelings of anxiousness and anger.

The circumstances are not hard to imagine. Using detailed audience surveillance, digital marketers will be able to track the media activities of their target audience in considerable detail. To ensure that they view targeted, perhaps even customized, commercials on the web or on TV, the marketers may offer audience members discounts to programming, music downloads, game networks, or a panoply of other subscription or pay-per-use activities. Some marketers may 'sponsor' particular evenings on the web, the home tube or both. They, or others, may even embed products into audiovisual material that they know from their surveillance activities that the consumer (or the consumer's parents) is thinking about purchasing. Some of the offers may be public; that is, marketers may make it clear that consumers who pass a litmus test will get more rewards than those who do not. Other offers may be private between the marketer and the individual.

Consider that Sally, a single young woman, has been searching the web for information towards a car purchase. Through the sponsorship of various audiovisual activities, Ford might have established a relationship with her so that the automaker knows her concerns about autos. Ford might also pay AOL, TiVo, and her cable TV provider (which in the future may be one company) to send information about Ford to all young women in their databases who have searched the web for cars. More boldly, Ford might pay the producers of popular television programs to have their central characters drive new Fords that change depending on the type of person or household that watches the show. Ford has decided that people like Sally would be interested in the Mustang, so that is the car she sees. Ford tells her in e-mail that the discount she will receive reflects Ford's conceptions of her style and spending ability.

Around the corner from Sally, a sixtyish widower named Jim gets none of these perks even though he might want them. Some advertisers and media firms pursue him, but these are not nearly as attractive as the ones that his son and grandchildren receive. Jim has read about the amazing discounts and sponsorships that some individuals receive, and he can't help but feel (even though he doesn't know for sure) that the ones he gets are less lucrative. Sometimes, he even lies about his age when on the web or answering a TV ad. Databases about him seem to be self-correcting, however, and he feels caught in a somewhat mysterious web of constructions about him that goes far beyond his age, gender, race, location, and even income.

Awareness of targeting activities that lead to different programs and product placements is making Jim nervous: he thinks he is being manipulated but is not sure by whom and when. When he views a news story about a local hospital's cancer centre, for example, he is not sure if it was produced in cooperation with the hospital to be sent to people like him. When his TV menu touts a special about the military, he is nervous that perhaps the databases think he is right-wing and pro-war.

Sally, too, worries increasingly about the media blandishments coming her way. She thinks it is a bit freaky that companies seem to know so much about her, and it sometimes worries her that the information might get into the wrong hands and ignite a flood of offers into her home. She too has read that marketers are privately sending different discount levels and sponsoring differently editing programs to different customers. She wonders how the suggestions on her TV menu would change if she were a lawyer, like her friend Sue, with a globetrotting reputation. And she can't help but think at times that if the trail she has left for marketers had been different, she would have opportunities that acquaintances have received but she has missed.

The point, of course, is that seemingly benign relationships in the new digital environment can quickly lead to feelings of discrimination, anger, and suspicion of institutions. Executives in charge of loyalty programs have already learned that some customers become quite angry or anxious to find that they have been categorized and treated differently from others. They have also learned that the key to managing such anger is to make the customer see tension-inducing rules as almost an interpersonal issue between company and customer. 'Failure' to get benefits or offers within the scheme would then be a private issue resulting from the rules of collaboration rather than one needing public remedy.

In the best of cases, customers try to show by their purchases that they deserve to be treated at a higher level of service. Some customers exit the 'relationship,' which because of low volume may suit the company (Turow 1997: 141–52). The approach could still blow up into public trouble, as when Victoria's Secret was accused of charging more in its west coast than its east coast catalogues (Demarrais 1997). Still, loyalty programmers believe that properly configured programs based on one-to-one interactions can keep legal discrimination in the interest of marketing a private matter. Moreover, they believe that inducing anxiety that causes unwanted customers to feel they don't belong to a consumption group is good business. Efficient marketing increasingly means 'managing' the customer roster – 'rewarding some, getting rid of others, improving the value of each of them' (Peppers and Rogers 1993: v–vi).

The upshot might well be a personal anxiety that pervades many levels of American life. The traditional mandate to keep up with the Joneses is being replaced with a twenty-first-century version in which the Wongs, Riveras, and Kowalskis are never sure how much the Joneses paid and whether media and marketers are treating them better or worse than everyone else (Turow 2006). Reflecting that anxiousness mixed with anger, one marketing executive (Ephron 2003: 30) even opined that 'brands seldom really give loyal customers the better deal.' He added: 'They're treated as hostages. I can open a BankOne Visa account and "save with a low introductory APR of 0%." But if I'm a cardholder in good standing, I pay 9.74%. AT&T sends others a check for $80 to switch, but I'm a customer so they send me a bill. I'm never certain I'm getting the right deal from the brands I'm loyally buying. Are you?'

As media-based loyalty programs gain traction, consumers will add to those suspicions concerns about the entertainment, information, and news they receive from their longtime providers. Even though they will not be able to make any sense of the firms' privacy policies, they will understand vaguely that the firms are continually collecting data about them. They will like the menus of content choices being offered to them, but they will also be aware that others – possibly even members of their own households – are being offered different program choices, or the same program choices with different kinds of content. With those differences based on the semi-secretive flow of data that the programmers and marketers take in about them, many viewers may not be sure that they are being treated fairly. Because everyone's programming menus are somewhat different, however, concerned consumers might not be sure what the best comparisons, or the right choices, are. Moreover, media firms and marketers are likely to continue the tack of managing customer anger by framing tension-inducing provisions as an issue between company and customer rather than a social issue.

This paper suggests that is wrong. A society in which media assumptions are based on the values of customer relationship marketing is one which will breed surveillance of consumers, and, consequently, consumer suspicions of individuals, groups, and institutions. Although the examples of Sally and Jim may seem extreme, they parallel illustrations in the trade press over the years. More important, they fit the strategic logic that underlies product placement, direct marketing, and loyalty programs. So far, the drawback to accelerating this logic through customized media, walled gardens, and interactive television has had less to do with corporate will than with the development of cost-efficient technologies. Given the strong interest of media and marketing firms, there is good reason to believe that these technologies will be implemented in the decades to come. Perhaps analysis of the strategies and where they originate can help concerned citizens understand the trajectories at work so they might try to impede them.

REFERENCES

Advanced Promotion Technology Vision News. 1995. 'Two-thirds of Major Retailers Are Using Database Marketing.' Jan.–Feb.: 3.

Advertising Age. 1982. 'Direct Marketing.' 19 July: M-27.

Barnauw, E. 1970. *The Image Empire*. New York: Oxford University Press.

CNN.com. 2003. 'Preferences.' http://www.cnn.com/services/preferences/

Crain, R. 2003. 'Urgency of Change Is Touted, but Marketers Favor Safe Bets.' *Advertising Age*. 24 Mar.: 28.

Cyr, D. 2003. 'Drug Money.' *Direct*. 1 Mar.: 3

Delay, R. 1980. 'Direct Marketing – 'Way beyond Catalogs.' *Advertising Age*. 30 Apr.: 188.

Demarrais, K. 1997. 'Scams Hurt Marketing Industries.' *The Record (Bergen County, NJ)*. 21 Sept.

Elkin, T. 2002. 'Seeking Payoff on the Web.' *Advertising Age*. 7 Oct.: 4

Emerson, R. 1962. 'Power-dependent Relations.' *American Sociological Review*. 27:31–40.

Ephron, E. 2003. 'CRM Must Be Real.' *Advertising Age*. 30 Mar.

Fitzgerald, K. 2002. 'Eager Sponsors Raise the Ante.' *Advertising Age*. 10 Jun.: S18.

Freeman, L. 1993. 'Direct Contact Key to Building Brands.' *Advertising Age* 25 Oct.: S-2.

Friedman, W. 2003. 'Panel Discussions: Marketing in a TiVo World.' *Advertising Age* 10 Feb.: 13.

Giddens, A. 1990. *The Consequences of Modernity.* Stanford: Stanford University Press.

Graham, P. 2002. A plan for Spam. http://www.paulgraham.com/spam.html.

Hansell, E., and S. Olsen. 2002. 'Spam: It's More Than Bulk e-mail.' CNET News.com, 8 Oct.

Hansell, S. 2003. 'AOL Providing Software to Customers to Block Pop-Ups.' *New York Times*, 12 Mar. c8.

Levey, R.H. 2003a. 'Live from Gartner CRM Summit 2003: Nectar Feeds Sainsburys Segmentation Efforts.' *Direct.* 5 Mar.: 1

– 2003b. 'The Trouble with Points Programs.' *Direct.* 1 Apr.: 5

MacDonald, F. 1983. 'The Clutter in the Mailbox 'Aint No Junk'.' *Advertising Age.* 17 Jan.: M-32.

Mitchell, A. 2003. 'Advertising Futures.' *Brand Strategy.* 4 Feb.: 40.

MMI Product Placement. 2003. http://www.mmiproductplacement.com/benefits.htm.

Oser, K. 2003. 'Holiday Cheer.' *Direct.* 3 Mar.: 3.

Pagnetti, J.A. 1983. 'Sales Sprout from the Seeds of Segmentation.' *Advertising Age.* 17 Jan.: M-9

Peppers, D., and M. Rogers. 1993. *The One to One Future.* New York: Doubleday.

Pettigrew, A. 1985. 'Examining Change in the Long-term Context of Culture and Politics.' In J.M. Pennings, ed., *Organizational Strategy and Change*, 268–318. San Francisco: Jossey Bass.

Progressive Grocer. 1995. 'Using Databases to Seek Out the Brand Loyal Shoppers.' (Feb.): S-10

Quinn, S. 2002. 'Creating a Salon Oasis Just for Teens.' *Plano Morning News.* 6 June.

Rapp, S., and T. Collins. 1990. *The Great Marketing Turnaround.* Englewood Cliffs, NJ: Prentice-Hall.

Sharkey, B. 1988. 'Do Moviegoers Recall the Products They See in Films?' *AdWeek.* 20 June: 1.

Taylor, C.P. 2003. 'Interactive Quarterly: Independent Agency; Avenue A; Proving That Expertise Pays, This Media Agency; Retained Clients and Increased Revenues.' *AdWeek.* 3 Feb.: 5.

Turow, J. 1997. *Breaking Up America.* Chicago: University of Chicago Press.

– 2005. 'Open to Exploitation: American Shoppers Online and Offline.' Report of the Annenberg Public Policy Center, June.
– 2006. *Niche Envy: Marketing Discrimination in the Digital Age*. Cambridge, MA: MIT Press.
Weiss, R. 2003. 'AOL Ramps Up Fight against Spam for Users.' *Computerworld* online, 20 Feb.

12 (En)Visioning the Televisual Audience: Revisiting Questions of Power in the Age of Interactive Television

SERRA TINIC

In a recent television advertisement for Shaw Communications Inc.,[1] an unseen individual responds to a knock at the door. He/she finds a smiling, baseball-capped fellow from the local cable operator 'checking' in' to ensure that the subscriber is enjoying the full benefits of his/her expanded digital television service. Our cheery cable representative – Digital Dave – is on the scene to remind our subscriber that digital television provides all customers with an expanded range of channel choices to the extent that he/she can watch virtually anything that fits her/his individual interests.

Let us briefly deconstruct this commercial moment. This is no anonymous cable television subscriber answering the door. Rather, through the positioning of the camera, we, as individual members of the cable television audience, are invited to see ourselves as the subjects of Digital Dave's exhortations. This particular subject positioning establishes a personalized relationship with 'Dave' – as opposed to an objectified business relationship between a telecommunications company and a subscriber – that promises to provide us with a veritable bounty of customized television offerings in which we, the viewers, will have control over our media repertoires. In effect, we are witnessing a transference of power. No longer are we slaves to the meager offerings of three or four network broadcasters. Instead, we as individuals – not preconstituted audience aggregates – are free to roam the digital cableways in search of programming that speaks to our own subjective desires. If we wish to add Shaw's Internet package to our digital cable

---

1 Shaw Communications is one of the two largest cable operators in Canada and dominates most of the western region of the country.

service we will further accelerate our advance into the driver's seat of televisual control. Or will we?

What is interesting about the Shaw advertisement is the extent to which it corresponds with the central themes underlying current academic debates about the consequences of media convergence in the age of digital communications: namely, the propensity to personalize, individualize, and customize forms of electronic communications technology that were once seen to be the pre-eminent domain of large corporate or government institutions. Beginning in the late 1980s, media convergence became a hot topic in mass communications research. Much of the dialogue heralded the coming era of integrated computer, fibre optic, telephonic, and broadcast technologies. With the introduction of the personal computer and increased accessibility to the Internet – and what would become the World Wide Web in the early 1990s – technology proponents painted a picture of an environment in which so-called ordinary viewers would be able to circumvent the controls of the broadcast industry and produce their own television schedules while operating every conceivable household appliance with their television remote controls. Much early prognosticating about a future world of 'interactive television' depicted a utopian vision of audience empowerment in a highly technologically deterministic manner (see Negroponte 1995). The assumption was that because these new technologies offered two-way communication, audiences would no longer be passive in their viewing, but active in creating new media content.

What such arguments overlooked were the economic, social, and political contexts in which convergence was taking place; in particular, the moves towards deregulating media ownership structures and weakening anti-trust laws under the Reagan administration. The consequences of deregulation were evidenced by the rush by telephone companies to buy into the cable industry and the conglomeration of media outlets into horizontally integrated companies with holdings in every aspect of entertainment and news production as well as online Internet services. Today, six to eight corporations (depending on the merger and acquisition plays at any given moment) own most of the world's media and all have a vested interest in how convergence unfolds.

As of a few years ago, we entered the first generation of media convergence with the public introduction of interactive television. We are now in a relatively better position to question some of the arguments concerning power in the new media age. In this chapter, I discuss the patterns that are evolving in how media companies are taking advantage of the potential of two-way communication. I argue against the type of technological determinism that marked early predictions about the future of televisual communications technologies. The first section contextualizes interactive television (ITV) and explains how it operates as a surveillance technology. This is followed by an analysis of the larger implications of this new media form with a particular emphasis on its potential as a socially fragmenting force. The central argument is that rather than witnessing a revolutionary new phenomenon of empowerment, which is often the description of interactive television, we are in fact seeing television – as a consumerist medium – fulfilling its industry logic as a marketing platform. In brief, this essay evokes a more dystopian view of the social implications of interactive television, accentuating its capacity to accomplish unprecedented levels of surveillance of the domestic sphere with the goal of rendering audiences visible to advertising interests in ways that were previously impossible.

In this respect, interactive television can be seen as a component of the larger technological apparatus that Oscar Gandy calls the panoptic sort, to the extent that it 'involves the collection, processing, and sharing of information about individuals and groups that is generated through their daily lives ... and is used to coordinate and control their access to the goods and services that define life in the modern capitalist economy' (Gandy 1993: 15). This paper attempts to explicate how the companies behind interactive television seek to exploit their access to such information about the private sphere with the hope of encouraging members of the household to become complicit in their own disciplinary subjectification as units of consumption. It is important to note here that, contrary to utopian depictions of the possibility of a new electronic public sphere, corporate ambitions regarding ITV contribute to processes of exclusion and fragmentation as the goal is to invite only select individuals to participate fully in the world of the new and improved medium. These are people who already occupy positions of economic, social, and cultural privilege – the preferred targets of manufacturers and their advertising interests. Herein, ITV epitomizes the ways in which Gandy's panoptic sort operates as the ultimate *difference machine* (15).

The Brave New World of ITV

What could sound crazier than someone saying 'I think my TV is observing me?' What could sound crazier than someone saying 'I think my TV is sending messages only meant for me?' (Burke 2000: 115)

At the most rudimentary level, interactive television is the integration of broadcasting and computer Internet capability through two-way communications via a set-top box connected by modem, cable, or satellite systems. At full capacity, ITV allows viewers to use e-mail, participate in Internet chat groups, play online video games, and/or shop while they watch television. One of its most significant features is that it allows audiences to personalize their television practices. Through the use of an electronic programming guide, viewers can set their systems to search and retrieve favourite shows or conduct searches for programs by certain directors or starring certain actors. All of this content can then be organized into a schedule customized for each individual to watch at his or her convenience rather than at the scheduled airing determined by a broadcast network.

The first step into the interactive field was the digital video recorder, now more commonly referred to as the personal video recorder (PVR), of which the two leading brands are TiVo and Replay. PVRs are set-top boxes that allow viewers to digitally record up to sixty hours of television programming. Sonic Blue, the company that developed Replay, recently increased its machine's capacity to 320 hours. Regardless of differences in storage capacity, both machines sell themselves on enabling viewers to create television 'wish lists' that allow them to create personalized television schedules, sometimes referred to as customized television 'channels.' Each machine is able to go one step further and actually anticipate viewers' preferences, based on their submitted profiles, and send back recordings of programs it 'thinks' the viewer would like. However, one of the dominant selling points of these PVRs has been the feature that allows viewers to skip past television commercials or avoid recording them altogether. Not surprisingly, these features have met with fury from the television industry as personalized schedules impede a television network's ability to offer a predicted audience to an advertiser at a set time. This is to say nothing of the commercial deletion feature, which makes the advertisement an uncertain venture regardless of when it is aired.

The apparent ability to circumvent the constraints of broadcast schedules and commercial appeals led to marketing discourse that paralleled that of the early work of researchers mentioned at the beginning of this chapter. Namely, that media convergence, in the form of ITV, had ushered in a new era of viewer empowerment in which individuals would be able to navigate the airwaves free from broadcasters' attempts to direct their attention to specific media content or product promotions. The promotion of an intangible bounty of social goods to be garnered from these new technologies is epitomized in a recent book by satellite and Internet marketer Phillip Swann (2000: 11): 'Companies are spending billions of dollars to develop new digital set-top boxes that will scratch your every itch and satisfy your every desire. These new boxes will make your life more convenient, more fun, and, in many ways, more enriching. They will take you on adventures of the mind, bring you closer to friends and loved ones, perhaps even give you a greater sense of control and self-esteem. They will change your life in ways you can barely imagine.'[2] Here, again, we are told of the ethereal wonders wrought by technological progress, but we are not given any indication of how such technologies are becoming structured according to the economic logic of capital. Indeed, if people were to follow Swann's polemic they could not be criticized for believing that the industries involved in ITV were actually charitable technocrats working to establish a liberatory electronic public sphere. However, a different story unfolds when we examine the current industry battles to control the structural environment of PVRs and ITV.

In an effort to regain control over the commercial television arena, the major motion picture studios and broadcast networks have brought their grievances against PVRs to the courtroom. Sonic Blue has been named the defendant in a number of lawsuits alleging the company violates copyright laws by allowing viewers to create personalized, time-shifted television menus. The plaintiffs further argue that people who watch television without commercials are in fact stealing. In the words of Jamie Kellner, chairman and CEO of Turner Broadcasting: 'It's theft. Your contract with the network when you get the show is you're going to watch the [commercial] spots' (Vikhman 2002). The other major PVR player, TiVo, has sought to avoid these problems by not activating the commercial-deletion option in its machine and, moreover, by developing what it calls 'advertainment' – ten-minute spots

2 Swann's book is a rather interesting piece of promotional literature for the ITV industry. Of the ten chapters and 126 pages extolling the wonders of interactive television, only three pages address the negative aspects of the surveillance potential of new media technologies. Here Swann admits that 'the television networks and ITV companies will know nearly everything about you' (123) and quotes a WebTV executive who acknowledges that the company has a 'department that does nothing but look at the information' (124). Swann implies, nonetheless, that these problems will be minimal as it will be in the companies' best interests to assure their customers of privacy protection. The argument corresponds closely with economic perspectives that assert that the market will always correct itself to achieve profits efficiently.

that promote movies (through trailers and actor interviews) and other products offered to targeted viewers (Smallman 2002). This is the point where we see the initial incursions into surveillance practices. Because of the two-way capabilities of the technology, these companies are able to monitor every move an individual makes with a remote control – what was selected, replayed, fast-forwarded, paused, and/or deleted. That information, in addition to records of audience wish lists, allows for profiling and research that can then be sold to advertising interests. In fact, this is exactly what TiVo did during the 2001 Superbowl. The company documented which plays were most frequently rewound and which commercials received the most viewing time and were replayed the most often. Pepsi Co. must have been happy to find that its Britney Spears commercials received more instant replays than the game itself, as TiVo discovered through monitoring a sample of 10,000 of its viewers. In fact, all of the results of this surveillant activity were sold to interested businesses through TiVo's marketing department ('Britney Rules' 2002). It is access to this type of audience information that makes Hollywood's claims against Sonic Blue of copyright infringement and commercial theft appear particularly disingenuous. Firstly, the standard videocassette recorder (VCR) allows for time-shifted viewing and fast forwarding through any television commercial; albeit not at the speed or automated level provided by PVRs. Second, the plaintiffs against Sonic Blue are asking (as a form of reward against damages) that the type of subscriber information sold by TiVo's manufacturers also be made available from Replay's owners (Wiley 2002). Sonic Blue's steadfast refusal to violate the privacy of its subscribers underlines that corporate policy – not innate technological capabilities – will, in the end, determine where power lies in the age of media convergence.

PVRs are, however, merely the first foray into the new media domain because they lack the Internet and computer interfaces that are a vital component of interactive television. ITV is truly where it all comes together – the ability, as mentioned earlier, to surf the Web while customizing your own television schedule, electronically chatting with friends and fan groups while simultaneously shopping or banking online, makes ITV participants completely transparent to those who want to sell them products and services. This ability to visualize or 'know' the audience has always been a paramount concern of television broadcasters and advertisers. The history of the relationship between audience measurement companies, such as A.C. Neilsen, and the broadcast networks is a series of attempts to know what types of people are watching which types of programming and when. These attempts began with early research methods such as phoning a sample of (preferred) households after specific shows were aired to ask people if they had watched the broadcast and retained any memories of content and corresponding advertisements. The limited information garnered through this episodic tracking was superseded by efforts to measure longitudinal viewing habits through the use of media diaries, whereby people were asked to track and record their weekly television viewing over the course of a month or two. The diary system proved particularly problematic as research companies determined that people lied about their viewing habits for fear of stigmatization. For example, rather than admitting to being an avid soap opera fan, an individual was likely to skew diary entries towards public broadcasting programs or news and educational shows. Consequently, concerted efforts were invested in developing audience surveillance technologies that could be placed, as unobtrusively as possible, in people's living rooms. An early example is Neilsen's 'people metre,' a device in which every member of a household would 'press-in' his/her identifying information when watching television. Measurement problems arose here as people would often forget, or neglect, to log in after turning on their television or they would leave the room without turning off their set, thus measuring the viewing patterns of an empty room. As a result, infrared and heat-seeking machines were developed to register when an actual body was sitting in front of the television. As anecdotal evidence has it, Neilsen measured the viewing repertoires of many household pets through this device.[3] In recent years Neilsen and other companies such as Arbitron have provided select households with packages that allow them to scan their groceries' barcodes and record television activities in the hopes of gleaning that most desired portrait: the link between the types of programs watched and actual consumption patterns.

These forms of audience measurement, and the entire television ratings system, are largely defensive mechanisms on the part of network executives to produce and schedule programs that they can sell to advertisers based upon serving up a *construction* of the audience for their products. In brief, it is a type of risk management that transforms

3 See Ang (1996) for an examination of some of these audience measurement techniques and for an excellent analysis of the problem with 'constructing' and 'measuring' audiences in general.

'the audience' into a commodity. Today, ITV solves many of the problems of audience measurement and goes beyond market researchers' wildest dreams to the extent that they can surreptitiously enter people's homes and watch them through the electronic hearth. Here is how it works. With the convergence of the Internet, personal computing, and broadcasting, software and hardware companies are developing desktop configurations for our television screens that invite us to 'log on' every time we choose to watch television. The enticement to do so is based on the ability to personalize our viewing through the use of the aforementioned electronic program guide (EPG). Once the individual is verified, his/her own EPG will process and provide him/her with the preferred content based on actual and anticipated wish lists. This single feature is paramount as the personalization of content opens the door to the advertiser's most desired strategy: one-to-one targeted marketing.

Logging on also allows the ITV subscriber to use e-mail and instant messaging to chat with other people watching the same program – all of which is observable to the service provider. Such electronic conversations provide instantaneous feedback about responses, or lack thereof, to any television program or advertisement. Moreover, just as the cookie system in web surfing is used to track the sites an individual visits on the Internet, such clickstream monitoring of ITV activity (and particularly that which involves online purchasing) provides the service provider with a composite profile of television surfing and consumption. The potential audience surveillance is seen as so promising that software developers have joined forces with market researchers and advertisers to develop both the systems and content that will allow for the embedding of the sales pitch in any television segment. Some of the largest companies, such as Synergy and Claritas, are developing both tele-surveys that develop psychographic audience profiles as well as online video games that construct scenarios in such a manner that an individual's personality and motivations – it is hoped – will be revealed with every manoeuvre made in the game (Burke 2000). Interactive advertising is also being market tested whereby icons appear during programs that allow viewers to click on an image and 'learn' more about a product. An example of this is Dove Cove, brought to us by the soap manufacturer Unilever, in which the viewer enters a virtual world designed to resemble a health spa or 'sanctuary' and then moves through rooms that, as one marketer explains, 'encapsulates them in the brand' (Burke 2000: 23). Throughout these 'visits,' viewers leave an electronic trail that links them identifiably with their subscriber information.

According to Josh Bernoff, an analyst with Forrester Research, the effectiveness of such interactive advertising far exceeds the current commercial-break model: 'With an interactive ad, when people respond to it, you have a name and an address and a lead that is 100 times more valuable than an impression to a marketer' (Cooper 2000). By way of example, Bernoff points to the success of a recent venture by a San Francisco company that aired an interactive Domino's Pizza commercial during a *Star Trek: The Next Generation* marathon. Of the estimated 1,000 ITV subscribers who saw the commercial, 220 'clicked' for further information and 140 homes ordered a pizza (Cooper 2000).

These, however, are early stages in the development of ITV. Industry members are currently working to exploit the possibilities of customized television to bring every individual their own personalized advertising and program content. The vision here is such that people living on the same street, and even in the same house, will be sent different advertisements, customized news broadcasts and, perhaps, even different endings to their favourite shows based on the psychographic and demographic data they have either intentionally or unknowingly sent to their service provider. The future scenario predicted by people working on ITV is one in which we will all be able to 'click' on the shirt of our favourite television character and have it immediately sent to our homes as the service will already have our credit card numbers and addresses stored on the central network (Roman 2002). Development of the technology proceeds apace as artificial intelligence and neural network software are also being incorporated so that our televisions will soon adapt to our lifestyles over time. Here, our options to use 'commercial-deletion' options will be circumvented as our consoles will enable personalized advertisements to follow us from one channel to another until we can no longer avoid them (Burke 2000). These interactive features are marketed to the public as ones of convenience and, somewhat misleadingly, as empowerment and content control. Little, however, is mentioned about how personal information is gathered or how it is used. Even less is told to the public about how we are encouraged to become complicit in our own transformation into all-consuming individuals. On the other hand, these processes are discussed at great length within the interactive marketing industries: 'In direct marketing, you make a spear, hunt down the consumer you want, and impale him. In brand advertising, you make a really, really sharp spear, chuck it into a crowd of consumers and hope it impales as many as possible. With online marketing you make a spear and invite the consumer to come and impale himself' (Rishad Tobaccowalla of Giant Step – interactive marketing in Burke 2000: 71).

Tobaccowalla's comments illustrate the marketing world's changing conceptualizations of the relationship between consumers and product promotion in the age of digital, interactive media forms. As McAllister and Turow (2002) underline, this contemporary emphasis on *synergistic marketing communication* – 'the use of multiple media channels and publicity methods to sell products, services, and ideas' – combined with digital media, should be examined within the 'overarching hallmark of the twentieth and twenty-first centuries: the commercialization of society' (507 and 508). Indeed, current strategies to accelerate ITV's capacity as a synergistic marketing medium sine qua non support McAllister and Turow's assertions about the reach and intensification of consumer culture into all facets of social life. Here, an examination of the central players in ITV provides us with an initial glimpse as to how this potential can be realized.

It will probably come as no surprise that America Online (AOL) and Microsoft Corp. are the major ITV actors in North America, with AOLTV and MSNtv (formerly WebTV and/Ultimate TV, respectively). Media mogul Rupert Murdoch is the central ITV force in Europe and much of Asia with his Sky TV and Star TV satellite services. AOLTV appears to have the current advantage in the North American market because of its existing Internet subscriber base and its takeover of Time Warner – a move that provides the company with one of the world's largest libraries of programming. In their efforts to capitalize on the commercial marketing potential of ITV, both AOLTV and MSNtv have been frustrated by the public's slow adoption rate of ITV. This fact has largely been attributed to the high cost of the requisite hardware in combination with subscription rates. MSNtv, for example, has found it difficult to pass the one million subscription mark and, according to interactive marketers, advertisers will not come fully on board until they can be guaranteed a threshold of two million viewers (Ford 2000). Consequently, North America's largest ITV companies have lobbied ardently to advance the switchover date of the industry standard from analogue to digital, which is scheduled for 2006 in the United States, 2008 in Canada, and anywhere between 2006 and 2010 in Britain (Roman 2002; Howe 2001). It is believed that once the necessary digital foundation is firmly established, people will become rapidly accustomed to the increased capacity of 'new' television and, from there, move comfortably into complete interactivity.

These lobby efforts have initiated an outcry on the part of privacy interest groups, including EPIC and the Center for Digital Democracy, which have called for immediate legislation of the new technologies in light of early practices by ITV companies. For example, every night MSNtv downloads the viewing practices of its subscribers for analysis and it is conceivable that the company is using this data as a pilot study for the future development of ITV and streamlined marketing strategies. According to the company's own public relations material, these new technologies 'enable faster and better decision-making through an innovative data warehouse that aggregates and stores information on all user activity [and provides] rich personalization and targeting of content and ads to consumers based on their television viewing and Web surfing histories and preferences' (EPIC 2001).

ITV companies have countered their critics and defended their practices by stating that they are in compliance with industry privacy regulations stipulated under the 1984 *Cable Communications Policy Act*, whereby cable operators must receive informed written consent from subscribers before they are able to collect or use any data that could identify any individual personally (Benner 2001). Herein, MSNtv and AOLTV emphasize that their subscribers must knowingly 'opt in' and agree to have their personal data used. This adherence to existing legislation that governs the cable industry – and not specifically satellite or other ITV modes of delivery – provides a measure of defence against those who would argue that broader policies must be developed to expand privacy protection beyond cable companies alone. In this regard, MSNtv and AOLTV are not, in fact, bound in any way by the *Cable Communications Policy Act*. And, despite their claims to respect privacy, these companies have sought to circumvent the issue through their assertion that they are only studying data provided by 'anonymous' individuals and general profile information provided by the subscribers who use the personalized features of their ITV service.

They are also continuously looking for incentives to encourage customers to opt in and assent to the collection of personal information. The most coercive practice in this regard is that of negative billing on the part of Rupert Murdoch's SKY TV, wherein subscribers are charged £248 if they refuse to activate the interactive features of their satellite service (Burke 2000).

MSNtv and AOLTV have taken an alternative route and chosen to entice viewers by pitching the features requiring surveillance as so attractive that viewers would be silly not to want to participate. The customized electronic programming guide is a case in point. Another means to opening the door to surveilling the private sphere is to target the group which is both one of the most desirable demographic groups as well as one of the most willing to surrender personal information: children and teenagers. Marketers are well aware that young people feel little compunction about divulging household details, particularly in exchange for coupons for toy discounts or free promotional offers (Burke 2000). If the price benefits exceed the costs of information exchange, parents will be more amenable to filling out surveys and thereby providing the requisite permission needed for children under the age of thirteen to participate in the data-mining process.

Microsoft, in particular, has shown a heightened awareness of the consumer power of children, especially in their ability to pressure their parents to purchase almost any conceivable product, and has engaged in broader marketing strategies to ensure that young people continue to be socialized in the world of television and interactive technologies. In 1997, the computer software giant introduced its ActiMate line of toys: dolls, based on characters from popular children's television programs, that are equipped with computer chips that allow them to interact with children while they watch the series on television. Interactive Barney, PBS's purple dinosaur and guru of love and kindness, launched the ActiMate line and was called the 'breakout hit' of the year by Consumer Reports (Behrens 1998). At its functional level, Interactive Barney is pitched as an educational toy and viewing buddy that will 'say' approximately seventeen phrases without its interactive software but increase its vocal and entertainment capacity substantially when the purchaser adds a television adapter pack (sold separately, of course). The adapter activates the embedded chip that allows Barney to interact with his television program and sing, play games, and say such things as: 'I like watching television with you.' However, Barney only remains interactive as long as the child remains within a specified radius of the television set. To exit the televisual realm is to lose what one person has intimated is a 'magical' relationship with the stuffed animal (Cassels in Behrens 1998).

In critical theories of consumer culture, goods have long been said to act as magical totems for a host of human bonds, from group belonging to love itself. In the case of the ActiMates, we see yet another corporate attempt to fetishize a product to the extent that children view not only Interactive Barney but also television as a friend. Thus, at the social level, Microsoft's ActiMates are a means to ensure that children not only grow up encircled by interactive media but do so in such a way that they are comfortably domesticated within a televisual apparatus that seduces them with the discourses of consumer happiness. If interactive television increases, exponentially, your 'power' to choose among a bounty of offerings from your televisual friend, why would you need to look elsewhere for personal fulfilment? While this is a heavy-handed statement, it is intended to provoke us to muse about the larger social and cultural implications of interactive television. ITV is a unique surveillant technology in that it enters into the heart of the private sphere: the home. As with all forms of surveillance, privacy is an issue of concern in the new media age. However, the emphasis on privacy issues, in both popular and policy discourses, detracts our attention from the broader societal and economic dimensions of ITV's potential impact; in particular, ITV's propensity to contribute to far greater social fragmentation along class, gender, and ethnic lines combined with the desire of its makers to intensify the use of personal information to solidify the foundations of consumer society. It is to these concerns that the concluding section turns.

Questions of Power in the Age of Interactive Television?

The boob tube is dead; the future is all about smart, personalized television ... The point is: This is an industry that's always been changed and impacted by new technologies that have consistently made it better for viewers and a more powerful economic engine for business.
          Stacy Jolna, Chief Programming Officer, TiVo (in O'Leary 2000).

Jolna's statement is telling in two respects. First it echoes the more utopian depictions of media convergence by those working on issues of new communications technologies: a win-win situation for audiences who will cast aside the shackles of the network (and, for that matter, neo-network) age. Second, it is somewhat contradictory as it points to simultaneous revolution and continuity in television; namely, that television has long been a powerful engine for business. The two need not be mutually exclusive but they are not easily reconcilable either. Business concerns, through ownership structures and advertising pressures, have consistently operated as constraints on the creative process of storytelling in the private television sector. Issues of social representation through the construction and depiction of race, class, gender, and sexuality in both news and entertainment programming have always been bound by the bottom line: the perceived need to play to imagined 'middle-of-the-road' audiences in ways that would not offend their prevailing sensibilities and particularly not those of advertisers.

Indeed, private television has long been referred to as the 'sponsor's medium,' a term harking back to the days when the new medium transplanted the structure of the radio series/serials in which product companies owned the programs aired. With almost unlimited script control, these organizations used programs as showcases for their goods with the hope that the content would correspond with the rhythms of emergent middle-class, suburban domestic life and consumption patterns. The soap opera is, of course, the ultimate example. The symbiotic relationship between corporate America and the television industry did not change drastically after the quiz show scandals of the 1950s forced the FCC to investigate standard operating procedures in the Big Three networks: ABC, NBC, CBS.[4]

The abolishment of the single-sponsor production system and the subsequent move to licensed network programming and the thirty-second commercial did not substantially alter the perception that television programs should still be produced with the advertisement in mind. Programs that promoted idealized portrayals that supported the ideological narratives of modernity still dominated prime time schedules in order to provide preferred audiences with images of the good life to which they were encouraged to aspire. Today, we only have to watch an episode of *Friends* to see the process at work. Where else but on television would we find a group of young, beautiful, white people who, despite being chronically under- or unemployed, lived in spacious New York City apartments and set the standard of fashion for dress and hairstyles in North America for over a decade? Indeed, programs that spoke to the experiences of those living in the margins of economic,

---

4 In the 1950s, quiz shows such as *Twenty-One* and *The $64,000 Question* were among the most popular programs on network television. Sponsors, such as Revlon cosmetics, saw the potential to link their products in idealized ways to preferred winning contestants – notably, people who they wanted their customers to aspire to emulate. Consequently, the sponsors pressured producers to rig the shows so that the select contestants would continue to win as long as they were seen as beneficial to product sales. Not surprisingly, contestants who failed to measure up as idealized representatives for the sponsor's products (attractiveness, class, and ethnicity all playing roles herein) were set up to lose. When the FCC began investigations into the 'quiz show scandals,' it was not because of concerns over representational issues. Rather, there was a concern that television was too central to public dialogue and political participation to be seen as a dishonest medium.

political, and cultural power were rarely popular with network executives seeking to sell airtime to advertisers who wanted to promote a 'happy, buying, mood' among targeted audiences who were already incorporated, to some extent, into those spheres of power. This is the synoptic phase of television – where the many watch the few – and the broadcast industry, in concert with marketing interests, has long desired to move towards the panoptic where they, the few, can watch us, the many. ITV brings the synoptic and the panoptic together in ways that may not operate in our best interests as citizens as opposed to consumers, which is how these industries would like to position us.

As Turow (1997) underlines, it is important to distinguish between 'society-making media' and 'segment-making media,' because form contributes to the structure of content and the degree to which any medium can attempt to create a public rather than a market. ITV is a segment-making media as its panoptic capabilities change the primary mode of address to audiences whereby we become an aggregate of individuals as a opposed to a public of citizens. This is a fundamental change in television: even in the days of traditional commercial programming, when audiences were segmented in the minds of marketers, they still experienced the televisual address of a public. Drawing on the work of John Dewey and his statement that society exists through communication, many media scholars examined television through a 'ritual' perspective where the medium was regarded as an electronic public sphere (albeit imperfect and marked by relations of power) or virtual space where viewers would encounter issues and people that they might otherwise never come to know (e.g., Carey 1989). Television began to lose much of its ritualistic capacity during the neo-network era, when the proliferation of cable and pay channels expanded the number of channels and executives and advertisers initiated narrowcasting strategies. However, the potential for any form of televisual public space vanishes completely in the age of interactive television where all programs and commercials can be customized to our individual interests and desires. It is the ultimate difference machine, if we again invoke Gandy's words, because it only speaks to those who serve its interests and excludes others who are already socially marginalized or deemed demographically undesirable. Moreover, by personalizing our televisual worlds, ITV ensures that we need never come into contact with those who differ from us. We never have to see images that diverge from our view of society. By isolating us, political and commercial interests can tell each individual what she or he wants to hear and collective debate and action could very well become the casualties of media convergence.

These, I argue, are the imperative issues to consider in examining ITV and they are the ones that most often become subsumed by the emphasis on privacy as currently conceptualized in many examinations of surveillant technologies. Privacy is, indeed, important. It is not difficult to imagine our interactive practices of television viewing and websurfing being placed under scrutiny to determine whether we are international security risks or credit or health risks and, thus, need to be monitored accordingly. However, on a quotidian level, ITV is evolving into an integrated marketing technology concerned not with who we are as embodied individuals, but rather with establishing classificatory categories that can be profiled for the purposes of selling products. MSNtv is not interested in Mary or Bob as people who hold abstract hopes and dreams about philosophical concerns but, rather, as Individual A and Individual B whose hopes and dreams can be translated and exploited for sale. In this regard, my thoughts on ITV correspond with Lyon's (2001: 17) depiction of the disappearing body within surveillant practices and technologies: 'The fiction that the inside of a home is a haven from outside demands and pressures is subverted by the ways in which electronic devices take data into and out of the house, sometimes without our knowledge. Even our bodies, often thought in modern times to be our "own," and thus private, become a source of surveillance data. Paradoxically, the embodied person is still not in view. Only the image or the trace counts. The disappearing body makes an exclusive focus on privacy less salient to surveillance.' It is for these reasons that the larger social implications of ITV become paramount when we examine the medium's surveillance components in relation to the marketing discourses that hail a revolutionary new technology.

As I argued at the outset of this chapter, ITV is best described as television that has fulfilled its logic as a medium of consumption rather than as a rupture in the fabric of the industry. As Kim (2001) astutely explains, every technology is subject to an organizing ideology that governs the way in which its industry leaders will guide its development; and interactive television, structurally, differs little from traditional television beyond increased program choice and the technology's capacity to monitor its audiences. This stands in stark relief to the arguments of a revolutionary moment for newly empowered audiences and brings into question the conceptualization of power itself. If we follow Foucault's reasoning that power and resistance are mutually constitutive, we can see that the resistance of the ITV viewer is pragmatically located within the realm of 'choice' (increased channel capacity and personalized viewing options), while the power to establish the parameters of decision making remains on the side of industries who produce the content and monitor the choices made for instrumental purposes: 'The proliferation of new technologies ... and the ever greater range of specialized audiences is creating an image world which seems to suggest that 'there is something for everyone's taste' – a delirium of consumer sovereignty and unlimited choice ... The discourse of choice is a core element of that legitimation. Seen this way, the figure of the "active audience" has nothing to do with "resistance," but everything to do with incorporation: the imperative of choice *interpellates* the audience as "active"!' (Ang 1996: 12). Choice of not unlimited options, indeed, is not synonymous with the control of the means and modes of production; and to the extent that ITV viewers remain receivers of programs, rather than producers of content, the celebration of a newly empowered audience in the making appears premature. Rather, power, when defined as the power of representation or the power of information gathering, remains in the hands of industry. We agree to become transparent in exchange for the appearance of choice. In other words, ITV – in its present configuration – remains business as usual only more so.

Is this, however, what people truly want? Have we seen the end of watercooler television shows or debates about the political bias in news programming? Would we be happier to dwell in our own private, customized televisual worlds of consumption? It is too soon to predict the future of ITV but it is important to initiate the dialogue on the role it will play in the social sphere. Early studies show that many people are reluctant to integrate their televisions with their PCs as the first is associated with entertainment and relaxation while the latter defines the world of work (Stipp 1998; Stewart 1999). However, members of the ITV industries have time on their side and can afford to be patient. The next generation that will drive the market is one which is increasingly comfortable with multi-tasking their technologies and can move fluidly between the information and entertainment aspects of the digital world. Interactivity is something they have grown to expect and following shortly behind will be their younger sisters and brothers who, perhaps, were 'friends' with Interactive Barney.

REFERENCES

Ang, I. 1996. *Living Room Wars: Rethinking Media Audiences for a Postmodern World*. London: Routledge.

Barnouw, E. 1978. *The Sponsor: Notes on a Modern Potentate*. New York: Oxford University Press.

Barthold, J. 2001. 'Someone to Watch over You.' *Telephony*. 16 Apr.: 83.

Behrens, S. 1998. 'We'll Look Back on This Old Barney: An Early Input-Output Gizmo You Could Hug.' *Current Online*. 19 Jan. http://www.current.org/tech/tech801b.html (accessed 19 May 2003).

Benner, J. 2001. 'Beware the Eye in ITV.' *Wired News*. 26 June. http://www.wired.com/news/privacy/0,1848,44801,00.html. (accessed 3 March 2003).

'Britney Rules.' 2002. *Victoria Times Colonist*. 5 Feb.: D5.

Burke, D. 2000. *Spy TV: Just Who is the Digital TV Revolution Overthrowing?* London: Slab-O-Concrete Publications.

Carey, J. 1989. *Communication as Culture: Essays on Media and Society*. Boston: Unwin Hyman.

Cooper, J. 2000. 'Inside the Box.' *Brandweek*. 8 May: C32

EPIC. 2001. 'Report Examines Interactive TV and Privacy.' *EPIC Alert*. 29 June. Washington: Author.

Ford, B. 2000. 'CBS, Microsoft Ink Interactive Pact.' *Ecommerce Times*. 8 Sept. http://www.newsfactor.com/perl/story/4236.html (accessed 19 May 2003).

Gandy Jr, O. 1993. *The Panoptic Sort: A Political Economy of Personal Information*. Boulder, CO: Westview Press.

Howe, J. 2001. 'Private Viewing.' *MediaWeek*. 9 Sept.: 11(33): IQ12.

Kim, P. 2001. 'New Media, Old Ideas: The Organizing Ideology of Interactive TV.' *Journal of Communication Inquiry* 25 (1): 72–88.

Lyon, D. 2001. *Surveillance Society: Monitoring Everyday Life*. Buckingham: Open University Press.

McAllister, M., and J. Turow. 2002. 'New Media and the Commercial Sphere: Two Intersecting Trends, Five Categories of Concern.' *Journal of Broadcasting and Electronic Media* 46 (4): 505–14.

Negroponte, N. 1995. *Being Digital*. New York: Vintage.

O'Leary, N. 2000. 'Smart TV.' *Print* 54 (6): 132–5.

Roman, K. 2002. 'Future TV.' *Ottawa Citizen*. 23 May: F2.

Simitis, S. 1987. 'Reviewing Privacy in an Information Society.' *University of Pennsylvania Law Review* 135:706–46.

Smallman, J. 2002. 'The Medium May Kill the Message.' *Financial Post*. 8 Sept.: FP12.

Stewart, J. 1999. 'Interactive Television at Home: Television Meets the

Internet.' In J. Jensen and C. Toscan eds., *Interactive Television: TV of the Future or the Future of TV?*, 231–68. Denmark: Aalborg University Press.

Stipp, H. 1998. 'Should TV Marry PC?' *American Demographics* 20 (7): 16–21.

Swann, P. 2000. *TVdotCom: The Future of Interactive Television*. New York: TV Books.

Toscan, C., and J. Jensen. 1999. 'Introduction.' In J. Jensen and C. Toscan, eds., *Interactive Television: TV of the Future or the Future of TV*, 11–23. Denmark: Aalborg University Press.

Turow, J. 1997. *Breaking Up America*. Chicago: University of Chicago Press.

Vikhman, F. 2002. 'Making 'Thieves' Out of TV Viewers: The Fight over DVRs Reaches the Courts.' *National Post*. 3 Aug.: B1.

Wiley, L. 2002. 'SONICblue Battles Hollywood over Use of Digital TV.' 28 May http://www.emedialive.com/r10/2002/news0702_02.html (accessed 3 March 2003).

Winseck, D. 1999.' Back to the Future: Telecommunications, Online Information Services and Convergence from 1840–1910.' *Media History* 5 (2): 137–57.

13 Cultures of Mania: Towards an Anthropology of Mood

EMILY MARTIN

Numbers are the product of counting. Quantities are the product of measurement. This means that numbers can conceivably be accurate because there is a discontinuity between each integer and the next. Between two and three there is a jump. In the case of quantity, there is no such jump; and because jump is missing in the world of quantity, it is impossible for any quantity to be exact. You can have exactly three tomatoes. You can never have exactly three gallons of water. Always quantity is approximate. (Bateson 1979)

In my research on the anthropology of science, I have been following the awarding of value to certain psychological conditions that, not long ago, were seen as simply pathological. These conditions are characterized by 'hyperactivity' and exuberant energy.[1] For one example, Attention Deficit Hyperactivity Disorder is often associated with the 'Edison effect,' and said to be responsible for the successful lives of people like Winston Churchill, Thomas Edison, Albert Einstein, and Bill Clinton. For another, manic depression, or bipolar disorder, is associated with the innovations and creativity of numerous politicians, actors, CEOs, and artists, such as Theodore Roosevelt, Robin Williams, Ted Turner,

---

1 This paper is based on several years' ethnographic fieldwork on the cultural meanings of mania and other 'hyper' conditions. The research took place in numerous support groups for manic depression and Attention Deficit Hyperactivity Disorder (ADHD) on the east and west coasts of the United States; in large and small rounds in a private tertiary care hospital in an east coast city; in a neurological lab; in worker training sessions; and in off-the-record interviews with middle-level employees of major pharmaceutical corporations. Support from the Spencer Foundation and the NSF is gratefully acknowledged.

or Vincent Van Gogh. When Ted Turner was named *GQ*'s man of the year, his manic depression was cited as part of what makes him the 'corporealized spirit of the age.'

The desirability of hyper or manic traits means they are frequently cited in advertisements for commodities like Armani perfumes for men and women, athletic shoes, luxury linens, and a host of other things. In jokes and cartoons, the mania of bipolar disorder has come to be emblematic of extravagant creativity. In a *New Yorker* cartoon, a couple are shown facing an abstract painting in a gallery. One says to the other: 'It's good but it doesn't say "Bipolar".' The humour turns on the current association between extraordinary creativity and the diagnosis of bipolar disorder. If the painting were more than merely good, it would say 'bipolar.'

In the mid-1990s I described in *Flexible Bodies* how a consulting firm was running training sessions for more than 20,000 employees of a Fortune 500 corporation based on outdoor experiential education (Martin 1994). Poised to cope with the massive downsizing of the mid-1990s, the training centred on gymnastic exercises on high wires intended to teach workers to adapt to change, tolerate fear, take risks and, in general, cultivate flexibility.

In my current research, similar firms are offering training in something a little more unbalanced. One firm, Leapfrog Innovations, has developed a training exercise called 'It's a Mad, Mad, Mad, Mad Hunt,' a 'manic scavenger hunt' in which teams of employees compete for points through embarrassing, risqué, or mischievous activities – like taking manikins from a story window – designed to push the bounds of propriety. When the teams come together to tabulate the points they have earned, the heightened excitement in the room is meant to be (and is experienced as) akin to the state of mania.

I am interested in how the domain of interior psychic states is being made visible in particular ways and how it is being selectively harnessed for culturally specific purposes. This paper concerns the mood chart, a small, mundane surveillance technology by which interior states, moods, are being monitored and optimized. People living under the description of manic depression are often encouraged to keep a 'mood chart.' An example is 'Amy's self-rated mood chart,' found in a popular contemporary handbook on managing bipolar disorder by Michael Miklowitz (2002). In general terms, the mood chart encourages rational management of a subjective domain. Such management is part of a historical process by which, according to Nikolas Rose, 'Psychiatry, psychology, and psychoanalysis ... constituted the domain of subjectivity as itself a possible object for rational management ... Desired objectives – authority, tranquility, sanity, virtue, efficiency, and so forth' became conceived 'as achievable through the systematic government of subjectivity.' For systematic government of this domain, 'one not only needs the terms to speak about and think about it, one also needs to be able to assess its condition. That is to say, one needs intelligence or information as to what is going on in the domain one is calculating about. Information can be of various forms: written reports, drawings, pictures, numbers, charts, graphs, statistics, and so forth' (Rose 1998). Peter Miller describes how in accounting schemes, the individual is enlisted in calculation of his inner states, which instils 'practices of individualization and responsibilization.' These practices involve the building of an identity at the individual level, both general (I am a self-regulating calculating, rational person) and specific (these are my mood patterns, and this is how I respond to specific medications) (Miller 1992). Information is brought out from the inside, subjective realm to the outside, social realm, for use there by family, doctors, the pharmaceutical industry, researchers, and teachers (Latour 1987).

How can we understand what happens when people reflect on their moods and fill out mood charts? Is the way they experience themselves – their subjectivity – changed by filling out the chart, or merely reflected in it? To gain some historical perspective on these questions, I turn to Emil Kraepelin's charts of moods in *Manic Depressive Insanity and Paranoia*, originally published in 1921 and reprinted in colour in the new lifetime edition of 2002.

The first chart included shows us that for Kraepelin there were a limited number of types of mood states within the general category manic depression, each marked with a special graphic or shade of colour: mania, hypomania, raving mania, manic stupour (all in shades of red), depression (light and heavy), depression with flight of ideas, depressive excitement (all in shades of blue) (Kraepelin 2002). But the many other examples of charts he included in the book demonstrate that Kraepelin recognized the immense variety of ways individuals with manic depression cycle through these mood states or a subset of them. He argued that efforts of others to describe subtypes of the illness characteristic of groups of individuals were futile: 'the multiplicity of the courses taken by manic-depressive insanity ... is absolutely inexhaustible. The cases reported only show that there can be no talk of even an approximate regularity in the course, as has formerly been frequently assumed on the ground of certain isolated observations' (149). For Kraepelin, the time scale was in years (numbered on the left side of the chart), subdivided only into months and portions of months (labelled across the top of the chart). So compact was the chart that the entire lifetime of a person could be contained on a single page. The condensed information in Kraepelin's charts was compiled by doctors familiar with patients in a clinical setting, over many years. Patients themselves had no role in noting down their symptoms; rather, doctors kept careful records of patients' health, from how much they weighed to how neatly they could write.

According to Kraepelin, though the outcome of manic depressive insanity was more benign than dementia praecox (later known as schizo-phrenia), ordinarily no cure would be possible for patients admitted to institutions: he said most were 'forever lost' (Kraepelin 1962). The only hope would come from prevention: in Kraepelin's view, at least one third of all mental illnesses had causes he regarded as preventable, such as alcoholism, syphilis, traumatic injury, or addiction to morphine and cocaine (1962: 151).

Kraepelin saw manic-depressive insanity as a disease which had a natural, inevitable progression. He regarded this disease as a 'natural disease entity' (Hoff 1995) which he and his associate discovered through clinical record keeping of 'countless' patients (Engstrom 1995). A spe-cial form was devised to categorize information on each patient, called the 'Zählkarten' (Roelcke 1997). Although masses of data were col-lected, there is no evidence that Kraepelin quantified these data to present in the form of graphs or charts summarizing many patients. Nor did he use mathematical forms of representation to show, say, the rise and fall of mood in individual patients over time. In his charts, what changed over time was only that subtypes of moods succeeded one another.

In spite of the gloomy prospects for his patients, Kraepelin's charts, as well as his descriptions, showed the possibility for patients to enjoy prolonged and frequent disease-free periods. The remissions were called by Kraepelin 'long lucid intervals' when patients were 'able to reenter the family, to employ themselves profitably, and to return to their profession' (Kraepelin 1981).

Observations about Contemporary Mood Charting

In contrast to Kraepelin's charts, contemporary mood charts are in-tended to be kept by the individual patient. Ubiquitous, they appear in popular books, magazines, brochures handed out at doctors' offices, and on the web. Their use is explicitly encouraged by many groups interested in manic depression as described below:

Researchers: The website for the Harvard Bipolar Research group provides a sample chart already filled in and a blank chart you can download and print out for your own use.

Consumer advocates: The National DMDA website provides a sample chart, a chart for downloading, and a page of detailed instructions.

Pharmaceutical companies: Lilly's consumer website provides a web page with a slider to register where you are in the range of moods, and a button to click if you want to see the changes in your moods over time as a graph.

The government: NIMH gives you a complex chart, with extensive instructions and the ability to feed your data into their research effort.

Teachers and parents: Various kinds of charts have been developed for use in homes and classrooms, using coloured markers on an erasable surface or plastic markers that can be moved around the 'mood tree.'

As in Kraepelin's charts, contemporary charts record a range of psychological states. But in contemporary charts, only simple, everyday feelings and behaviours are listed (anger, sadness, irritability, tiredness, hunger, etc.), rather than technically defined complexes of traits (such as Kraepelin's 'manic stupour'). This simplification adds transparency to the workings of the chart for a non-expert, who only needs to reflect inwardly and report what is there. The traits charted, it is assumed, are knowable by the individual directly. Do I feel sadness? Elation? Do I experience energy? Fatigue? The charts do not call out for observations made by a trained external observer of the sort used by Kraepelin.

Compared to Kraepelin's charts, the contemporary mood chart has undergone a certain elaboration. In Kraepelin's case, an individual's entire life span could be described on one page; commonly today, a page contains the details of only a single day. Each day, in turn, can be divided into periods of hours and minutes and each quality or activity can be registered practically by the minute.

In sharp contrast to earlier charts, contemporary charts invariably contain a place to record the means of ameliorating mood disorders. All charts I have seen have a section, often occupying nearly half of the chart, for recording what medicines the person takes as well as how much and when they take them throughout the day. Material explaining the utility of the charts makes it plain that the careful plotting of medications in relation to symptoms is done so that adjustments to medication can be made in order to alter particular aspects of the mood pattern. A software company has devised a program, Mood Monitor®, which is sold to doctors so they can make it available to their patients on the web. When patients fill out the chart in their homes, the doctor can keep tabs on their condition remotely, and can calculate summary data that allow mood to be juxtaposed to medications, and medications adjusted as necessary. One testimonial from the company's website reads: 'I have used Mood Monitor® in my clinical practice and am very impressed. I have long found that Bipolar patients are unable to accurately measure their moods and other parameters between sessions. With Mood Monitor®, I can see at a glance how their moods are varying as well as how the patient is sleeping. This allows me to make adjustments in medications with more confidence.' Even without a doctor's intervention, people frequently say that representing their moods on a chart over time allows them to see more clearly exactly what differences medications make: 'I've found that minor changes in medication can make big changes in how I felt, so tracking dosages was useful. And embarrassing as it was, tracking when I didn't take medicines was useful.' In a sense, patients are being turned into machines for biofeedback as they carefully plot their moods' changing relationship to medications. The assumption in commentaries on mood charts is generally that patients are honest record keepers to the best of their ability, though the reality is undoubtedly more complex. Patients are frequently urged to take notes on themselves daily or even more often, to avoid lapses of memory.

Although the contemporary chart seems to offer more hope of improvement through medication than Kraepelin's, one effect of the detailed moment by moment scrutiny is to emphasize the abnormal. There is no reason why people should not mark their moods and other states as lying right along the middle axis of the chart, along 'normal,' but this seldom happens. What counts as 'normal' can occupy as little as one point on the chart, as on the Harvard bipolar Research Program web logo. In my fieldwork, support groups not uncommonly followed the practice of beginning each meeting by going around the room for a brief introduction. Each person would state his or her first name and then, using a kind of oral mood chart, give a number on a scale from –5 to +5, indicating a range of moods from very depressed to very manic.

'I'm Jan and I'm +3'; 'I'm Dave and I'm –2' and so on. Most meetings were attended by people with a diagnosis of manic depression, but one time a regular member brought along her husband, who does not have manic depression. He listened as each person gave his or her name and score. When it was his turn he said, uncertainly, 'I'm Brad and I guess I must be zero.' This practice illustrates the way people participate in the measuring of a psychological quality within themselves through a process of abstraction. Each person can occupy a different point on the scale, but everyone is located on the same scale, and no one but the visitor considers him or herself 'normal.'

Possibilities for Resistance

The argument has been made that it is virtually impossible to resist accounting schemes of this kind: in Peter Miller's view, when there is resistance to one such scheme, the new form usually takes the original as reference point. One regime is overturned in favour of another that remedies the shortcomings of the first. 'In the most extreme instances, where a particular regime of calculation is "successfully" overturned, it is generally in favor of a new regime whose advantages are argued for and understood in relation to the alleged deficiencies of the preceding one' (1992: 74). Miller's point is that accounting schemes have a way of focusing our attention on the efficiency and accuracy with which they measure something rather than on the appropriateness of measuring that something at all. Once accounting schemes for moods are in place, our tendency, the argument goes, is to try to improve how well they work, not to question the validity of measuring moods on an abstract scale in the first place.

In his classic description of modernity, Weber made clear how the precision, speed, and efficient standards associated with rationalization extended ever more widely across institutions and ever more deeply into them. Nonetheless he believed the process of rationalization could never be complete. Weber commented that 'the calculation of consistent rationalism has not easily come out even with nothing left over' (Weber 1958, 1946: 281). He meant that no abstract scheme could completely capture the intricate detail of any aspect of human life. Only by means of a fiction, one that holds in suspension everything that is left out of an accounting scheme such as a mood chart, could such a chart be imagined to be a complete representation of what people experience as 'moods.' The 'left overs' that an accounting scheme does not capture often stimulate activities that go far beyond remedying the shortcoming of the scheme. For example, as the publishing capacities of the web have opened a new door to individual creativity, people are beginning to design their own mood charts on the web. These charts do much more than reinscribe the calculations of the mood chart in a more efficient or effective form. They may change the definition of what 'moods' include, or alter the dimensions on which information about them is recorded.

For example, a chart on a website published by a man who identifies himself as Jinnah includes large amounts of information about his particular life, undercutting the depersonalized and abstract qualities of most charts. The more specific the information, the less readily it can be reduced to a number and compared to other individuals. Jinnah's chart also separates measures of mood from measures of functionality, opening multiple axes on which a person can compare different aspects of his or her condition. The additional axis could increase the surveillance of the chart over Jinnah's life, but at the same time it opens the possibility of challenging the way standard DSM categories group traits. In the DSM, moods on either end of the manic-depression scale are abnormal. By charting functionality as well as moods Jinnah discovers that he can be functional while his moods are abnormal, thus opening a possible rejection of the DSM's presumptions. However, he does not interpret it this way. Instead, he concludes he is never normal: 'I have used the charts to show my family that when they thought I was normal (i.e. functional), I wasn't emotionally stable. It came as quite a shock to them often because they couldn't detect anything wrong with me … [The chart allowed me] to realize I had no periods of normality.'

In principle, there is no reason a mood chart could not increase a person's sense that they had normal periods. Kraepelin's charts, for example, portrayed numerous blank spaces that represented the months or years the patient was in remission and presumably normal. Contemporary charts could easily designate a band of space above and below 'zero' on the vertical axis and label it 'normal' or an equivalent. In practice, the area of charts designated as 'stable' or 'okay' is narrow and sometimes allowed only one point on the scale. By means of this simple feature of the chart, a person's sense that he or she generally occupies only 'abnormal' space may be encouraged. Of course, broader cultural tendencies that are increasing the number and extent of mental and physical states that we regard as pathological would be involved here as well.

Hyperrationality as Resistance

I would argue that the problem of finding loci for resistance may be misconstrued in the case of 'mental illness.' What is at issue in this case is whether the person can be rational, possess reason. Where rationality itself is at issue, doing the accounting defined as self-management might be considered a strike for one's capacity to be a rational person. Charting one's moods could be seen as demonstrating one's rationality – and this in itself could constitute a form of resistance to being categorized as irrational.

From this point of view we can reconsider Amy's mood chart. What is being measured here? What is the something that goes up and down, or gets a numerical designation? Moods? Feelings? Energy? Will? As we saw, in Kraepelin's charts, different types of moods succeeded one another in an infinite variety of ways. In his charts there was no single entity that was measured in degrees of difference on a single scale. In contemporary charts, the something that goes up and down is the same commensurable stuff. Whatever this stuff is imagined to be, it comes from a private, individual, and internal space. The chart converts specific internal experiences into abstractions through numeric measurement, but it also makes these experiences social along the way. In an analogous process, when a specific form of labour is transformed through abstraction from something with use value into something with exchange value, it also becomes social. Marx explained how the specific labour of tailoring a coat could become equivalent to the very different specific labour of weaving linen. First, the concrete labour of tailoring becomes 'directly identified with undifferentiated human labour,' which is measured by labour time. This makes tailoring 'identical with any other sort of labour,' including the labour of weaving linen. Although tailoring, like all labour that produces commodities, is 'the labour of private individuals,' 'yet, at the same time, it ranks as labour directly social in its character ... The labour of private individuals takes the form of its opposite, labour directly social in its form' (Marx 1967: 64).

When one is categorized as mentally ill and hence outside the realm of the fully human, having one's private experience count as part of the social holds great significance. The individual uniqueness of experience might be lost in the homogenizing process of abstraction, but in return the private moods of an individual take the form of their opposite, moods 'directly social' in their form. This might not be a bad bargain when compared to the cost of being considered less than fully human.

Mood charts make private feelings social by placing them on a scale which is putatively applicable to everyone. One's personal mood states become commensurable with the moods of others, even though one's moods may be more extreme in their ups and downs. What makes this process social is that all moods can be charted, and hence all moods can be compared. Judging from my fieldwork, group interactions involving mood charts (which might provide more robustly social interactions) are rare, except for the brief introductory remarks common in support groups. Nonetheless, when the alternative is the stigmatized state of being mentally ill, a state that throws one's rationality itself into question, even a weak form of sociality might seem valuable.

The kind of social inclusion mood charts provide is minimal. It is also based on a form of abstraction from the unique individual case to broad categories in which particulars are submerged. To return to Marx, one of his main concerns in the case of labour was the way such abstracting processes enabled certain forms of power and control over labourers and the means of production. This is a major issue in the case of the pharmaceutical industry as well.

Beyond the question of who is included in human sociality, what kinds of populations emerge from these record-keeping practices? Keeping a mood chart is said to be part of 'mood hygiene.' In a popular book about manic depression, Mondimore includes a picture of Hygeia, the goddess of hygiene, who, in this context, stands for 'Practices and habits that promote good control of mood symptoms in persons with bipolar disorder.' 'Preventive measures can be [important] for improving symptom control in bipolar disorder' (Mondimore 1999). This is a powerful image, because it is well known how efficacious hygienic practices (cleaning water, sweeping houses, and washing bodies) was in the reduction of mortality and morbidity that took place in the early twentieth century. But this is also an odd image. Hygiene reduces physical disease better the more *pathogens* are reduced or eliminated. What is the hygiene of moods meant to reduce or eliminate? Are the surges and dips above and below the line of 'normal' meant to be reduced? If this were to happen, would emotions, feelings, and sentiments in general be reduced? Would it be more 'hygienic' if they were reduced to almost nothing? In Mondimore's book, the answer to this question is left unspecified.

To see what 'mood hygiene' could be imagined to improve, I would refer to other parts of my research, in which I show that the category 'mania' embraces two kinds of psychological states, mood and motivation. This doubleness of mania allows us to see how it can be considered a disordered mood that Hygeia might want to sweep out and also a form of behaviour that is so desirable that it is trained for and inculcated in the workplace. For example, all the mood charts I described include links between moods, high or low, and productivity, high or low. In Jinnah's diary, the link is explicit. Needless to say, elevated mood, even including some degree of mania, is considered an asset to productivity. Psychiatrists and psychopharmacologists explicitly speak of optimizing mania to enhance productivity.

In contrast, depression is more often seen as simply a liability. In the mass media, and some of the research that lies behind its stories, 'moods,' especially depression, are often associated with lack of productivity or inability to work. There has been a spate of articles in print media, the web, and on television looking at links between moods like depression and the inability to leave the welfare rolls by means of finding a job. In a recent *60 Minutes* story on welfare and depression, Lesley Stahl begins:

> One reason there are still 5 million people on welfare is that a huge number of them are depressed, not just suffering from a case of the blues, but seriously medically depressed. It's an epidemic of depression among America's poor.'

> Dr. Kessler of Harvard's School of Public Health estimates between a third and a half of people still on welfare are clinically depressed.

> Dr. Carl Bell of a mental clinic in Chicago: 'The state of Illinois, bless their heart, finally figured out that maybe the people who were going to be left on welfare were people with psychiatric disorders, and so maybe somebody ought to be here screening for that and referring people for treatment.'

> Stahl: So people come in for welfare, for their checks, to make their applications, and if somebody is there that perhaps spots these symptoms ...

> Dr. Bell: You can screen them out – everybody can get a very simple screening form – find out who's got what, and then treat them. (*60 Minutes*, 10 Nov. 2002)

In the *60 Minutes* scenario, depression is not only located in individuals but also in social groups at the bottom of the social hierarchy. The implication is that in individuals thought capable of self-maximization, who have the resources to optimize themselves with drugs, it would be optimum for depression to simply disappear. Ubiquitous pharmaceutical advertisements in the United States frequently imply that depression can be eradicated, and that its eradication would be a good thing. On behalf of Prozac and Serafem, Lilly urges you to 'Get your life back' (from the depression that has taken it away) and after treatment to remove the depression, declares 'Welcome Back!' On behalf of Zoloft, Pfizer exhorts, 'When you know more about what is wrong, you can help make it right.' Taking Zoloft will correct the 'chemical imbalance of serotonin in the brain,' which, it is implied, is the physical signature of depression. In medical taxonomies such as DSM-IV, depression takes a number of distinct forms: unipolar depression is a form of mood disorder characterized by sustained low mood; bipolar disorder is a form of mood disorder in which depression alternates with mania. In the public discourse and in many pharmaceutical advertisements, depression is often disconnected from these more complex medical categories and treated as a separable mental state that can be ameliorated or even eradicated with the help of drugs.

With depression eliminated, the way seems open to cultivate a kind of manic energy stripped of its non-functionality. Indeed, explicit programs to help achieve this end are abroad in the public domain. Through many best-selling books and television programs, Barbara Sher advocates a program of self-improvement in which moods are considered a distraction. Moods are important to identify so they do not get in the way of the real goal: building motivation. 'You can't ignore your emotions. They're strong and primitive and must be dealt with' (Sher 1997: 54). But moods and emotions are to be identified and swept away (as if by Hygeia's broom) so that your 'hidden motivators' and 'untapped energy sources' can be unleashed, even when, her website states, 'you are in a lousy mood.'

Mood charts are mostly intended for people who feel afflicted by their moods, and need to know more about them in order to control them by practising mood hygiene. The old-fashioned word 'hygiene' should be a tip-off to us that *selections* are being made among mood states. The fantasy is that – as depression withers away altogether, even among those on welfare – the mania of manic depression can be tamed or optimized. In Marx's account of labour, abstraction allows qualitatively different kinds of work to be compared on the same scale and brings measures of productivity into clearer view. In the case of moods, the abstractions of the mood chart make different moods comparable, and hence make the states of having moods more visible. With this vantage point, it is easier to imagine (if not actually bring into being) a population that is less 'moody' altogether. It remains to be seen whether moods actually diminish in a significant way and whether, in their place, heightened motivation arises to encourage the production of economic value.

REFERENCES

Bateson, G. 1979. *Mind and Nature: A Necessary Unity*. New York: E.P. Dutton.

Engstrom, E.J. 1995. 'Kraepelin: Social Section.' In G.E. Berrios and R. Porter, eds., *A History of Clinical Psychiatry: The Origin and History of Psychiatric Disorders*, 292–301. New York: New York University Press.

Hoff, P. 1995. 'Kraepelin: Clinical Sections, I.' In G.E. Berrios and R. Porter, eds., *A History of Clinical Psychiatry*, 261–79. New York: New York University Press.

Kraepelin, E. 1962. *One Hundred Years of Psychiatry*. Cambridge, MA: Harvard University Press.

– 1981. *Clinical Psychiatry*. Delmar, NY: Scholars' Facsimiles & Reprints.

– 2002. *Manic-depressive Insanity and Paranoia*. Bristol, England: Thoemmes Press.

Latour, B. 1987. *Science in Action: How to Follow Scientists and Engineers through Society*. Cambridge, MA: Harvard University Press.

Martin, E. 1994. *Flexible Bodies: Tracking Immunity in America from the Days of Polio to the Age of AIDS*. Boston, MA: Beacon Press.

Marx, K. 1967. *Capital, Vol. 1.*, ed. F. Engels. 1887; repr. New York: International Publishers.

Miklowitz, D.J. 2002. *The Bipolar Disorder Survival Guide: What You and Your Family Need to Know.* New York: Guilford Press.

Miller, P. 1992. 'Accounting and Objectivity: The Invention of Calculating Selves and Calculable Spaces.' *Annals of Scholarship* 9 (1–2): 61–86.

Mondimore, F. 1999. *Bipolar Disorder: A Guide for Parents and Families*. Baltimore: Johns Hopkins University Press.

Roelcke, V. 1997. 'Biologizing Social Facts: An Early 20th Century Debate on Kraepelin's Concepts of Culture, Neurasthenia, and Degeneration.' *Culture, Medicine and Psychiatry* 21 (4): 383–403.

Rose, N. 1998. *Inventing Our Selves: Psychology, Power and Personhood*. Cambridge: Cambridge University Press.

Sher, B. 1997. *Live the Life You Love*. New York: Random House.

Weber, M. 1958. *Max Weber: Essays in Sociology*. 1946; repr. New York: Oxford University Press.

14 Surveillant Internet Technologies and the Growth in Information Capitalism: Spams and Public Trust in the Information Society

DAVID S. WALL

In July 1993, *The New Yorker* published Peter Steiner's celebrated cartoon of two dogs furtively watching a computer screen and one saying to the other: '[o]n the internet no one knows that you're a dog' (69 (20): 61). The cartoon gained instant popularity by symbolizing optimistically the free commons of cyberspace. Now consigned to the dustbin of history, it is quite clear in the cold light of the twenty-first century that not only does the technology detect that you are a dog, but the same data flows will indicate what breed you are; your fur colour and type; whether you prefer Eukanuba™, Pedigree Chum™, or some other proprietary brand of dog food; how old you are; and whether your kennel cough injections are up to date. From this information, other technology will trawl through networks of databases to combine your information with that from other, similar breeds of dog to evaluate your probable state of health and anticipate the financial and personal burden that you will make upon your owners – not to mention any other risks to them or society that you are likely to pose, such as potential adverse changes in your temperament brought about by hormonal changes due to your age. Of great significance here is the fact that this information about your canine characteristics has a monetary value which can be exchanged in the informational marketplace.

One of the hallmarks of the new industrialism that underpins the information age is the valorizing of information and the subsequent growth of a new economy based upon information capital(ism), which is the accumulation of profit and wealth arising from the exchange and exploitation of informational sites of value.

This chapter explores how the 'surveillant' qualities of Internet technologies have facilitated the growth in information capital(ism). It also illustrates how the medium is shaping the message, because just as this new economy has merged with, and spans across, formal (legitimate) economies to create entirely new and beneficial business opportunities, the same processes that have given rise to it have also generated the opportunities for new forms of harmful and intrusive behaviours. These behaviours are endangering the establishment of public trust in the technology of the information society.

This chapter begins by discussing the emergence of the information age and the rise of information society. It then looks at the 'surveillant' technologies of the Internet which make possible the accumulation and exploitation of valuable personal information. It next addresses the construction of the industry in spams, or unsolicited bulk e-mails (UBEs), as they are often called, to illustrate the economy of the trade in information capital. Finally, it assesses the new opportunities for offending which arise from the appropriation of information capital and which are characteristically different from 'traditional' forms of criminal or harmful activity. Consideration is given to the implications of these developments for trust in social relations. The analysis and discussion contributes to a broadening of the debate over the 'new politics of surveillance' which emerged during the past decade and was further shaped by the political aftermath of 9/11 (see Lyon 2001a; Levi and Wall 2003: 179). The findings are largely drawn from research funded by a Home Office innovative research award (Wall 2002a) (see postscript).

Capital in an Information Society

The most prominent thinker on the topic of the information age is Manuel Castells, who published his analysis in a trilogy of works under the series title *The Information Age: Economy, Society and Culture* (Castells 2000a, 2000b, and later in 2000c). In his much contested argument (see Van Dijk 1999), Castells claims that the 'information age' has transformed the relationships of production/consumption, power, and experience (2000c: 5). He goes on to observe that a new economy has emerged which has three distinctive characteristics (2000c: 10). First, it is based upon *informational* productivity, which is: 'the capacity of generating knowledge and processing/managing information.' Second, it is *global*: its 'core, strategic activities, have the capacity to work as a unit on a planetary scale in real time or chosen time.' Third, it is *networked*, giving rise to a new era of economic organization in the form of networked enterprise. As a consequence, the main unit of production has shifted from the organization to the project. Castells contends that while the organization remains the legal unit of capital accumulation, it is nevertheless just one node in a global network of financial flows (2000c: 10; also see discussion in Mendelson and Pillai 1999).

In short, the information age argument centres on a shift in values from more tangible to less tangible forms of wealth; from things to ideas and information which have a global currency. Advancing Castells's argument, if the organization remains the legal unit of capital accumulation, then it can also be the illegal unit of accumulation. The global networked informational economy is also the source of harmful, acquisitive behaviour.

The New Technologies of Surveillance via the Internet

The main generators of new 'information' are the surveillant Internet technologies illustrated below. Followers of the debates over surveillance will be forgiven for the rather confusing conceptual variations in breadth and difference that are often employed, especially between theoretical discourse and practical applications, but also between different populations – either personally or en masse. Bowden and Akdeniz's (1999: 96–7) useful demarcation between mass surveillance and personal surveillance is employed here to demonstrate the qualitative differences between different technologies.

One-to-one personal surveillance on the Internet usually takes place via discrete software known as Spyware. The classic, though illicit, spyware package is the Cult of the Dead Cow's *Back Orifice* program, a 'remote administration tool' that gives system administration – type privileges to a remote user through the computer's Internet link. It allows remote users to gain access to the target's computer and surveil the contents of the computer and monitor its operation. Back Orifice is illegal because it contravenes the United Kingdom's *Computer Misuse Act 1990* and the equivalents in other jurisdictions. There are, however, also a range of commercial spyware packages such as Spectorsoft's *Spector Pro 4* which, once installed on a computer, can record e-mails, transactions in chat rooms and instant messaging, access to websites, keystroke recording, and VCR-like snapshot recording. The manufacturers boast that: 'Spector pro automatically records everything your spouse, children and employees do online' and the many customer endorsements included on the website glow:

[Jason] '[t]he first night I installed Spector I caught my girlfriend talking with many different men, and forced her to make some decisions about our relationship'

[S. Bata] 'Spector Pro saved my child's life on the internet'

[Scott] 'My office manager was spending 2.5 hours every day on the internet. Couldn't [previously] find out why she wasn't getting her work done ...'

(http://www.spectorsoft.com/products/SpectorPro_Windows/customers.html – cited 5 March 2003)

Three more types of automated personal surveillance software on the Internet are also significant. The first is the web search software, or 'spam spider bots,' which works its way through websites surveilling and 'hoovering' up information about individuals – typically web addresses which are then compiled into e-mail lists to be sold to spammers (see below). The second is the 'cookie' (see Cookie Central), a small string of personal data generated by a web server in response to embedded HTML information in a website that the user has visited. The cookie is then stored in the user's computer ready for access by the website when the user next visits the site. The primary function of a cookie is to simulate a continuous connection to a website and therefore assist the user to navigate the site with as little obstruction as possible. However, in carrying out its intended purpose it also performs a clandestine and automatic function of transmitting personal information to the web server once a specific web page has been selected. It thereby enables the owners of the website to surveil the users in order to obtain important data about them, for example, about the frequency of use of a particular website and what is being accessed on it. The third is the software that facilitates the decentralized peer-to-peer technologies, such as Gnutella, which depend upon participants allowing their 'swop' directories to be surveilled by other participants.

Mass surveillance, in contrast to the above, typically involves the scrutiny of complete population groups to identify patterns of behaviour through computer profiling. Three prominent examples can be given of this form of surveillance. The first is the use of third-party cookies by marketing companies such as Doubleclick and Engage (see Ward 2000) to compile information for the profiling of users and matching of e-mail addresses with individuals, mainly for advertising purposes. The second example is the ECHELON network, a joint USA/UK government-run interception system that surveils large numbers of 'transmissions and uses computers to identify and extract messages of interest from the bulk of unwanted ones.' It thus captures information about potential terrorist or other threats to the national infrastructure and also any critical commercial intelligence that might affect national interests (Bernal 2000). Whereas ECHELON listens in to communications networks, the third example of a mass surveillance system, the U.S. Total (subsequently Terrorist) Information Awareness network seeks to join up a range of databases and employ 'dataveillance' (Clarke 1994) techniques through data search and recognition pattern technologies to uncover terrorist threats to the national infrastructure (DARPA 2002).[1] Each of the three systems described above enable the 'few' to surveil mass populations and contribute to what Haggerty and Ericson refer to as a 'surveillant assemblage' – a term describing the relationship between heterogeneous surveillance technologies that '"work" together as a functional entity,' but do not have any other unity (2000: 605).

However, while Bowden and Akdeniz's distinction between personal and mass surveillance delineates the scope of surveillance, it still leaves outstanding the question of who is surveilling who and why. In answering this question it is constructive to draw upon Foucault's much-used conceptualization of Bentham's Panopticon prison design, which allows prison officers to see prisoners without being seen (Foucault 1983: 223). Although Foucault never had electronic communications in mind, he nevertheless signifies the power relations present in the construction of the panopticon as well as the direction of that relationship, which is characterized by the 'powerful' being able to see the few without themselves being seen (Foucault 1983). Simply put, public knowledge of both systems of mass surveillance has the potential to temper the behaviour of the populations being surveilled, although the shaping of behaviour is currently limited: most internet users are not yet aware of the transparency of their actions and the volume of surveillable 'traffic' data they currently generate in their Internet transactions.

The Internet panopticon must not be conceived as a simple, unidirectional power-relationship – even if the panopticon is 'super' (Poster 1995), 'virtual' (Engberg 1996), or 'electronic' (Lyon 1994: chap. 4) – because network technologies facilitate levels and types of connectivity

---

1 Although this program has now been abandoned by the U.S. government in the face of public pressure, comparable programs are being developed in other governmental departments. More importantly, the logic of the Total Information Awareness Network has already becomes deeply embedded in official thinking about national security.

not previously provided by communications technologies. In particular, network technologies generate both multidirectional and multiple information flows (Wall 2003) – the former is discussed here. Thus, 'the many can also watch the few' (Mathieson 1997: 215), as the Internet also possesses a synoptic quality which shapes behaviour patterns and potentially exercises a reverse panoptical power relationship.

Four basic, everyday examples of synoptical surveillance using the Internet are described below. The first is found in the City of Ontario, California, which, under Megan's Law (1996), displays on its official city and police force web pages the locations of known sex offenders within the city. The second is located in Newfoundland, which, like many government authorities, networks its roadside cameras to show weather and road conditions and traffic volume, but also to let local motorists see that they are being seen. The third example is the car rental company Easycar, which publishes the surveillance pictures of car rental defaulters on the Internet to publicly name and shame them into returning the overdue cars. The final example offered here are the many lists of individuals wanted by law enforcement agencies throughout the world. Perhaps the most famous of these is the FBI's 'Top Ten Most Wanted Fugitives,' but most other police agencies now publish their 'Most Wanted' on their own Internet sites. Although the practice predates the Internet, the 'Most Wanted' information is now publicly available on a global scale. Indeed, Officer.Com publishes a worldwide compilation of 'Most Wanted' lists (mostly in English language) and The World's Most Wanted, Inc. provides a searchable list of fugitives and unsolved crimes. The common feature of each of these examples is that the many can surveil the few from the comfort of their own computer screen, wherever it is located.

What is particularly distinctive about Internet information flows, however, is not that they can be *synoptic* as well as *panoptic*, but that they are simultaneously multidirectional. Multidirectionality gives the Internet and its technologies an 'idolatrous dream of omniperception' and a 'minacious twinkle in the electronic eye' (Lyon 2001b: 147), creating distinctive surveillant assemblages (Haggerty and Ericson 2000). Just as the surveillers surveil their populations then, in principle, the surveilled can surveil the surveillers. This surveillant mutuality has profound implications for debates over Internet governance, especially the potential for new regimes of governmental accountability. To assist with this function is a range of software tracking tools and facilities, often built into commercial operating systems, which enable individuals to gain information about who is sending them messages and about the registered owners of the domain names of the websites they access. Of course, Internet service providers (ISPs) retain records of their through traffic, either for their own use or because of legal requirements, thus contributing to 'the disappearance of disappearance' (Haggerty and Ericson 2000: 619) and increasing the efficiency of the Internet as a surveillance tool.

Since every Internet transaction leaves a trace in the ISP's traffic data, the sex offenders in the City of Ontario, the motorists of Newfoundland, the Spyware surveilled employees, spouses, and children, and the 'Most Wanted' could in principle and with the right resources and software counter-surveil those who are looking at them. Indeed, it is currently the case that security software such as Norton Internet Security will not only provide users with full details of illicit surveillers, it will also display the location of those surveillers on a map and, through WHOIS.Com, also identify the ISP and prepare the relevant identification information needed in order to make a formal complaint to the surveiller's ISP. Clearly, the balance between the exercise of the two sources of power is rarely even, but the concepts of panoptic and synoptic power are increasingly important tools in the analysis of governance of online behaviours.

The point of this section is not, however, to debate surveillance – contributing authors elsewhere in this collection have already done this, to much greater effect. Rather, it is to illustrate the surveillant practices inherent in Internet technology which produce or compile information that has a marketable value.

The Construction of the Spam Industry

The workings of the ubiquitous spam industry provide a very useful and contemporary example of the production of information and the market for information capital. Spamming is of great interest because its relatively minor and non-sensational impact upon each individual recipient has meant that UBEs are rarely discussed as anything other than a nuisance or an incivility. Spamming nevertheless affects all online users and the collective and harmful impacts of UBEs to commerce and online society as a whole are very considerable. Not only do the majority of spams constitute pure cybercrimes (see Wall 2002b: 193) in that they perform small-impact, multiple victimizations which are individually negligible, though collectively significant, they also represent an experience that most Internet users share. Therefore, it is important that UBEs should not be seen simply as a series of electronic mail shots, and they should certainly not be confused with terrestrial junk mail, which is actively encouraged because it economically supports the postal service. Rather, they are initiators of a potential interactivity between offender and victim which results in the latter losing out. Importantly, UBEs are solely the products of information age technologies; they are distributed through networks, they have a global reach, they are informational, and if you eliminate the Internet, they would not exist.

In the broader perspective they also represent a shift in the criminal division of labour in that new information technology places considerable productive power – some would say destructive power – in the hands of a few 'empowered small agents' (see Rathmell 1998: 2; Pease 2001: 22), the civil and criminal law enforcer's worst nightmare. This power falls outside the traditional, capitalist, power framework to temper the growth of public trust in new technology and threaten the commercial viability of the information society. Within the context of this chapter, spams are an 'indispensable prism through which social structure and process may be seen' (Lyon 2001b, quoting Abrams 1982: 192).

Spamming is so called because UBEs are: '... something that keeps repeating ... to great annoyance,' as in Monty Python's 'Spam Song,' wherein the term 'spam' was sung repeatedly by a group of Vikings in a restaurant (*Compuserve* 1997; Edwards 2000: 309; Khong 2001). Unsolicited bulk e-mails (UBEs) are e-mail advertisements that contain announcements about, and invitations to participate in, nefarious ways to earn money, obtain products and services free of charge, win prizes, spy upon others, and obtain improvements to health or well-being through revolutionary ways to lose weight, replace lost hair, increase one's sexual prowess, or cure cancer.

While, there are some vague arguments in favour of UBEs based upon the promotion of legitimate commercial activity and also upholding rights to free expression, the demerits outweigh the merits as UBEs generally degrade the quality of life of Internet users and invade the sanctity of an individual's privacy. UBEs rarely appear to live up to their promises, often carrying unpleasant payloads in the form of either potent deceptions, harmful computer viruses and worms, or Trojan Horses which enable third parties to gain clandestine access to the user's computer. They not only choke up Internet bandwidth and slow down access rates, reducing efficiency, they also waste the time of Internet service providers and individual users, who must manage UBEs and deal with the computer viruses that they often carry. Rather worrying is the prediction, supported by empirical observations (Wall 2002a; 2003) and other commentaries (Yaukey 2001), that the problem of indiscriminate UBEs is likely to continue to increase during the coming years.

The spamming industry in fact comprises two quite different sets of enterprises. The first is the compilation and production of bulk e-mail lists which are then sold on to e-mail marketers (spammers). The second is the impact of the content of unsolicited bulk e-mails upon recipients once they have been distributed by spammers to e-mail addresses on those lists.

E-mail List Compilation

The production of bulk e-mail lists is largely, though not exclusively, organized around the wholesale acquisition of online users' private information. There are two main strategies for compiling bulk e-mail lists. The first is to encourage voluntary subscriptions to e-mail lists, which is the current legal position in EU countries under the EU 'Directive on privacy and electronic communications' (2002/58/EC). Article 13 of the directive outlaws UBEs and enforces the principle that individuals have to voluntarily 'opt in' to mail lists in order to receive bulk e-mails – this is also the position in other jurisdictions (CAUCE) (see postscript). Subscription may be proactive, as in the case of specific interest groups. For example, the United Kingdom's Joint Information Systems Committee (JISC) hosts a facility for academics to subscribe to mail lists that will keep them informed of important issues and developments in their subject or interest areas. Subscriptions may also be reactive: users may be encouraged, even coerced, into providing personal information as part of the subscription procedure to a 'free' online service, in effect selling their personal information in return for a service. The second strategy for obtaining information is to employ surveilling website 'spam spider bots' (described above) to scour the websites for e-mail addresses.

Individually, e-mail addresses have no perceivable individual financial value, but when collated with 10, 20, 40, 80 million or so other e-mail addresses they accumulate value. UBEs are usually sent to e-mail addresses that have been supplied on a CD-ROM and sold to the spammer by a bulk e-mail compiler. The following are two advertisements for CD-ROMs containing lists of e-mails:

MULTILEVEL MARKETING OPPORTUNITIES: E-mail Addresses 407 MILLION in a 4-disk set
** Complete package only $99.95!! **

WE WILL SEND Successfully E-mails 1.Million ADDRESSES =Only $99.95!! Nowhere else on the Internet is it possible to deliver your e-mail ad to so many People at such a low cost.100% DELIVERABLE Want to give it a try? Fill out the Form below and fax it back:

The majority of e-mail addresses on the CD-ROMS tend to be unconfirmed or unprofiled and therefore do not provide the spammer with their intended responses. Indeed, some of the major victims of the spam list compilation part of the industry are themselves spammers who have been duped into buying expensive CD-ROMS of unvalidated and useless e-mail addresses. Having said that, not many responses may be required in order to recoup the start-up costs and make a profit.

Confirmed or validated e-mail addresses have a much higher value and their worth is greatest when they are profiled by subject or owner characteristics, because like advertisements, UBEs that contain information of direct relevance to the recipient are most likely to obtain a positive response and result in a successful transaction.

We specialize in successful, targeted opt-in e-mail marketing. MORE THAN 164 CATEGORIES UNDUPLICATED E-mail Addresses!! ** business Opportunity seekers, MLM, Gambling, Adult, Auctions, Golf, Auto, Fitness Health, Investments, Sports, Psychics, Opt-in Etc..

A 'spoof spam' strategy is frequently employed by bulk e-mail list compilers to 'validate' e-mail addresses and obtain additional information to assist in profiling. Indeed, the findings of an ongoing longitudinal study of UBEs and their content suggests that only a small proportion, possibly as few as 10 per cent of all UBEs, appeared to be genuine attempts by vendors to inform the online public about their wares, or to provide useful information (Wall 2003). The remaining 90 per cent or so lack reasonable commercial or informational plausibility, suggesting that the majority of UBEs either seek simply to elicit a response, are incredibly naive, or deliberately intend to deceive the recipient (Wall 2003). A rough guesstimate is that about a third of all e-mails are 'spoof spams.' Three distinct spoofing tactics can be identified.

The first tactic is to send out a blank UBE requesting an automatic response from the recipient's computer when opened. The second is to include subject matter that is so offensive or ludicrous that it incites the recipient to 'flame' the sender, sending back an angry reply. The third tactic is to include within the content of the UBE an invitation to 'deregister' from the mail list. Some of these invitations also constitute 'remove.com' scams that are discussed below, whereby the recipient innocently responds and subsequently becomes ensnared in a scam.

In each case the spammer obtains confirmation that the e-mail address is valid and may also receive some important information about the recipient that includes personal details, occupational information, and business information. E-mail replies frequently reveal much information about the sender, such as where they work: for example, 'ac.uk' denotes that they work in a UK university and gov.uk in a UK government office, while 'nameofbusiness'.com reveals that they are in business. E-mail signatures reveal even more specific personal information.

The Contents of Unsolicited Bulk E-mails

Table 14.1 provides a proportional breakdown of the contents of UBEs received at one account during the first two years of the longitudinal study into UBEs mentioned earlier. It is followed by a more detailed description of their content which illustrates the complex and multiple information flows that UBEs generate. It also lends weight to the implausibility argument made earlier, and outlines the types of risk recipients face. Though not a precise match, the categories and proportions roughly equate with Brightmail's 2002 survey entitled *Slamming Spam* (Brightmail 2002).

Income-generating claims contain invitations to the recipient, supported by unsubstantiated claims, to take up or participate in lucrative business opportunities. Examples include the following: a) investment reports and schemes; b) lucrative business opportunities such as pyramid selling schemes; c) earning money by working at home; d) 'Pump and Dump' investment scams; e) e-mailed Nigerian Advanced Fee scams; and f) invitations to develop websites and traffic for financial gains.

Pornography and materials with sexual content. Examples include the following: a) straightforward invitations to gain access to a website containing sexually explicit materials; b) invitations to join a group which is involved in sharing images and pictures about specific sexual

Table 14.1. UBE content

Content	%
Income-generating claims	28
Pornography and materials with sexual content	16
Offers of free or discounted products, goods, and services	15
Product adverts/Information	11
Health cures/Snake oil remedies	11
Loans, credit options, or repair of credit ratings	9
Surveillance/Scares /Urban legends	5
Opportunities to win something, online gambling options	3
Other	2
Total	100

activities; and c) invitations to webmasters to increase their business traffic by including invitations to obtain access to sexually oriented materials on their sites. Many of these UBEs contain entrapment marketing scams (see below).

Offers of free or discounted products, goods, and services (including free vacations). For recipients to be eligible for these offers, they usually have to provide something in return, such as money, a pledge (via a credit card), or information about themselves, their family, their work, or their lifestyle. Enticements include the following: a) free trial periods for products or services – for example, mobile phones, pagers, satellite TV (it was up to the recipient to withdraw from the service – as long as the recipient first signed up to the service); b) free products, such as mobile phones, pagers, or satellite TV, if the recipient signs up to the service for a specified period of time; c) UBEs which seek to exploit import tax or VAT differences between jurisdictions by selling items such as cheap cigarettes, alcohol, and so forth; and d) UBEs which sell goods across borders, from jurisdictions in which the goods are legal to those where the goods are either illicit or restricted, such is the case with prescription medicines.

Advertisements /information about products and services. Some of these advertisements are genuine, others are blatantly deceptive. Examples include advertisements for the following: a) office supplies, especially print cartridges; b) greatly discounted computing and other equipment; c) medical supplies and equipment; d) branded goods at greatly discounted prices; e) qualifications from unknown educational institutions; f) Internet auction scams, whereby an advertisement containing information about the auction is spammed; and g) bulk e-mail lists.

Health cures/snake oil remedies. Spammers who send out UBEs advertising health cures or snake oil remedies seek to prey on vulnerable groups like the sick, the elderly, and the poor. Examples include offers of: a) miracle diets; b) anti-aging lotions and potions; c) the illegal provision of prescription medicines; d) expensive non-prescription medicines (such as Viagra) at greatly discounted prices; e) hair loss or similar remedies; f) various body enhancement lotions or potions to effect breast, penis, or muscle enlargements or fat reduction; g) operations to effect the above; and h) cures for cancer and other serious illnesses.

Offers and invitations to take up loans or credit options, or to repair credit ratings. Examples include the following propositions: a) instant and unlimited loans or credit facilities or instant mortgages, often without the need for credit checks or security; b) the repair of bad credit ratings; c) credit cards with zero or very low interest; and d) offers which purport to target and engage with people whose financial life, for various reasons, exists outside the centrally run credit-rated driven banking system.

Surveillance information, software, and devices. This category is hard to disaggregate from the mischief section found below. The two are included together in table 1 because it is hard to tell whether the information and products are genuine or not. Examples include: a) scare stories about the ability of others to surveil their Internet use to coerce recipients into buying materials – books, software etc. – about how to combat Internet surveillance and find out what other people know about them; b) encouraging recipients to submit their online access details, purportedly to find out what others know about them; c) recommending a web-based service for testing the recipient's own computer security; and d) encouraging recipients to purchase spyware that allegedly equips them to undertake Internet surveillance upon others (see earlier descriptions).

Hoaxes/ urban Legends, mischief collections. Examples include: a) UBEs that appear to be informative and tell stories that perpetuate various urban legends; b) hoax virus announcements or 'gullibility viruses' which seek to convince recipients into believing that they have accidentally received a virus and then provide instructions on to how to remedy the problem, deceiving them into removing a system file from their computer; c) messages which appear to be from friends, colleagues, or other plausible sources that deceive the recipients into opening an attachment which contains a virus or a worm; d) chain letters which sometimes suggest severe consequences to the recipient if they do not comply; such letters may also engage the recipient's sympathy with a particular minority group or cause, for example, female single parents, or women in general (the Sisterhood Chain Letter scam); e) e-mail-based victim-donation scams that emerged on the Internet soon after the events of 9/11; f) invitations to donate funds to obscure religious-based activities or organizations; and g) links to hoax websites, such as Convict.Net which originally started as a spoof site, but was so heavily subscribed by former convicts that it eventually became reality.

Opportunities to win something, online gambling options. Examples include: a) notification that the recipient has won a competition and must contact the sender so that the prize can be claimed, or that they must provide some information before the prize can be received; b) offers to enter a competition if information or money is provided; and c) free lines of credit in new trial gambling websites. Mostly these are disguised forms of entrapment marketing (see above).

Risk and the Appropriation or Manipulation of Information Capital

The burning question about spams is what impact do they have upon the recipient? Of course the first thing that springs to mind from the typology above is that there is nothing really new about them; in fact, they follow tried and tested fraudulent routines. Not only do most criminal and civil justice systems already have legal remedies for these deceptions, they are also no stranger to the occupational cultures and investigative skill sets of criminal justice professionals, or the experience of the general public to reject them. But the content categorizations and the detailed descriptions of the activities under them also illustrate quite vividly the extent to which they rely upon the manipulation of information. They also indicate some significant differences in their scope, span, and penetration of potential victim populations.

There are currently three main forms of deception and entrapment to be wary of. The first is *direct engagement with victims* through UBEs on a one-to-one basis. Individuals, duped by the invitation or proposal delivered by UBEs that do not simply seek address validation, reply and become embroiled in an exploitative relationship with a fraudster who could be located almost anywhere. The inevitable outcomes are 'scams,' frauds, and deceptions, described earlier in the content analysis. Inducements include get rich quick schemes, pornography sites, free goods, snake oil remedies/health cures, credit repairs, surveillance software, online gambling, and so forth.

The second is *mechanized entrapment*, a sort of automated entrapment marketing through embarrassment / blackmail typically linked to reminding recipients of their prior access to pornography sites. A popular method is to entice potential victims into subscribing to thirty-day or three-month free trial offers upon supply of their credit card details. The recipient receives free access to the service for the prescribed period of time but has then to take the initiative to withdraw from receipt of the service, otherwise it will be automatically assumed that they wish to continue and billing will commence. Once entrapped into receiving a service, recipients who seek to disengage from it are often given mild indications, even warnings, that if they cease accepting the service information about their subscription could be made public. At the least, they may be requested to enter into land mail correspondence with the service provider to confirm their intent to withdraw, for which they will receive a letter to their home which clearly states the source (a pornography site). At the worst, there are a number of urban myths (which are hard to corroborate) to the effect that service providers have reimbursed fees to complainants in the form of personal cheques bearing the provider's name and in some cases graphic sexual imagery in the background print, thus reducing the likelihood that the complainant will submit it for payment.

The third type is the *automated remove.com scam*. Fed up with receiving nonsensical UBEs, recipients are encouraged to click onto a service that promises to remove them from bulk e-mail lists. In order to proceed with the removal service which allegedly ensures their continued privacy, recipients are asked to provide the provider of the service with important information about their identity. Usually a small 'administration' charge is requested, paid via a credit card, which is often recurring. This fact is often hidden in the extensive terms and conditions. In practice, recipients are rarely supplied with the service that they seek, but neither can they confirm conclusively that the service delivers what it promises. Therefore, they continue to pay the very small recurring charge, typically about $2.95 every three months or so.

The risks from the content of many UBEs to online users are potentially horrendous. Furthermore, the fact that English-language UBEs are continuing to double annually despite more restrictive legal regulation demonstrates the extent to which the spam industry circumnavigates or even exploits the transjurisdictional capabilities of Internet technology. This suggests strongly that spamming will continue to thrive despite the law and therefore calls into question the effectiveness of legal interventions against spam, such as Article 13 of the EU directive which prohibits unsolicited e-mail throughout the EU member states. Arguably, if we need more law, it should be enacted within the countries of origin from where most UBEs originate (in the Far East), or the remaining states in America that currently have no anti-spam legislation.

It is wrong to simply link risk assessment with the extent or growth in the prevalence of UBEs. There is considerable evidence to show that UBEs pose less risk than is often assumed. Although the effectiveness of the law may be limited in a transjurisdictional environment, it is still likely to have a tempering effect upon the distribution of UBEs by legitimate enterprise, especially following the growth in case law in common law countries. More encouraging is evidence of an increasing amount of prophylactic activity against UBEs. Internet users are ceasing to regard UBEs as a threat and are viewing them more as an incivility which can be coped with, usually by deleting the offending UBE (see Wall 2003). The other major development in the 'war against spam' is the increasing technological sophistication of spam filters.

However, while UBEs are a declining threat to the Internet savvy, we still know very little about the way that UBEs, their content, and various payloads impact upon the more vulnerable groups within society: the poor, single parents, those in remote locations, those with learning difficulties, the disadvantaged, and 'newbies' or the newly online and other groups who lack experience of the Internet and are unable to judge between the plausible and less plausible invitations. The newly retired are particularly susceptible to fraudsters because they are anxious to invest their retirement lump sums, the size of which they have rarely had in their possession before, to support their needs in later life.

My analysis of the spam industry shows it to be based upon the informational products derived from the surveillance of online populations and which exploit the panoptic and synoptic capabilities of the Internet to manipulate and deceive the same populations. While the behaviour patterns may seem familiar, spams are, as mentioned earlier and elsewhere (Wall 2003), 'true cybercrimes,' in that they are wholly dependent upon the Internet, which is an efficient vehicle for enabling offenders to engage in a broad range of small-impact multiple victimizations on a global scale (Wall 2002b: 195). This latter point is very important because the individual victimizations are often minimal and, if even reported to police, do not warrant the expenditure of police resources (Wall 2002b: 195). Yet the cumulative impact can be quite considerable.

I have only focused here upon e-mail-based systems which use fairly basic Internet communications technologies. Newer forms of technology, such as Peer-to-Peer (P2P) technology and prospective (future) technologies based upon 'location' and 'ambience' (IPTS 2003: 60-97) fully exploit the Internet's surveillant characteristics. Distributed, as opposed to centrally indexed, P2P software creates networks of users and enables the contents of participants' 'swop' directories to be surveiled by other participants. Developed for and popularized by MP3 music file exchange, P2P is now a common basis for all sorts of data exchange and many commercial variants exist. Some enable cash-based transactions, others 'swop' or barter economies, some require the individual to trust the technology completely.

Conclusions: Trust in the Internet or Trussed by IT

By exploring the production of information capital arising from the shift in the location of value from the tangible to the intangible, from physical objects to the virtual means by which ideas are expressed, this chapter has revealed some of the background to the emerging information economy. As the symbols and images of material production from the declining old economy are reduced into 'information,' only to then be reconstructed as simulation within web-based production, not the least in computer games, the signs of the real become reconfigured in cyberspace as a new 'real,' the hyper-real (Baudrillard 1994). Significant to the earlier discussion is the additional observation that this same 'information,' in the form of familiar symbols and images, is now being used as a tempting lure to attract attention and initiate the (often automatic) elicititation of further information about people which is then manipulated and sold.

For the information economy to thrive and flourish the new technologies must engender public trust. A major impediment to this trust is the entrepreneurial manipulation of multidirectional (surveillant) information flows over the Internet to capture all available information and exploit its value, whether legally or illegally, and seemingly without regard to the broader impact on trust building. The boundaries between the two positions are not yet – and may never be – firmly set, so it is not surprising that many of the spams discussed earlier are not what they might initially seem. They include advertisements that are really designed to solicit only a response, fraudulent solicitations, misleading and spoof advertisements, and services that seek to entrap. Unfortunately, genuine attempts to promote e-commerce, provide important information, or share knowledge tend to become lost among this 'white noise' of the Internet. Moreover, like the last scene of Orwell's *Animal Farm*, where the animals cease to be distinguishable from the humans, it is especially hard to distinguish between legitimate and illegitimate business operations in the compilation and sale of information. Although the indications are that most spammers tend to be fairly small independent operators, it is impossible to establish whether or not information acquired by the corporate sector is actually used for the purposes it was collected.

What is exposed here is the tension between the need for users to share information in order to participate on the Internet while maintaining the individual's right to control his or her most private information. If such control is a desirable principle, which it surely must be as it underpins trust building, then the prognosis is not altogether good. Participants in online activities are displaying more than a casual willingness to agree to Internet service provider and individual website terms and conditions, notably with regard to the acceptance of cookies, in return for a material, or informational, advantage to themselves. In the process they effectively forego their online privacy rights (see Haggerty and Ericson 2000; Poster 1995). Not only does this trend suggest that privacy is becoming a tradeable commodity rather than a moral right – a trend that is central to the new politics of surveillance – it also impedes the establishment and maintenance of trust.

Even if individuals were more concerned about their own information, someone else may still be selling it. The 'super-panopticon' of databases (Poster 1995) that currently contain our most private information are a veritable goldmine of potentially valuable data (see chapter 15 in this volume). Poster argues that they are so multi-authored that no specific creator could claim authorship, and morally they are no single person's property because they are composed of everyone's data, yet they belong in law to the institutions that compile them and as such are their property: 'the database is a discourse of pure writing that directly amplifies the power of its owner/user ... they are politicised' (Poster 1995: 85–6). While the owners of these databases may take great pains to assure the public of the integrity of their security, the chain of trust is only as strong as its weakest link, and there have already been a number of topical examples whereby employees of database owners, in some cases government bodies, have data mined the records of celebrities to sell their personal information for profit. Examples of this breach of trust include employees of the United Kingdom's Inland Revenue service who were caught selling information from the tax returns of celebrities (BBC News Online 2003b). Another example is the case of U.S. police officers passing on to websites such as The Smoking Gun custody photographs of the rich and famous in their darkest moments, presumably in exchange for some form of reward.

For the mass media, this prized information achieves the double whammy of creating scandalous stories that reveal the intimate secrets of celebrities while also creating equally scandalous stories that expose the very weaknesses in the system that they, the mass media, initially sought to exploit. The result is a further impediment to trust building. The problem for individual victims is that once the genie of private information is 'out of the bottle,' it cannot be retreived.

Ironically, there is perhaps some hope in that the very qualities of the Internet which threaten trust building by preventing the establishment of control over personal information – namely, that the transparency of multidirectional information flow which allows the few to watch the many at the same time as the many can watch the few – can be used to create accountability for actions and impose rules of fair usage. Perhaps this truly is the age of an infotopic transparent (information) society (Brin 1998), in which technology 'may embolden and empower a new kind of citizenship' (Brin 2002). Such a society restores within the individual, either alone or in partnership, the responsibility to resolve conflicts. It signals a brave new world borne out of a frustration with the old. As Marge Simpson once observed, 'You know, the courts might not work any more, but as long as everybody is videotaping everyone else, justice will be done' ('Homer Bad Man,' episode 2F06 of *The Simpsons*, 1994).

On the other hand, as Brin himself ponders, there is always the creeping possibility that the practical realities of the transparent society may produce exactly the opposite effect and mask an opaque Orwellian dystopia. This is a nightmare scenario, described satirically by Adams (1998), in which 'new technology will allow the police to solve 100 percent of all crimes. The bad news is that we'll realize 100 percent of the population are criminals, including the police.'

Postscript

Since this chapter was written in mid-2003 there have been two major changes in the spamming world which reinforce the arguments put forward in it. The first has been a massive escalation of spamming activity through the growth of botnets – networks of (Zombie) personal computers that have been infected by remote administration viruses. The botnets not only extend the reach of the 'cybercriminal' by facilitating the launching of spamming or distributed denial of service attacks, but they are also very valuable informational artefacts that can be hired out, sold, or traded. Botnets represent a step-change in the automation of offender-victim engagement online through the reorganization of criminal labour. In varying configurations, depending upon the purpose of the criminal venture, spammers may employ hackers who then employ virus writers. The second has been the introduction of anti-spam legislation in the European Union, the United Kingdom, and the United States. On 11 December 2003, the United Kingdom introduced compulsory opt-in legislation in the form of the Privacy and Electronic Communications (EC Directive) Regulations 2003 (SI/2003/2426). These regulations brought into effect Article 13 of EU Directive 2002/58/EC on privacy and electronic communications, which had been passed in July of the previous year. The U.S. Federal (opt-out) legislation 'Controlling the Assault of Non-Solicited Pornography and Marketing Act of 2003,' or the 'CAN-SPAM Act of 2003' (S. 877) was passed by Congress in early December 2003 and came into effect on 1 January 2004. The U.S. legislation sits on top of a variegated patchwork of state legislation. Some of this legislation is very strong – for example, in California – while in other states it is either weak or non-existent. Law has had a marked indirect impact upon spamming activity (see Wall forthcoming, chaps. 4 and 7).

CASES

Compuserve. 1997. *CompuServe Inc. v. Cyber Promotions, Inc. and Sanford Wallace,* 962 F.Supp. 1015 (S.D. Ohio 3 Feb. 1997).

LEGISLATION

Megan's Law. *1996. Sec. 2. Release of information and clarification of public nature of information.* H.R. 2137 http://www.ojp.usdoj.gov/vawo/laws megan.htm (accessed 4 Mar. 2003).

REFERENCES

Abrams, P. 1982. *Historical Sociology*. Shepton Mallet: Open Books.

Adams, S. 1998. *The Dilbert Future: Thriving on Business Stupidity in the 21st Century*. London: HarperBusiness.

Baudrillard, J. 1994. *Simulcra and Simulation*. Ann Arbor: University of Michigan Press.

BBC News Online. 2003a. 'Net "Naming and Shaming" Wins Support.' *BBC News Online*. 7 Jan. http://news.bbc.co.uk/1/hi/technology/2634809.stm (accessed 7 Mar. 2003).

– 2003b. 'Tax Records "For Sale" Scandal.' *BBC News Online*. 16 Jan. http://news.bbc.co.uk/1/hi/business/2662491.stm (accessed 6 Mar. 2003).

Bernal, F. 2000. 'Big Brother Capabilities in an Online World – State Surveillance in the Internet – State Surveillance Schemes.' http://www.bernal.co.uk/capitulo3.htm (accessed 5 Mar. 2003).

Bowden, C., and Y. Akdeniz. 1999. 'Cryptography and Democracy: Dilemmas of Freedom.' In J. Cooper, ed., *Liberating Cyberspace: Civil Liberties, Human Rights, and the Internet*, 81–125. London: Pluto Press.

Brightmail. 2002. *Slamming Spam*. San Francisco: Brightmail.

Brin, D. 2002. 'Citizen Gain.' In *Telephony*, 4 Feb. http://telephonyonline.com/ar/telecom_citizen_gain/index.htm (accessed 24 July 2003).

– 1998. *The Transparent Society: Will Technology Force Us to Choose Between Privacy and Freedom*. London: Addison-Wesley.

Castells, M. 1997. *The Information Age: Economy, Society, and Culture, Volume 2: The Power of Identity*. Oxford: Blackwell Publishers.

– 2000a. *The Information Age: Economy, Society, and Culture, Volume 1: The Rise of the Network Society*. 2nd ed. Oxford: Blackwell Publishers.

– 2000b. *The Information Age: Economy, Society, and Culture, Volume 3: End of Millennium*. 2nd ed. Oxford: Blackwell Publishers.

– 2000c. 'Materials for an Explanatory Theory of the Network Society.' *British Journal of Sociology* 51 (1): 5–24.

Clarke, R. 1994. 'Dataveillance: Delivering "1984."' In L. Green and R. Guinery, eds., *Framing Technology: Society, Choice and Change*, 117–30. Sydney: Allen & Unwin

DARPA. 2002. *Defense Advanced Research Project Agency's Information Awareness Office and Total Information Awareness Project*. http://www.darpa.mil/iao/iaotia.pdf (accessed 5 Mar. 2003)

Edwards, L. 2000. 'Canning the Spam: Is There a Case for the Legal Control of Junk Electronic Mail?' In L. Edwards and C. Wealde, eds., *Law and the Internet: A Framework For Electronic Commerce*, 2nd ed, 309–40. Oxford: Hart Publishing.

Engberg, D. 1996. 'The Virtual Panopticon.' In *Impact of New Media Technologies*. Fall. http://is.gseis.ucla.edu/impact/f96/Projects/dengberg/ (accessed 8 Mar. 2003).

Foucault, M. 1983. 'Afterword: The Subject and Power.' In H. Dreyfus and P. Rainbow, eds., *Michel Foucault: Beyond Structuralism and Hermeneutics*, 2nd ed., 208–26. Chicago: University of Chicago Press.

Haggerty, K., and R. Ericson. 2000. 'The Surveillant Assemblage.' *British Journal of Sociology* 51 (4): 605–22.

IPTS. 2003. *'Security and Privacy for the Citizen in the Post-September 11 Digital Age: A Prospective Overview.'* Report by the Institute for Prospective Technological Studies, Joint Research Committee, Seville, to the European Parliament Committee on Citizen's Freedoms and Rights, Justice and Home Affairs., European Commission, July.

Khong, W.K. 2001. 'Spam Law for the Internet.' *Journal of Information, Law and Technology* 3 http://elj.warwick.ac.uk/jilt/01-3/khong.html/ (accessed 23 Jan. 2002).

Levi, M., and D.S. Wall. 2003. 'Crime and Security in the Aftermath of September 11: Security, Privacy and Law Enforcement Issues Relating to Emerging Information Communication Technologies.' In IPTS, 105–12, 178–204.

Lyon, D. 1994. *The Electronic Eye: The Rise of Surveillance Society*. Minneapolis: University of Minnesota Press.

– 2001a. 'Surveillance after September 11.' *Sociological Research Online* 6 (3) http://www.socresonline.org.uk/6/3/lyon.html.

– 2001b. *Surveillance Society*. Buckingham: Open University Press.

Mathieson, T. 1997. 'The Viewer Society: Foucault's Panopticon Revisited.' *Theoretical Criminology* 1: 215–34.

Mendelson, H., and R. Pillai. 1999. 'Information Age Organisation, Dynamics and Performance.' *Journal of Economic Behaviour and Organization* 38 (3): 253–81.

Pease, K. 2001. 'Crime Futures and Foresight: Challenging Criminal Behaviour in the Information Age. In D.S. Wall, ed., *Crime and the Internet*, 18–28. London: Routledge.

Poster, M. 1995. *Second Media Age*. Cambridge: Polity Press.

Rathmell, A. 1998. 'The World of Open Sources (2). Information Warfare and Hacking.' Paper delivered to the 1998 International Conference for Criminal Intelligence Analysts, 'Meeting the Challenge from Serious Criminality,' Manchester.

Van Dijk, J. 1999. 'The One-Dimensional Network Society of Manuel Castells: A Review Essay.' *The Chronicle*. http://www.thechronicle.demon.co.uk/ archive/castells.htm (accessed 6 Mar. 2003).

Wall, D.S. 2002a. 'DOT.CONS: Internet Related Frauds and Deceptions upon

Individuals within the UK.' Final Report to the Home Office. March (un-
published).

– 2002b. 'Insecurity and the Policing of Cyberspace.' In A. Crawford, ed.,
Crime and Insecurity, 186–210. Cullompton: Willan.

– 2003. 'Mapping Out Cybercrimes in a Cyberspatial Surveillant Assemblage.'
In Webster, F. and K Ball, eds., *The Intensification of Surveillance: Crime
Terrorism and Warfare in The Information Age*, 112–36. London: Pluto Press.

– Forthcoming. *Cybercrimes: The Transformation of Crime in the Information Age*.
Cambridge: Polity.

Ward, M. 2000. 'Web Gets Wise to Who You Are.' *BBC News Online*. 15 Aug.
http://news.bbc.co.uk/1/hi/sci/tech/879890.stm (accessed 6 Mar. 2003).

Yaukey, J. 2001. 'Common Sense Can Help You Cope with Spam.' *USA Today*
19 Dec. http://www.usatoday.com/life/cyber/ccarch/2001/12/19/
yaukey.htm (accessed 6 Mar. 2003).

WEBSITES

CAUCE. http://www.cauce.org (accessed 6 Mar. 2003).

City of Ontario, California. http://www.ci.ontario.ca.us/index.cfm/22/2284
(accessed 4 Mar. 2003).

Cookie Central. http://www.cookiecentral.com (accessed 7 Mar. 2003).

Cult of the Dead Cow. http://www.cultdeadcow.com/ (accessed 5 Mar. 2003).
(for a useful description of Back Orifice see http://www.nwinternet.com/
~pchelp/bo/bo.html).

Doubleclick. http://www.doubleclick.com/us (cited 7 Mar. 2003).

Engage. http://www.engage.com (accessed 7 Mar. 2003).

FBI Top Ten Most Wanted Fugitives. http://www.fbi.gov/mostwant/
terrorists/fugitives.htm. (accessed 6 Mar. 2003); see also Most Wanted
Terrorists. http://www.fbi.gov/mostwant/terrorists/fugitives.htm
(accessed 6 Mar. 2003).

Gnutella. http://www.gnutella.com (accessed 7 Mar. 2003).

Joint Information Systems Committee. http://www.jisc.ac.uk (accessed 9 Mar.
2003).

Newfoundland. http://www.roads.gov.nf.ca/cameras (accessed 4 Mar. 2003).

Officer.Com. http://www.officer.com/wanted.htm (accessed 6 Mar. 2003).

Smoking Gun. http://www.thesmokinggun.com/mugshots/index.html
(accessed 7 Mar. 2003).

Spectorsoft. http://www.spectorsoft.com (accessed 5 Mar. 2003).

World's Most Wanted, Inc. http://www.mostwanted.org/Search.html
(accessed 6 Mar. 2003).

15 Data Mining, Surveillance, and Discrimination in the Post-9/11 Environment

OSCAR GANDY, JR

Public awareness of the role that data mining and other forms of remote surveillance have come to play in the management of the concerns of daily life has increased rather dramatically in the United States following the immediate, and often quite hostile, response to the plans announced for the creation of a Total Information Awareness (TIA) program in the Department of Defense (Mittelstadt 2002). No small part of the concern being expressed was based on an assessment of the character of the program's head, the same retired Admiral John Poindexter who had been convicted of lying to Congress and other efforts towards derailing Congressional investigations of official misconduct in the Iran-Contra Affair (Moscoso and Edmonson 2003).

Although concerns about the use of data-mining techniques by corporations for the purpose of segmentation, targeting, and other forms of 'relationship management' have been expressed in the past (Danna and Gandy 2002), it was primarily in the wake of announcements about proposed TIA activities that similarities and differences between the uses, users, and risks associated with the technique became the focus of so much public attention (Leahy, Feingold, and Cantwell 2003; Markle Foundation 2002).

This chapter seeks to bring a critical light to bear on the ways in which the use of data mining and associated techniques of knowledge production seem likely to transform social relations within both the public and private spheres. After a brief introduction to the technology and its uses within business and government organizations we will turn our attention to the social concerns that are raised by such increased, widespread, and presumptively legitimate uses of computer-enhanced discriminatory techniques. The final section will focus on the limitations to the development of responsive public policies that are inherent in the traditionally narrow framing of privacy legislation towards the protection of what has been termed 'individually identifiable information' (Hatch 2001; Lyon 2003; Pollack 2001; Skok 2000).

The Goals of Data Mining

Data mining is a process that has as its goal the transformation of raw data into information that can be utilized as strategic intelligence within the context of an organization's identifiable goals. At its core, data mining is concerned with prediction. Data-mining efforts are directed towards the identification of behaviour and status markers that serve as reliable indicators of a probable future. Because the development of data-mining applications is dominated by concerns about risk minimization or avoidance, the most relevant behaviours are those that can be assessed in terms of relative risk. The use of a data-mining resource to identify the best prospects – the 20 per cent of the customers who are likely to provide 80 per cent of the profits – can still be understood from a risk perspective by noting that selection of the best also means avoidance of persons and relations that are less desirable, or risky.

The identification of the best prospects or the highest risks necessarily involves some agreement about standards of value. In the business environment, an individual who is a 'good customer' under one system of valuation may be of considerably less value under another. Decisions within the realm of consumer credit provide some excellent examples (Leyshon and Thrift 1999). A customer who pays off her credit cards in full each month would surely be identified as a good 'credit' risk, in terms of the likelihood of default. That same customer would not be so highly valued in terms of her contribution to the bank's bottom line, as she would not incur any interest charges. A riskier customer, one who is late, and may even miss a payment on occasion, may actually be more valuable because of the interest and penalties he accrues (Adams, Hand, and Till 2001).

The traditional challenge for data miners is to determine which customers are more valuable, and therefore worth keeping. A somewhat different challenge is the determination of that point at which a previously 'good' customer becomes less desirable.

In one set of analyses based on some 90,000 customers, cluster analysis was used to divide the sample of customers into two classes. Pattern-searching techniques were then used to identify what the analysts characterize as 'interesting or unexpected observations' (Adams, Hand, and Till 2001: 1020). With this technique the analysts identified a class of customers who began as 'good,' and then rather suddenly switched to missing payments, which the analysts characterized as 'bad behaviour' under the first approach to valuation. The provision of 'warnings' about the likelihood of default by individuals is precisely the kind of predictive services that credit agencies have begun to offer their customers. Clearly, these behavioural analyses differ in important ways from the sorts of credit-scoring analyses that are relied upon in deciding whether to issue a card in the first place (Leyshon and Thrift 1999).

A great many terms that describe data-mining applications are broadly familiar, and while some may be unique to a particular organization or administrative specialty, fundamentally they refer to the same underlying task (Bowker and Star 1999). For example, although the identification of individual terrorists is of paramount importance to those who are involved in the pursuit of 'homeland security,' the assignment of individuals to classes, categories, or groups on the basis of an analysis of seemingly independent, and in some cases, anonymous, activities has become a routine administrative activity (Computer Science and Telecommunications Board 2000; Hobart and Schiffman 1998).

Data-mining applications in the area of security abound. Concern about the security of computers and other critical infrastructural resources has led to an accelerated push for the development of techniques to detect and prevent intrusion into sensitive systems (Computer Science and Telecommunication Board 2002). In the detection of attacks on networked computer systems, the challenge is one of identifying deviations from the normal flow of traffic to a server. Unlike the evaluation of traffic accidents or other discrete events that generate a reliable record, a great many attacks on computer systems go undetected, and as a result, analytical models that depend upon those distinctions are of limited use. It is here that the discovery of patterns within data that 'call attention to themselves' enables analysts eventually to arrive at distinctions between 'probes,' outright attacks, and legitimate variations in network traffic over time (Rothleder 2000).

Consumer profiling involves the attempt to assign individuals to one or more groups or segments on the basis of attributes that they share, or are assumed to share to some degree. The nature of consumer profiling has been explored in considerable detail by the Federal Trade Commission (2000) with regard to its use by online businesses. Online profiling refers primarily to the techniques used by companies with an online presence to deliver banner advertisements and editorial copy to individuals.

A variety of techniques, many of which are hidden from, or not disclosed, to website visitors are used to gather transaction-generated details about them (Pollack 2001). Among the most frequently used techniques are 'cookies' or relatively durable markers that are placed on the visitor's hard drive. These cookies make it possible for firms, or their representatives, to 'track' individuals as they move to different sites around the web (Helling 1998). As testimony revealed to the FTC, data gathered from this 'clickstream' is often combined with additional information acquired from third-party sources. 'This enhanced data allows the advertising networks to draw a variety of inferences about individual customer's interests and preferences. The result is a detailed profile that attempts to predict the individual consumer's tastes, needs, and purchasing habits and enables the advertising companies' computers to make split-second decisions about how to deliver ads directly targeted to the consumer's specific interests' (Federal Trade Commission 2000: 5).

Some users of customer-profiling systems report dramatic improvements in the productivity of their websites. In some cases, the source of the improvement was a fairly rough index of consumer interests based on little more than a zip code (Stevens 2001). In other cases, profiling systems are continually updated through the incorporation of multiple streams of information.

The provision of targeted advertising is not the only use of data-mining techniques within the networked environment. The profiling of visitors to corporate websites can also influence business decisions about the cost and quality of the services to be offered to that visitor (Danna and Gandy 2002). The use of data-mining techniques to facilitate this kind of discrimination in e-commerce transactions is referred to as 'weblining' (Stepanek 2000) because of its similarity to familiar but illegal forms of racial discrimination known as 'redlining.' Its name is derived from a common practice in financial service markets of drawing red lines around neighbourhoods where loans would not to be made or property insurance would not be underwritten (Nier 1999).

Firms that use data-mining in support of 'customer relationship management' (CRM) are interested in differentiating among existing and potential customers in terms of their estimated total value. Reportedly, there are very few segments of the market that mine customer data more actively than firms in the gaming industry. According to one source, 'Casino operators gather and store data from loyalty programs, direct mail databases, socioeconomic data, and credit card issuers, among other sources.' Harrah's Entertainment is an exemplary user of data-mining for CRM. Harrah's gathers data from its nearly five million 'loyalty' program members and uses transaction-generated information from its loyalty cards as well as slot machine data and information derived from food and beverage sales (CFO Research Services 2002: 11).

Data-mining Techniques

Data mining, or knowledge discovery in databases, as the insiders would have it known, is a small but rapidly expanding specialty within the field of applied mathematics that seeks to derive meaningful intelligence from the analysis of patterns within sets of data, including the virtual mountains of data that are generated by a host of routine interactions enabled by networked computers (Fulda 2000).

Data mining is both descriptive and predictive, and although evaluation is not always explicit, the process of identification nearly always includes some form of evaluative assessment (Gandy 2000). Descriptive analysis may involve the assignment of 'objects' or entities into pre-existing categories in accordance with well-developed rules for classification (Bowker and Star 1999). Objects of interest may also be classified into categories that are the products of the analysis reflecting the application of statistical or other criteria for the determination of boundaries (Norusis 1994).

At the descriptive level, analyses common to data-mining can provide condensed summaries of massive amounts of information. We readily understand common statistical measures, such as means or averages, as summaries. We also understand correlation coefficients as summary statements about relationships between objects, or the attributes of objects. At a multivariate level, we might understand factor scores as a summary statement about relations between a large number of attributes and a large number of objects that might be persons or groups of persons. We are perhaps less familiar with summaries at the level of discourse.

Data-mining systems applied to the analysis of text can produce summaries of varying length (Mani, Guilder, Clifton, and Concepcion 2000). Not long ago, the MITRE Corporation announced the development of a data-mining tool for text processing that is capable of processing a large number of news stories, identifying the key 'players' in the news, and generating a concise summary. In one of the examples provided in a promotional article, 60,000 news stories were condensed into an eighty-six-word summary (Mani et al. 2000: 3).

At the descriptive level, the process of discovery is often aided through the use of techniques that facilitate the visualization of patterns and relationships in data. These patterns may allow distinctions to be made between persons, behaviours, and outcomes on the basis of relations between the attributes of each. Increasingly these relationships may be discerned in spatial terms with the aid of geographic information systems (Baker and Baker 1999; Phillips and Curry 2003). Online analytical processing tools (OLAP) make it easier for subject area specialists who are not necessarily familiar with advanced statistical techniques to pursue an analysis of complex multivariate relationships merely by manipulating objects on screen, or responding to queries about alternative models or conditions (Roberts-Witt 2001).

Clustering techniques are frequently used to assign objects of analysis to groups or classes on the basis of criteria that can be adjusted on the basis of theory and experience. While classification into categories or groups may be the ultimate goal of these analytical efforts, a preliminary stage in the process may emphasize the discovery of patterns of association, or covariation. Unlike traditional statistical methods where specific tests are used to evaluate a hypothesis, statistics are used within data-mining efforts to generate the hypotheses that might be explored through other means, or through subsequent analyses of enhanced data bases (Carbone 2000).

It is also likely that the associative links which emerge through analysis will make even more sense when they are examined over time. Thus, insurers may decide to classify automobile accident claims as problematic on the basis of a determination of the temporal relationship between the calls the insured made to his doctor and those he made to his attorney.

As an area of applied statistics, data-mining involves the application of increasingly sophisticated techniques to larger and larger pools of data by more and more decision makers within organizations. This diffusion of analytical capacity has been enabled in part by increases in the diffusion of relatively inexpensive computing power to desktop and even to laptop computers (Darling 1997; Lawson 2002). Commonplace statistical resources used for prediction (or explanation in very polite company) include multiple regression and discriminant analysis.

More sophisticated techniques involve the use of so-called neural networks that can process data autonomously to generate clusters, and with sequential adjustment, develop fine-tuned predictive algorithms. Neural networks are said to provide advantages over traditional statistical techniques like multiple regression because assumptions about relations between data are not as restrictive. Unlike multiple regressions, neural network techniques are able to include a large number of variables that comes closer to reflecting both the complexity and the uncertainty that characterize routine business decisions (Garver 2002).

Neural networks or a related approach known as decision tree analysis may be used to identify the variables, or attributes of objects that are most important or influential in the prediction of some outcome. A relatively mundane example should suffice to illustrate the use of these techniques in the business environment.

Success in the competitive fast food environment is based in part upon superior knowledge about what governs consumers' decisions, and about what aspects of the experience produces a level of satisfaction high enough to increase the probability of repeat sales. Determining the relative importance of different attributes of pizza delivered to the home was the task at hand in one case study (Garver 2002). Semantic differential scaling techniques were used to capture customer ratings of aspects of the experience, including price, temperature of the pizza, and the 'friendliness' of the person delivering the product. While the neural network analysis and regression techniques both identified 'taste' as the most important factor, 'amount' only emerged as relatively important in the neural network analysis (Garver 2002).

Of course, only an experiment in which the model's parameters can be adjusted and tested against actual consumer behaviour would determine whether the hypotheses generated by the analyses win empirical support.

Data and the Real World

It is important to keep in mind that data-mining produces statements about the relationships between data as well as statements about the relationships between phenomena in the real world. Some analytical approaches common to data-mining efforts use insights derived from observed relationships between data to make statements about phenomena that have not been, and perhaps cannot be, observed directly (Gigerenzer et al. 1989; Smith 1998). Data mining is said to differ from ordinary information retrieval in that the information which is sought does not exist explicitly within the database, but must be 'discovered' (Fulda 2000: 106). In an important sense, the relationships between measured or 'observed' events come to stand as a 'proxy' for some unmeasured relationship or influence (Baard 2002).

For example, factor analysis programs generate estimates of the correlation, or association, between measured variables and unmeasured factors or influences. In the area of media studies, numerous examples can be found of factor analytic models that draw inferences about unmeasured tastes and preferences within population segments on the basis of the structural relationships that can be observed between the programs that hundreds of people report viewing with varying degrees of regularity (Frank and Greenberg 1980; Webster, Phalen, and Lichty 2000).

A fairly sophisticated application of this technique has been described by Predictive Networks in relation to their 'Digital Silhouettes' product (Predictive Networks 2001). In addition to traditional demographic indicators that include gender, occupation, and race, the 'anonymous predictive models of individuals' includes ninety categories of inferred 'content affinities.' The 'Digital Silhouette' technology can be seen as an example of a product or service that continually updates the profiles of individual web users. The technology is promoted as a resource for advertisers interested in reaching a particular type of consumer. This ideal type is assumed to share common affinities and demographics, and on the basis of the client's specification, according to the promoters, 'Predictive Networks selects the Digital Silhouettes most likely to respond to the ad' (Predictive Networks 2001: 3). As in many such applications, the accuracy or precision with which individuals are described by these profiles is said to improve over time. The brochure claims that as 'the user continues to surf the web, his or her Digital Silhouette becomes a refined composite of the demographic and affinity distributions of all the web sites he or she has visited' (Predictive Networks 2001: 3).

It is important to note here that each individual's profile is shaped not only by her own behavioural attributes, but also by the program's characterization of the profiles of all the other individuals within the database who visit particular websites. As a result, these profiles are fundamentally relational, or comparative, rather than *individual* identities (Gandy 2000).

The application of data-mining techniques for predictive purposes uses data gathered in the past in order to generate descriptions about events that may occur in the future. In the context of ordinary business applications, such predictions may be concerned about determining which individuals will respond affirmatively to a particular offer in the context of a particular marketing campaign (Adams, Hand, and Till 2001).

Applications in the realm of national security are a bit more specialized, but the underlying assumptions, resources, and techniques do not vary substantially, and both may be concerned with the development of behavioural profiles.

The Sources of Data

At the heart of public concerns about data-mining is the extent to which the technology invites the integration of data from previously independent data record systems. Concerns about the development of centralized government data banks expressed during early periods of concern about an emerging 'surveillance society' (Flaherty 1989) have been replaced by concerns about the sorts of 'virtual databases' enabled by high speed networking and distributed processing capacity (Solove 2001). Although different approximations of 'fair information practices' have sought to restrict the secondary use of information to the purposes for which it was initially gathered, there is an extremely limited record of success in this regard (Sovern 1999).

Heightened concern has been expressed about the enhanced capacity of government agencies to create virtual databases because of their recently increased access to public and private record systems. Both government agencies and private businesses have continued to utilize massive amounts of data from a host of sources, limited primarily by well-placed concerns about data quality and standardization (Gandy 2001).

The expansion in the pool of available data is being affected rather dramatically by changes in both technology and public policy. The growth in the number of databases under the administrative control of government agencies reflects more than the growth of government oversight and responsibility. It also reflects the impact of changes in the cost structure of data capture, storage, and transmission.

Record systems expand because the systems that enable the capture, storage, and retrieval of information have become more efficient, less expensive, and less demanding of specialized training. The fact that the data stored within these systems has also been determined to have economic value beyond its primary function plays a less important role in determining whether and how the information will be stored by government agencies.

However, a somewhat different logic may explain the rapid growth in the size and scope of private databases. Indeed, it may be the case that the need to meet government demands for information related to taxation and regulatory oversight explains much of the initial impetus for the development of computerized record-keeping systems within the private sector. The discovery that there was additional economic and strategic value in transaction-generated information (TGI) is of fairly recent vintage. Hundreds of articles are being published each year designed to remind business managers, especially those involved in electronic commerce, of the value to be derived from mining the 'gold' within their own record systems, and dozens more compare various techniques and analytical products in terms of their efficiency and ease of use (Greening 2000).

The direct marketing industry, a primary consumer of TGI about individuals, is said to have evolved into its present form in response to the enhanced potential for target marketing that was made possible by the establishment of postal zip codes in the 1960s (Harper 1986). This form of marketing has since come to support a segment of the business services industry that includes list vendors, consolidators, and other data managers that bill customers and clients in excess of $3 billion each year (Solove 2001: 1408).

Many database vendors also provide clients with access to lists derived from government databases, including those derived from voter registration files. One vendor, Aristotle International, offered political campaign managers lists of voters enhanced with additional information about ethnicity, voting frequency, motor vehicle ownership, and estimates of household income in addition to data about whether the registrants owned, or rented their homes (Aristotle 2002; Pressman 2000).

A major competitor to Aristotle offered an online service that allowed clients to specify the sorts of voters they wanted to reach. The vendors of a product called VISION (Voter Information System Integrated Online Network), made what has emerged as an obligatory nod in the direction of concern about privacy before going on to describe the nature and benefits of the services they provide. Privacy concerns are supposed to be reduced because VISION clients are invited to specify the criterion they want used in generating a list of prospects, rather than making requests for information about individually identified persons.

In one scenario provided in their online brochure (I-Centrix 2002), a client might request a list of all males over forty-five who are registered Democrats, who have an income above $70,000, who have voted in past Presidential elections, and who also happen to live in the 3rd Electoral District of Troy, New York. If the customer is willing to pay the established price for the 125 voters who would be produced by this multivariate sort, pre-printed mailing labels as well as phone numbers might be supplied. The VISION service claimed to have on offer an additional three hundred variables that might be used to define the preferred class of voters to be targeted by a specialized appeal (I-Centrix 2002: 2).

Of course, there are no guarantees that the information acquired from database vendors or consolidators will be accurate. On occasion, the consequences of almost certain errors can be quite substantial. In the case of ChoicePoint, a database company that claims to be primarily in the 'credential verification' business, we find an organization that combines identification with classifications along a number of critical dimensions. Unfortunately for the nation, and perhaps for the world, a ChoicePoint subsidiary provided credential verification services for the county election boards in Florida. Many of their characterizations of American citizens were grossly incorrect. As a result, an estimated eight thousand potential voters, many of them African Americans, were listed as convicted felons, and thereby denied the right to vote in a hotly contested national election. While not all of those who were excluded on the basis of this classification error were likely to have voted for Al Gore, they are far less likely to have supported George W. Bush. In any event, their numbers far exceeded the margin of victory that history has been forced to record (Coppola 2002).

ChoicePoint is now emerging as one of the largest private forensic laboratories in the country. Derek Smith, ChoicePoint's chief executive, seems driven to add DNA and other biometric identifiers to the list of commercially available data about individuals they will use in their identification and certification services (Coppola 2002).

Although we might not ordinarily consider the providers of search engines as being very similar to the list vendors that charge fees for providing access to those lists, it is clear that search engines also provide access to the information in databases. The developers of search engines have invested in the development of quite sophisticated algorithms for searching and indexing the rapidly expanding caches of information stored on servers around the globe. Google is readily identified as a leader within this very dynamic industry. Its search engines support in excess of 150 million searches per day (Fraim 2002).

Following its acquisition of Deja.com, a major innovator in the exploitation of Usenet newsgroup archives, hundreds of millions of personally identified comments, critiques, and representations became available for search, analysis, and commercial exploitation by Google (Google 2001). Fraim (2002) suggests that Google's demonstrated ability to segment searches into national and linguistic groups can easily be expanded into more meaningful 'communities' that might include corporations, universities, and neighbourhoods defined by zip codes. The fact that Google uses 'cookie' technology in ways that link searches to searchers has raised serious privacy concerns, even though Google claims that they currently limit the identification of searchers to broad aggregates (Olsen and Kane 2003). Nevertheless, analysis of the word and subject rankings that could be based on the millions of searches initiated within and between these 'aggregates' has strategic potential that is not lost on data miners within business and government.

Businesses are particularly concerned about the ways in which their firms and products are 'positioned' in relation to other firms at any moment in time. They are concerned about their status and visibility not only among the population at large, but among particular segments of the population. The analysis of searches and commentary can provide them with continual assessments of their relative position.

Policy makers are always interested in knowing about the salience of particular concerns among the electorate. Like corporate strategists, they are concerned about the salience of issues and policy options among elite segments of the population that are more likely to respond to what they see as threats or opportunities (Ferguson 2000). These assessments are increasingly easy to produce with the aid of text-mining software. Indeed, an analysis of elite or expert discussions and debates ongoing in Usenet newsgroups is precisely the sort of application that text-mining products like MITRE's 'Top Cat' have been designed to support (Mani, Guilder, Clifton, and Concepcion 2000).

The U.S. *Patriot Act* (2001) provided the U.S. government with substantially improved access to data within record systems in government agencies that are not generally concerned with law enforcement or national security. The guidelines issued less than a year later for the FBI reflected the increased ease with which government agencies can expect to gather data from public and private sources (Center for Democracy and Technology 2002). Where previous guidelines restricted agency investigations to cases where there is at a minimum some suspicion of criminal activity, the new guidelines reflect a greater appreciation of the ways in which data-mining techniques can be used 'as the basis for generating the suspicion of criminal conduct in the first place' (Center 2002: 1). Under the new guidelines, the FBI appears to be authorized to conduct 'fishing expeditions' on their own or with the support of commercial firms that are better equipped, and have far more experience in data-mining.

Although there are signs that some private corporations as well as librarians (Hull 2002) are resisting government attempts to gain access to the TGI in their files, there is also evidence that many corporations are voluntarily providing government agencies with access to information out of some heightened sense of patriotism. Of course, on occasion, this patriotic urge seems to coincide with corporate self-interest.

Reputedly acknowledged as the world's largest retailer, eBay.com has also been identified as an organization that is quite willing to share information about the transactions of its more than 60 million registered users if security personnel at the company have their suspicions aroused by the behaviour of buyers or sellers, or by the inquiries of government officials. Such arousal is likely because security personnel at eBay routinely use their own data-mining resources to continually search for patterns of suspicious behaviour that might indicate a risk of fraud.

Although eBay's 'director of law enforcement and compliance' reported that the company receives approximately two hundred requests for information about users each month, he noted that the overwhelming majority of these requests were informal, rather than official subpoenas or court orders (Dror 2003). While eBay doesn't provide unhindered access to its entire corporate database, which reportedly contains all the TGI the company has produced since 1995, a request from a law enforcement officer is likely to be rewarded with a complete history of a specific target's interactions with the eBay servers. This report may include the contributions that registrants may have made to the company's many discussion groups.

In those cases where there were explicit legal barriers to the sharing of financial information (Janger and Schwarts 2002), such as information that eBay gained access to through its acquisition of the PayPal transaction clearance service, security officials at eBay were reportedly quite willing to provide law enforcement with sufficient 'hints' about how to produce the appropriate court order (Dror 2003).

TIA and Data-mining Unleashed

Of course, eBay is not alone in its willingness to provide government investigators with information about customers that they believe might be useful in the identification of terrorists or other miscreants (Baard 2002; Wylie 2003). As one astute observer noted, 'It didn't take long for data-management companies to realize that if their software could find links in customer buying patterns and improve retailers' inventory decisions, perhaps it could find links among the government's vast terrorism-related intelligence warehouses and enhance the government's ability to prevent the next attack' (Franklin 2002).

Many entrepreneurs rushed to offer their services to government, and hundreds more responded to requests for proposals designed to meet the information needs of the Defense Advanced Research Projects Agency (DARPA) Total Information Awareness (TIA) program (Electronic Privacy Information Center 2003). According to the initial solicitation, 'DARPA is soliciting innovative research proposals in the area of information technologies that will aid in the detection, classification, identification, and tracking of potential foreign terrorists, wherever they may be, to understand their intentions, and to develop options to prevent their terrorist acts' (Defense Advanced Research Projects Agency [DARPA] 2002).

At the core of the research initiative was the development of 'ultra large all-source information repositories' that were later defined as 'large scale databases covering comprehensive information about all potential terrorist threats.' What DARPA was in search of was 'innovative technologies needed to architect, populate, and exploit such a database for combating terrorism.' After noting that DARPA already had 'on-going research programs aimed at language translation, information extraction from text' and other techniques, the agency made it clear that it was inviting 'new ideas for novel information sources and methods that amplify terrorist signatures and enable appropriate response.'

Because the successful proposals were classified, it was not possible to determine precisely what the twenty-six early 'winners' had promised to deliver. It is easy, however, to imagine the sorts of data-mining resources that will emerge from the Palo Alto Research Center (PARC), given its title: 'Knowledge-based tracking of content change in growing collections of text documents' (EPIC 2003).

Although the U.S. Senate managed to call a moratorium on the use of data-mining by the departments of Defense and Homeland Security (Gross 2003) because of the unprecedented level of intrusion into the lives of average citizens that the program seemed to invite (Mack 2002), it is also clear that the underlying rationale for expanding the use of the technology has many supporters. Indeed, while a recent report on national security from the Markle Foundation emphasized the need to balance the value of strategic intelligence against the risks to civil liberties, there was little room for doubt that the members of the Task Force endorsed the widespread use of data-mining in support of national security (Markle Foundation 2002).

The Markle group (Markle Foundation 2002: 45) agreed that there was a need for 'an information collection and analysis strategy that will allow us to detect and prevent dangers from these unanticipated sources.' Specifically, the working group identified the sorts of tools that might be developed and deployed in an effort to anticipate, and thereby prevent, some 'terrorist' action in the future. In their view, theoretically driven scenarios could be assessed in relation to empirically generated probabilities that have been derived from data-mining applications. Indeed, from their perspective one 'advantage of computer-generated scenarios is that very large numbers could be created without the unintended constraints of human definitions of "likely" or "plausible," thus decreasing the chances of surprise through novelty' (Markle Foundation 2002: 52).

We are of course familiar with the strategic advantages that can be derived from the relaxation of traditional boundaries around the possible or imaginable that are based on naive concepts of reasonableness. On the other hand, it is important for us to evaluate the risks that flow from the design of intervention strategies that might be pursued if government and corporate planners are no longer constrained by traditional conceptions of 'fairness,' 'justice,' 'equality,' or 'legitimacy.' I have discussed these outcomes elsewhere with regard to myriad ills that accompany the use of TGI to facilitate discrimination (Danna and Gandy 2002; Gandy 2000, 2001). Among the most serious of these concerns are those related to the ways in which discrimination in information markets reinforces disparities in the level and impact of participation in the public sphere by segments of the population. These disparities are likely to become sharper still because of the ways in which the strategic resources provided by data-mining may also be used to shape the 'preferences' for information that individuals come to recognize and express as their own (Zarsky 2002/3).

The Limits of Privacy Policy

I have argued in the past that the traditional policy framework generally relied upon to protect individual and collective interests in privacy are hampered by an emphasis on the use of 'individually identified information' (Gandy 1993). Individuals are placed at risk of discrimination by virtue of their membership in groups, rather than specifically on the basis of their individual identities (Gandy 1996, 2000). This 'categorical vulnerability' (Gandy 2001) is especially problematic in the context of a degraded public sphere because individuals may not be aware of the groups to which they have been assigned by various data mining analyses (Bruening and Schwartz 2002). Indeed, it is the tendency among those who engage in data mining and related approaches to the development of behavioural profiles to keep those profiles secret that limits the utility of the traditional privacy torts. Because of the assumption that information about individuals is made public on a widespread and routine basis, the interests of individuals in such information is minimized at the same time that corporate and government interest in the strategic intelligence generated through analysis of such data are protected by security classification and intellectual property regimes (Belgum 1999; Hatch 2001).

In the case of corporate actors, short-term strategic advantage is realized in part by the ability of competitors to treat such classificatory schemes as trade secrets (Montana 2001). For somewhat different reasons, government agencies that develop profiles that they believe will be useful in identifying smugglers, terrorists, or other subjects of interest tend to resist attempts by the public or civil libertarians to gain access to the details about the composition of these profiles.

Critics have suggested that the use of profiles as alternatives to traditional markers of stigmatized identity (Loury 2002) makes it possible for an increase in discriminatory acts that would be declared illegal if they relied solely or primarily on the use of 'suspect categories' like race, gender, or national origin (Belgum 1999). Because these meaningful categories are submerged or replaced by coefficients assigned to variables or explanatory factors, the names that have been assigned to these factors are less likely to attract the heightened scrutiny of the courts.

It should be clear, however, that these profiles, or 'signatures' of group membership (McCarthy 2000), have much in common with traditional stereotypes or schema that have been developed over time for subordinate groups. However, in the case of racial stereotypes, we have evidence that the targets of stereotypic thinking have a high degree of awareness of the ways in which they have been characterized and evaluated by those higher in the social hierarchy (Sigelman and Tuch 1997). It seems unlikely, however, that very many of the targets of data mining know much about the ways in which they are being classified. They are largely unaware of the ways in which they are the victims or the beneficiaries of such remote sensing.

In the absence of knowledge about the groups to which they have been assigned, individuals are less likely to engage in political organization or in the development of social movement organizations (Eskridge 2001). This is the case despite the suggestion that computer networking would make such organizational efforts less difficult to develop and sustain (Hula 1999; Kreimer 2001).

I suspect that it will be difficult to mount a sustained challenge to the use of data-mining and other discriminatory techniques that are at the heart of what I refer to as 'the panoptic sort' (Gandy 1993). The best we can hope for is transparency – that citizens and consumers are routinely and effectively informed about the nature and extent to which their options and opportunities are being shaped by the application of statistical techniques.

It seems unreasonable to expect that those who use these techniques will be the best sources of public awareness of the consequences of their use. Therefore it will be particularly important for journalists, scholars, and advocates of informed choice to be able to convey a sense of the array of individual and collective risks that flow from the use of data-mining and other discriminatory techniques.

REFERENCES

Adams, N.M., D.J. Hand, and R.J. Till. 2001. 'Mining for Classes and Patterns in Behavioural Data.' *Journal of the Operational Research Society* 52:1017–24.

Aristotle International, Inc. 2002. US voter lists. Aristotle International, Inc. http://www.aristotleonline.com/voterlists.html.

Baard, E. 2002. 'Your Grocery List Could Spark a Terror Probe: Buying Trouble' *Village Voice*. 30 July. http://www.oas.villagevoice.com/issues/0230/baard.html.

Baker, K., and S. Baker. 1999. 'Divide and Conquer.' *Journal of Business Strategy* 20:16–18.

Belgum, K.D. 1999. 'Who Leads at Half-time?: Three Conflicting Visions of Internet Privacy Policy.' *Richmond Journal of Law and Technology* 6:1.

Bowker, G.C., and S. Leigh Star. 1999. *Sorting Things Out: Classification and Its Consequences.* Cambridge, MA: MIT Press.

Bruening, P., and A. Schwartz. 2002. 'Privacy and Online Politics: Is Online Profiling Doing More Harm Than Good for Citizens in Our Political System?' Washington, DC: George Washington University Graduate School of Political Management. http://www.democracyonline.org.

Carbone, P. 2000. 'What Is the Origin of Data Mining?' *The Edge* 4:2–12.

Center for Democracy and Technology. 2002. *CDT's Analysis of New FBI guidelines: Center for Democracy and Technology.* http://www.cdt.org/wiretap/020530guidelines.shmtl.

CFO Research Services. 2002. *Mining the Value in CRM Data: A CFO Perspective.* Boston, MA: CFO Publishing Corporation. http://www.cfo.com.

Computer Science and Telecommunications Board. 2000. *Summary of a Workshop on Information Technology Research for Federal Statistics.* Washington, DC: National Academy Press.

– 2002. *Cyber-security Today and Tomorrow: Pay Now or Pay Later.* Washington, DC: National Academy Press.

Coppola, V. 2002. 'Derek Smith's Brave New World.' In *Georgia Trend* ChoicePoint, Inc. http://www.choicepoint.net/news/gatrend.html.

Danna, A., and O.H. Gandy Jr. 2002. 'All That Glitters Is Not Gold: Digging Beneath the Surface of Data mining.' *Journal of Business Ethics* 40:373–86.

Darling, C.B. 1997. 'Datamining for the Masses.' *Datamation.* (Feb.): 52–4.

Defense Advanced Research Projects Agency (DARPA). 2002. *BAA 02–08 Information Awareness Proposer Information Pamphlet.* Washington, DC: U.S. Department of Defense.

Dror, Y. 2003. 'Big Brother Is Watching You – And Documenting: EBay, Ever Anxious to Up Profits, Bends over Backwards to Provide Data to Law Enforcement Officials.' Haaretz.com. http://www.haaretz.com.

Electronic Privacy Information Center (EPIC). 2003. *EPIC Analysis of Total Information Awareness Contractor Documents.* Washington, DC: Electronic Privacy Information Center. http://www.epic.org/privacy/profiling/tia/doc_analysis.html.

Eskridge, W.N. Jr. 2001. 'Channelling: Identity-based Social Movements and Public Law.' *University of Pennsylvania Law Review* 150:419–525.

Federal Trade Commission, U.S. 2000. *Online Profiling: A Report to Congress.* Washington, DC: U.S. Federal Trade Commission, Bureau of Consumer Protection.

Ferguson, S.D. 2000. *Researching the Public Opinion Environment: Theories and Methods.* Thousand Oaks, CA: Sage.

Flaherty, D.H. 1989. *Protecting Privacy in Surveillance Societies: The Federal Republic of Germany, Sweden, France, Canada, and the United States.* Chapel Hill: University of North Carolina Press.

Fraim, J. 2002. 'Electronic Symbols: Internet words and Culture.' June 2002. http://www.firstmonday.org/issues/issue7_6/fraim/index.html.

Frank, R.E., and M.G. Greenberg. 1980. *The Public's Use of Television.* Beverley Hills, CA: Sage.

Franklin, D. 2002. 'Data Miners: New Software Instantly Connects Key Bits of Data That Once Eluded Teams of Researchers.' *Time.com,* 15 Dec. http://www.time.com/globalbusiness/printout/0,8816,400017,00.html.

Fulda, J. 2000. 'Data mining and Privacy.' *Albany Law Journal of Science and Technology* 11:105–13.

Gandy, O.H. Jr. 1993. *The Panoptic Sort: A Political Economy of Personal Information.* Boulder, CO: Westview Press.

– 1996. 'Legitimate Business Interest: No End in Sight? An Inquiry into the Status of Privacy in Cyberspace.' *University of Chicago Legal Forum 1996*: 77–137.

– 2000. 'Exploring Identity and Identification in Cyberspace.' *Notre Dame Journal of Law, Ethics and Public Policy* 14:1085–1111.

– 2001. 'Dividing Practices: Segmentation and Targeting in the Emerging Public Sphere.' In W.L. Bennet and R.M. Entman, eds., *Mediated Politics: Communication in the Future of Democracy*, 141–59. New York: Cambridge University Press.

Garver, M.S. 2002. 'Try New Data-mining Techniques.' *Marketing News* (16 Sept.): 31–3

Gigerenzer, G., Z. Swijtink, T. Porter, L. Daston, J. Beatty, and L. Kruger. 1989. *The Empire of Chance: How Probability Changed Science and Everyday Life.* New York: Cambridge University Press.

Google, Inc. 2001. 'Google Acquires Usenet Discussion Service and Significant Assets from Deja.com.' http://www.google.com/press/pressrel/pressrelease48.html

Greening, D.R. 2000. 'Data mining on the Web. There's Gold in That Mountain of Data.' *New Architect.* http://www.webtechniques.com/archives/2000/01/greening.

Gross, G. 2003. 'US Senate Blocks Government Data-mining Plan.' *InfoWorld Daily News.* 24 Jan. Washington, DC. http://www.infoworld.com/article/03/01/24/030124hntia_1.html.

Harper, R. 1986. *Mailing List Strategies*. New York: McGraw-Hill.

Hatch, M. 2001. 'Electronic Commerce in the 21st Century: The Privatization of Big Brother: Protecting Sensitive Personal Information from Commercial Interests in the 21st Century.' *William Mitchell Law Review* 27:1457–1502.

Helling, B. 1998. 'Web-site Sensitivity to Privacy Concerns: Collecting Personally Identifiable Information and Passing Persistent Cookies.' *First Monday*. http://www.firstmonday.dk/issues/issue3_2/helling/index.html.

Hobart, M.E., and Z.S. Schiffman. 1998. *Information Ages: Literacy, Numeracy, and the Computer Revolution*. Baltimore: Johns Hopkins University Press.

Hula, K.W. 1999. *Lobbying Together: Interest Group Coalitions in Legislative Politics*. Washington, DC: Georgetown University Press.

Hull, D. 2002. 'Libraries Face Privacy Test: Anti-terrorism Law Used to Demand Records.' *Mercury News*, 18 Oct. http://www.bayarea.com/mid/mercurynews/news/local/4312790/html.

I-Centrix. 2002. 'Voter Information System Integrated Online Network.' I-Centrix http://www.i-centrix.com/vision.html.

Janger, E.J., and P.M. Schwartz. 2002. 'The Gramm-Leach-Bliley Act, Information Privacy, and the Limits of Default Rules.' *Minnesota Law Review* 86:1219–61.

Kreimer, S.F. 2001. 'Technologies of Protest: Insurgent Social Movements and the First Amendment in the Era of the Internet.' *University of Pennsylvania Law Review* 150:119–71.

Lawson, N.J. 2002. 'The Emperor's New Clothes: Data mining, the Real Key to the Kingdom.' *Assessment Journal* (May/June): 49.

Leahy, P., R. Feingold, and M. Cantwell. 2003. Letter to Attorney General John Ashcroft. 10 Jan. U.S. Senate, Committee on the Judiciary.

Leyshon, A., and N. Thrift. 1999. 'Lists Come Alive: Electronic Systems of Knowledge and the Rise of Credit Scoring in Retail Banking.' *Economy and Society* 28: 434–66.

Loury, G.C. 2002. *The Anatomy of Racial Inequality*. Cambridge, MA: Harvard University Press.

Lyon, D. 2003. 'Surveillance as Social Sorting: Computer Codes and Mobile Bodies.' In D. Lyon, ed., *Surveillance as Social Sorting: Privacy, Risk and Digital Discrimination*, 13–30. New York: Routledge.

Mack, G. 2002. *Total Information Awareness Program (TIA) System Description Document (SDD)*, 1-150. Washington, DC: Defense Advanced Projects Research Agency

Mani, I., L. Van Guilder, C. Clifton, and K. Concepcion. 2000. 'Text Mining by Filter Composition.' *The Edge* 4: 1–4.

Markle Foundation, Task Force on National Security in the Information Age.

2002. *Protecting America's Freedom in the Information Age.* New York: Markle Foundation.

McCarthy, J. 2000. 'Phenomenal Data mining.' *Communications of the ACM* 43: 75

Mittelstadt, M. 2002. '"Orwellian" Data-mining Program Draws Fire from All Sides.' *Dallas Morning News,* 16 Dec. (accessed online, Lexis/Nexis, 13 Feb. 2003).

Montana, J. 2001. 'Data mining: A Slippery Slope.' *Information Management Journal* 35:50.

Moscoso, E., and G. Edmonson. 2003. 'Pentagon "Data mining" Program Fans PrivacyFears.' *Cox News Service.* 18 Jan. (accessed online, Lexis/Nexis, 13 Feb. 2003).

Nier, C.L. 1999. 'Perpetuation of Segregation: Toward a New Historical and Legal Interpretation of Redlining under the Fair Housing Act.' *John Marshall Law Review* 32:617–65.

Norusis, M.J. 1994. *SPSS Advanced Statistics 6.1.* Chicago, IL: *SPSS.*

Olsen, S., and M. Kane. 2003. 'Google Triggers Privacy Concerns.' *CNET News.com.* 3 Mar. http://znet.com/2100-1104-990724.html.

Phillips, D., and M. Curry. 2003. 'Privacy and the Phenetic Urge: Geodemographics and the Changing Spatiality of Local Practice.' In D. Lyon, ed., *Surveillance as Social Sorting: Privacy, Risk and Digital Discrimination,* 137–52. New York: Routledge

Pollack, M. 2001. 'Opt-in Government: Using the Internet to Empower Choice.' *Catholic University Law Review* 50:653–702.

Predictive Networks, Inc. 2001. 'Digital Silhouettes.' http://www.predictivenetworks.com

Pressman, A. 2000. 'Voters for Sale.' *The Industry Standard.* http://www.thestandard.com/article/0,1902,19864,00.html.

Roberts-Witt, S.L. 2001. 'Gold Diggers.' *PC Magazine.* http://www.pcmag.com/print_article/0,3048a%253D5356,00.asp

Rothleder, N. 2000. 'Data mining for IntrusionDetection.' *The Edge* 4:10–11.

Sigelman, L., and S.A. Tuch. 1997. 'Metastereotypes: Blacks' Perceptions of Whites' Stereotypes of Blacks.' *Public Opinion Quarterly* 61:87–101.

Skok, G. 2000. 'Establishing a Legitimate Expectation of Privacy in Clickstream Data.' *Michigan Telecommunication and Technology Law Review* 6:61–88.

Smith, B.C. 1998. *On the Origin of Objects.* Cambridge, MA: MIT Press.

Solove, D.J. 2001. 'Privacy and Power: Computer Databases and Metaphors for Information Privacy.' *Stanford Law Review* 43:1393–1462.

Sovern, J. 1999. 'Opting In, Opting Out, or No Options at All: The Fight for Control of Personal Information.' *Washington Law Review* 74:1033–1118.

Stepanek, M. 2000. 'Weblining.' *Business Week E.Biz.* 3 Apr.: EB26–33.

Stevens, L. 2001. 'It Sharpens Data mining's Focus – Instead of Building Data mining Applications with No Clear Goal, Companies Are Setting Priorities Up Front to Maximize ROI.' *Internet Week* (6 Aug.): 29.

U.S. *Patriot Act.* 2001. Public Law No. 107-56, 115 Stat. 272.

Webster, J.G., P.F. Phalen, and L.W. Lichty. 2000. *Ratings Analysis: The Theory and Practice of Audience Research.* Mahwah, NJ: Lawrence Erlbaum Associates.

Wylie, M. 2003. 'Homeland Security; With Marketers, Not Much Remains Private; Companies Own Data That Government Wants to Identify Terrorists.' *San Diego Union-Tribune* (5 Jan.): A1.

Zarsky, T.Z. 2002/3. '"Mine Your Own Business!": Making the Case for the Implications of the Data Mining of Personal Information in the Forum of Public Opinion.' *Yale Journal of Law & Technology* 5:1–55.

Contributors

William Bogard
Department of Sociology
Whitman College

J.P. Brodeur
International Centre for Comparative Criminology
Université de Montreal

Christopher Dandeker
Department of War Studies
King's College London

Aaron Doyle
Department of Sociology and Anthropology
Carleton University

Richard V. Ericson
Centre of Criminology
University of Toronto

Oscar Gandy Jr.
Annenberg School for Communication
University of Pennsylvania

John Gilliom
Department of Political Science
Ohio University

Kevin D. Haggerty
Department of Sociology
University of Alberta

Stéphane Leman-Langlois
International Centre for Comparative Criminology
Université de Montréal

David Lyon
Department of Sociology
Queen's University

Gary T. Marx
Professor Emeritus of Sociology
Massachusetts Institute of Technology

Emily Martin
Department of Anthropology
New York University

Serra Tinic
Department of Sociology
University of Alberta

Joseph Turow
Annenberg School for Communication
University of Pennsylvania

David S. Wall
Centre for Criminal Justice Studies
University of Leeds

Reg Whitaker
Department of Political Science
University of Victoria